Progress in Mathematics
Volume 156

Yves Guivarc'h
Lizhen Ji
J.C. Taylor

Compactifications of Symmetric Spaces

Birkhäuser
Boston • Basel • Berlin

Yves Guivarc'h
IRMAR
UFR Mathématiques
Université de Rennes-I
Rennes, France

Lizhen Ji
Department of Mathematics
University of Michigan
Ann Arbor, MI

J. C. Taylor
Department of Mathematics and Statistics
McGill University
Montreal, Quebec, Canada

Library of Congress Cataloging-in-Publication Data
Guivarc'h, Yves
Compactifications of symmetric spaces / Yves Guivarc'h, Lizhen Ji, John C. Taylor.
p. cm. -- (Progress in mathematics ; v. 156)
Includes bibliographical references (p.) and index.
ISBN-13: 978-1-4612-7542-8 e-ISBN-13: 978-1-4612-2452-5
DOI: 10.1007/978-1-4612-2452-5
1. Symmetric spaces. 2. Compactifications. I. Ji, Lizhen, 1964- . II. Taylor, J. C. (John Christopher), 1936- . III. Compactifications of Symmetric Spaces. IV. Series: Progress in Mathematics (Boston, Mass.); vol. 156
QA670.G85 1997 97-27755
516.3'62--dc21 CIP

AMS Subject Classification: 53C35, 22E30, 57T15, 60J45, 31C35, 43A85, 60J50

Printed on acid-free paper

Birkhäuser

Softcover reprint of the hardcover 1st edition 1998

Typeset by the Authors in $\mathcal{AMS}$-TEX

9 8 7 6 5 4 3 2 1

WE DEDICATE THIS BOOK TO OUR FAMILIES,
ESPECIALLY OUR WIVES,
AND TO ALL OUR TEACHERS

CONTENTS

PREFACE

Compactifications of symmetric spaces of non-compact type have been studied from various points of view. One of the main purposes of this monograph is to determine the Martin compactifications explicitly in terms of the geometry of the symmetric spaces, in particular, in terms of the spherical Tits building. In so doing, we also obtain new characterizations of the Satake–Furstenberg compactifications and of the Karpelevič compactification of the symmetric spaces.

The book is based on two independent preprints by Guivarc'h and Taylor and by Ji. While the main results of this work have been announced in [G17], it also contains new results obtained later. This accounts, in part, for a certain non-linearity in the exposition, which is also a reflection of the fact that some aspects of this project are still being investigated.

It is our pleasure to thank all the people who have helped us during this project. We would like to thank J.P. Anker and M. Babillot for very important information about the Green function. The first author is grateful to G.A. Margulis for bringing the question of computing the Martin compactifications to his attention. He would also like to emphasize the strong influence of Furstenberg's ideas on his work. L. Ji, the second author, would like to thank Professor S.T. Yau for imparting his philosophy of hands-on learning, and to thank Dr. S. Stafford for early conversations about Martin compactifications of Riemannian manifolds which were the starting point for the second author. He would also like to thank S. Helgason, M. Goresky, A. Korányi, G. Prasad and R. Spatzier for helpful conversations during various stages of this project. The third author, J. C. Taylor, would like to thank A. Korányi for introducing him to the study of symmetric spaces, and to thank his colleagues for their cooperation in bringing this project to completion. In addition, he would like to acknowledge his debt to M. Brelot who encouraged him to study the Martin compactification early in his career. Thanks are also due to M. Guillemer for help in typing and for invaluable help preparing the final draft of the manuscript. Finally, we thank A. Borel for many helpful comments and pointing out some typographical errors.

The authors would also like to especially thank their families, in particular their wives, for patience, support, and help.

Y. Guivarc'h was a member of CNRS during this project, L. Ji was partially supported by the National Science Foundation throughout this project, and J.C. Taylor was materially supported by the National Sciences and Engineering Council of Canada.

Compactifications of Symmetric Spaces

CHAPTER I

INTRODUCTION

Let X denote an open metrizable C^∞ manifold without boundary. To each Riemannian metric on X, there corresponds a number of invariant objects. Two obvious invariants are the Laplace–Beltrami operator L and the volume measure dx. The operator $-L$ acting on $L^2(X)$ is a non-negative operator and has a non-negative lower bound λ_0 to its spectrum. It is known (cf. Sullivan [S4], Taylor [T3, p. 131]) that, for $\lambda \leq \lambda_0$, the operator $L + \lambda Id$ has positive global solutions.

A natural class of compactifications of X are the Martin compactifications $X \cup \partial X(\lambda)$ of X associated with the operator $L + \lambda Id$ for $\lambda \leq \lambda_0$. The Martin boundary $\partial X(\lambda)$ parametrizes a set of generators of the cone of positive solutions of $Lu + \lambda u = 0$, and a subset corresponds to a set of minimal solutions that form a basis of the cone, i.e., the extremal elements. For example, if the strong form of Liouville's theorem holds for the Riemannian manifold, i.e., every positive harmonic function is constant, then the Martin compactification associated with L is the one-point compactification. The topology of the Martin compactification for L determines the asymptotic behavior of the Brownian motion on X. The Martin compactification is the natural setting for the study of the boundary behavior of the positive eigenfunctions. Furthermore, these compactifications and their boundaries are invariants of the Riemannian manifold.

The only extensive class of Riemannian manifolds for which these compactifications have been identified are the simply connected ones with "pinched" negative sectional curvature (see Anderson and Schoen [A4], Ancona [A2], [A3]). In this class, the Martin compactifications, for $\lambda < \lambda_0$, all coincide with the conic compactification $X \cup X(\infty)$ and every point of the sphere at infinity $X(\infty)$ gives rise to a minimal λ-eigenfunction.

In the case of non-positive and non-pinched curvature, the situation is essentially more complicated. This was observed first by Dynkin in the special case of the symmetric spaces of Hermitian matrices with determinant one [D4]. In this case, the Martin boundary contains an open set of non-minimal eigenfunctions.

In this book, these Martin compactifications will be determined explicitly and geometrically for all symmetric spaces X of non-compact type. It is shown that the Martin boundary $\partial X(\lambda_0)$ is isomorphic to the maximal Satake–Furstenberg compactification and has a cell complex structure that is dual to the spherical Tits building of X.

This duality between the Martin boundary $\partial X(\lambda_0)$ and the Tits building of X could be looked upon as a relation between classical and quantum theories: the Tits building can be defined in terms of geodesics while the

Martin boundary parametrizes a set of generators of positive solutions of $Lu+\lambda_0 u = 0, \lambda \leq \lambda_0$ (and can also be defined in terms of the "ground state" of L). For $\lambda < \lambda_0$, the Martin boundary $\partial X(\lambda)$ is fibered over the spaces $\partial X(\lambda_0)$ and $X(\infty)$. It inherits from these spaces a mixed cellular structure, the cells of which are products of cells in $\partial X(\lambda_0)$ and the simplices in $X(\infty)$. There are a finite number of types of cells, up to G-equivalence, and each type is determined by the structures of the geodesics converging to it and corresponds to a well-defined mode of convergence to the boundary, as well as to a well-defined type of λ-eigenfunction. This is closely related to the geometric point of view of scattering theory of Melrose [M6], that deals with the parametrization of certain non-square integrable solutions of $Lu+\lambda u = 0, \lambda \geq \lambda_0$, in terms of a geometric boundary (for more discussions about this relation see Ji [J1]).

This dual cell complex structure of the Martin boundary $\partial X(\lambda_0)$ comes from the combinatorial structure of the flats in X via the polyhedral compactification of a flat, and our result gives strong evidence to support the philosophy of Tits [T7, p. 217] concerning compactifications of a Lie group G and its symmetric space X: "The 'most natural' choice for the 'space at infinity' of G or X is 'often' closely related to the spherical Tits building of G". (For applications of the rational spherical Tits building to compactifications of locally symmetric spaces, see [J3].)

The study of eigenfunctions of the Laplacian L on X is closely related to that of eigenfunctions of convolution equations on G defined by probability measures (see Furstenberg [F3] and [F4], Corollary 12.6 and Proposition 13.3). This approach gives insight into the final results and, moreover, is suitable for generalizations to other spaces. Another important feature of this book is the detailed discussion of convolution equations and their associated Martin compactifications. The convolution results can be generalized to other groups and spaces, for example, reductive groups defined over p-adic fields and the Bruhat–Tits buildings associated with them. Moreover, they offer a complementary approach to the main problems considered in this book and put them in a conceptual framework that is suitable for further extensions. This is a subject in itself. The information presented here is only an introduction to this area, that is a natural extension of the main topic of this book.

Statement of some of the main new results

For simplicity, here one restricts to the case of symmetric spaces and the Laplacian. The Martin compactification $X \cup \partial X(\lambda)$ of X for $L + \lambda Id$ is obtained by adding to X an ideal boundary (constructed exactly as in the original Euclidean situation discussed by Martin [M5]), whose points are positive solutions of $Lu + \lambda u = 0$ that are limits of the normalized Green function. (For example, in the case of the Euclidean unit disc one obtains the Poisson kernel, a family of functions parametrized by the unit circle.)

Thus the Martin compactification is determined by the asymptotic behavior of the Green function at infinity. It is an important problem to determine the limit functions, called Martin kernels, to identify the Martin compactification in terms of the geometry of the manifold, and to study any additional structure on the Martin boundary inherited from the manifold.

Let G be a connected semisimple Lie group of non-compact type with finite center and K a maximal compact subgroup of G. Then the Killing form defines an invariant metric on $X = G/K$, and X is a Riemannian symmetric space of non-compact type. Then, some of the main new results of this book are summarized in the following statement.

Main Theorem.

(1) *At the bottom of the spectrum* λ_0, *the Martin compactification* $X \cup \partial X(\lambda_0)$ *is isomorphic to the maximal Satake–Furstenberg compactification* $\overline{X}^{SF}$ *of* X. *Further, the Martin boundary* $\partial X(\lambda_0)$ *has a cell complex structure dual to the spherical Tits building of* X.

(2) *Let* $X(\infty)$ *be the set of equivalence classes of geodesics in* X, *called the sphere at infinity. Let* $\overline{X}^{c} = X \cup X(\infty)$ *be the conic compactification. Then, for any* $\lambda < \lambda_0$, *the Martin compactification* $X \cup \partial X(\lambda)$ *is the smallest compactification* $\overline{X}^{c} \vee \overline{X}^{SF}$ *that dominates both* $\overline{X}^{SF}$ *and* $\overline{X}^{c}$.

(3) *Let* $\overline{X}^{K}$ *be the Karpelevič compactification. Then* $\overline{X}^{K}$ *dominates* $X \cup \partial X(\lambda)$, *and these compactifications are isomorphic if and only if either rank*$(X) = 1$ *or rank*$(X) = 2$ *and* $\lambda < \lambda_0$.

The Martin kernels are written down explicitly in Proposition 7.26 and Theorems 8.2 and 13.23. Part (1) is proved in Theorems 7.33 and 9.18 and Corollary 14.22, Part (2) is proved in Theorem 8.21, and Part (3) in Theorem 8.26.

Part (1) of this main theorem (without the information on the Tits building) was proved for the polydisc by Guivarc'h–Taylor in [G16] (see also [T3]) and the direct identification of the other Martin compactifications, i.e., Part (2), for the product of two hyperbolic spaces was given by Giulini–Woess in [G4].

The spherical Tits building of X is a simplicial complex with one simplex for each proper parabolic subgroup of G, and the inclusion relation between the simplexes is opposite to the inclusion relation between the parabolic subgroups. The underlying topological space of this simplicial complex can be canonically identified with the sphere at infinity $X(\infty)$. In other words, the spherical Tits building gives a simplicial complex structure on $X(\infty)$ (see Proposition 3.20). The types of cells correspond to the various types of geodesics and, hence, to the horospheres [B4].

The dual cell complex structure of $\partial X(\lambda_0)$ comes from new geometric realizations of $\overline{X}^{SF}$, that are basic constructions in this book and for which the convergent sequences are explicitly given (see Chapters III and IX). Briefly, a maximal totally geodesic flat submanifold in X is called a flat. The Weyl chambers of a flat determine a polyhedral decomposition of the flat, and its dual cell complex forms the boundary of a polyhedral compactification of the flat.[1] These polyhedral compactifications of flats can then be rotated by K to define a compactification of X. This compactification is called the dual cell compactification and is homeomorphic to $\overline{X}^{SF}$. Its boundary inherits a cell complex structure that is dual to the spherical Tits building of X. In fact, the spherical Tits building of X can also be described in terms of the Weyl chamber decompositions of the flats in X, and the cell complex structure on the boundary of the dual cell compactification is in a natural duality with the Tits building on $X(\infty)$. (See Chapters III and IX for details.)

The reason for the mixed structure of $X \cup \partial X(\lambda)$, when $\lambda < \lambda_0$, can already be seen in the splitting into two parts of the geometric formula (see Theorem 13.23) that gives the minimal Martin kernels and also in the asymptotic form of the Green function.

Characterizations of the compactification $\overline{X}^{SF}$

The Satake–Furstenberg compactifications $\overline{X}^{SF}$ were originally defined by Satake [S1] by embedding X into the space of positive definite Hermitian matrices of determinant 1. Furstenberg [F3] compactified X by embedding it in the space of probability measures on the (maximal) Furstenberg boundary. Moore [M8] showed that Furstenberg's procedure defined a family of compactifications that coincides with the family of Satake compactifications. This result of Moore plays an important role in several places in this book as it is useful to realize the maximal Satake–Furstenberg compactification as a space of measures. Besides the geometric polyhedral construction mentioned above, several other characterizations of $\overline{X}^{SF}$, the maximal Satake–Furstenberg compactification, are presented. Each approach has its own advantage and sheds light on the others.

Satake's approach reduces the problem of compactifying a general symmetric space to that of compactifying the most basic symmetric space: the space $\mathrm{SL}(n, \mathbb{C})/\mathrm{SU}(n)$ of positive definite Hermitian matrices of determinant one. A compactification of X is obtained by embedding X in $\mathrm{SL}(n, \mathbb{C})/\mathrm{SU}(n)$. Such an embedding X is equivalent to realizing the Lie algebra of G as a subalgebra of $\mathfrak{sl}(n, \mathbb{C})$, a procedure well-known in Lie algebra theory (see Chapter IV).

One advantage of the Furstenberg compactifications is that the Furstenberg boundaries appear naturally in the boundary of the compactifications as the set of unit point masses or delta measures because Furstenberg's

[1]The polyhedral compactification of a flat is closely related to Atiyah's convexity theorem of the moment map of a Hamiltonian action of a torus. In fact, using the main theorem above and Atiyah's convexity theorem, one can show that both the Satake–Furstenberg compactifications and the Martin compactifications are topological balls (see [J2]).

approach is to embed X into the space of probability measures on the boundary.

Since the maximal Furstenberg boundary can be identified with the Poisson boundary of X, the boundary behavior of bounded harmonic functions on X can be studied in the compactification $\overline{X}^{SF}$ (cf. Koranyi [K5]). This approach of using measures could be generalized to other spaces, in particular, non-symmetric rank one Hadamard manifolds.

The compactification $\overline{X}^{SF}$ can also be realized by embedding X into the space $\mathcal{S}$ of closed subgroups of G as follows: $gK \in X = G/K \to gKg^{-1} \in \mathcal{S}$. The closure $\overline{\mathcal{S}_0}$ of the image $\mathcal{S}_0$ of X in $\mathcal{S}$ is a compactification homeomorphic to $\overline{X}^{SF}$ (see Chapter IX). For every group $D \in \partial\overline{\mathcal{S}_0}$, the boundary of X in $\mathcal{S}$, there is a unique positive solution u of $Lu + \lambda_0 u = 0$ with $u(o) = 1$, where $o = K \in X$, that is invariant under the group. This solution h^D is a Martin kernel and determines a unique point on $\partial X(\lambda_0)$, and every Martin kernel arises this way. This gives an explicit identification between $\overline{X}^{SF}$ and $X \cup \partial X(\lambda_0)$. Furthermore, in this realization the cellular structure of $\overline{X}^{SF}$ can clearly be seen.

Another realization of $\overline{X}^{SF}$ is related to the following construction of the conic compactification $X \cup X(\infty)$. The limits of geodesic spheres passing through $o = K \in X$ (or any other point) are called geodesic horospheres. Every point $z \in X(\infty)$ determines a unique family of parallel geodesics and a unique geodesic horosphere passing through o. This horosphere is a level surface of the horofunction determined by z and could also be regarded as the geodesic sphere with center z and infinite radius. Therefore, $X(\infty)$ can be identified with the set of geodesic horospheres containing o, and the topology of $X \cup X(\infty)$ can also be characterized as follows: a sequence (y_n) in X, converging to infinity, converges if and only if the geodesic sphere with center y_n and containing o converges to a geodesic horosphere.

Inspired by the above construction, following Karpelevič [K3], one calls the set $gKg^{-1} \cdot o$ in X a group sphere, i.e., group spheres are orbits in X of maximal compact subgroups of G. Let $\mathcal{C}(X)$ denote the space of closed subsets of X equipped with the topology of Hausdorff convergence on compact subsets. Then the map $g \cdot o = gK \in X = G/K \to gKg^{-1} \cdot o \in \mathcal{C}(X)$ defines an embedding, and the closure of X in $\mathcal{C}(X)$ is a compactification of X homeomorphic to $\overline{X}^{SF}$ (see Theorem 9.21).[2] The ideal points of this compactification are limits of group spheres, called group horospheres, in comparison with the geodesic horospheres above. Using this identification, one gets a dynamical description of the topology of $\overline{X}^{SF}$. Starting from the basepoint $o = K \in X$, one has a collapsed group sphere, i.e., the basepoint o. As one moves away from the basepoint, the group spheres expand and one shifts them towards the direction one is moving in so that they pass through the basepoint o. On approaching infinity, these group spheres

[2]We thank Koranyi for suggesting this realization.

blow out in the direction of approach to infinity and become group horospheres. (These group horospheres are the closed sets of the form $D \cdot o$, where $D \in \partial \overline{S_0}$.) This description is similar to the picture Dynkin mentioned in his ICM talk [D3, p. 20]: The space of positive definite matrices of determinant one can be realized as the space of ellipsoids of volume one. Near infinity, the ellipsoids flatten and the lengths of some axes go to infinity. This situation is very easy to visualize in the case of the hyperbolic disc, as explained at the end of § 9.4. From the point of view of harmonic analysis, one can consider the "ground state" of L, i.e., the unique spherical function Φ for which $L\Phi + \lambda_0\Phi = 0$. Let $g \in G$ and define $S_g\Phi$ by the formula $S_g\Phi(x) = \frac{\Phi(g^{-1}\cdot x)}{\Phi(g^{-1}\cdot o)}$. If m_g is the representing measure of $S_g\Phi$ on the maximal Furstenberg boundary $\mathcal{F}$, then the map $g \cdot o \to m_g$ embeds X into $\mathcal{M}_1(\mathcal{F})$ and the closure of this image is again a realization of $\overline{X}^{SF}$ (see Theorem 10.9).

THE KARPELEVIČ COMPACTIFICATION $\overline{X}^K$

The Karpelevič compactification $\overline{X}^K$ was introduced by Karpelevič [K3, pp. 117–148] after his detailed study of the geometry of geodesics[3] and used by him to parametrize the minimal Martin boundary.[4]

While the original definition of the Karpelevič compactification uses induction on the rank of the symmetric space and is very complicated, here a non-inductive simple description is given. More precisely, by examining the sequences of a flat that converge in $\overline{X}^K$, a non-inductive characterization of this closure is given. After the nature of the Karpelevič topology, restricted to the flat, is clarified, an intrinsic description is given of the closure of a flat as isomorphic to a compactification of the flat determined by its polyhedral structure. Finally, it is pointed out that the Karpelevič compactification $\overline{X}^K$ can be defined by fitting together the Karpelevič compactifications of the flats in exactly the same way that the dual cell compactification is obtained from the polyhedral compactifications of the flats (see Chapter V for details). The Karpelevič compactification is fibered over $\overline{X}^c \vee \overline{X}^{SF}$ and the fibration is non-trivial in general. For a more precise description of this fibration, see Lemma 5.26, Proposition 5.27, and Corollary 5.28.

FIBERS OF MAPS BETWEEN THE COMPACTIFICATIONS

By the second part of the Main Theorem above, for $\lambda < \lambda_0$, there exist continuous surjective maps from $X \cup \partial X(\lambda)$ to $\overline{X}^c = X \cup X(\infty)$ and $\overline{X}^{SF}$

[3]The study of geodesics in [K3] plays an important role in the geometry, in particular, rigidity theory, of Hadamard manifolds and their quotients (see [B4] and the references there). The relation between geodesics and compactifications of bounded symmetric domains was also studied in [P3, Chapter 2, §9]

[4]Part (3) of the Main Theorem shows that the Karpelevič boundary is strictly bigger than the Martin boundary for spaces of higher rank.

that restrict to the identity on X. In this introduction, for simplicity, they will be denoted by p_c and p_{SF}, respectively.

A point in $\xi \in \partial X(\lambda)$ is said to be minimal if the Martin kernel K_ξ is minimal in the sense that any positive solution of $Lu + \lambda u = 0$ bounded by K_ξ is a multiple of K_ξ. The minimal Martin boundary $\partial_e X(\lambda)$ is defined to be the set of minimal points. It follows from the Main Theorem that the closure of the subset $\{\xi \in \partial X(\lambda) \mid \xi \text{ is the only point in the fiber } p_c^{-1}(p_c(\xi))\}$ in $\partial X(\lambda)$ is the minimal Martin boundary. The minimal Martin boundary is also the inverse image under p_{SF} of the Furstenberg boundary considered as the unique compact G-orbit in $\overline{X}^{SF}$.

The minimal Martin boundary was identified as a set by Karpelevič [K3] (see Chapter 13 below and Guivarch [G14] for a simpler proof). Note that to parametrize the minimal Martin boundary, both of the compactifications $\overline{X}^{SF}$ and $X \cup X(\infty)$ are needed.

The projection p_c fails to be injective only when the rank of X is greater than one, where the rank of X is the dimension of a maximal totally geodesic flat submanifold in X. In this higher rank case, $X \cup \partial X(\lambda)$ is a blow-up of $X \cup X(\infty)$. This could be explained in terms of the Tits building. If the rank of X is greater than 1, then K does not act transitively on the set $X(\infty)$ of equivalence classes of geodesics and, hence, there are different types of geodesics. This classification of geodesics into different types also gives rise to the Tits building structure on $X(\infty)$. Therefore, one needs to blow up $X(\infty)$ according to the Tits building structure or the types of the geodesics. In other words, different ideal boundary components correspond to different types of classes of geodesics. In fact, for any $z \in X(\infty)$, the fiber $p_c^{-1}(z)$ is the maximal Satake–Furstenberg compactification of a symmetric space of lower dimension (see Propositions 3.45 and 8.25). This symmetric space is called a boundary symmetric space and is defined by the parabolic subgroup of the unique simplex in the Tits building that contains z as an interior point (see Chapter III).

When the rank of X is greater than 1, the projection p_{SF} also fails to be injective. Any point $y \in \partial\overline{X}^{SF}$ is contained in the interior of a unique boundary symmetric space. This boundary symmetric space corresponds to a unique simplex of the spherical Tits building of X as in the previous paragraph. Then the fiber $p_{SF}^{-1}(y)$ can be canonically identified with the corresponding closed simplex, as follows from the proof of Theorem 8.21 (see also Proposition 3.46).

In terms of the realization of $\overline{X}^{SF}$ obtained by embedding X into the space S of closed subgroups of G, the above description of the fibers can be explained as follows. A point $\xi \in \partial X(\lambda)$ can be represented as an ordered pair $(z, D) \in X(\infty) \times \partial\overline{S_o}$ with $D \subset P_z$ the parabolic subgroup that is the stabilizer of z (see Proposition 13.31). Moreover, from the same result, the corresponding eigenfunction is $h_\xi(x) = h^D(x)e^{-\sqrt{\lambda_0-\lambda}\,\bar{d}(x,z)}$ where $e^{-\bar{d}(x,z)}$ is the Busemann cocycle and h^D is a barycenter of the square root of the Poisson kernel.

Clearly, ξ is minimal if this barycenter is trivial, in which case D is a compact extension of a maximal unipotent subgroup of G. It follows that the minimal Martin boundary $\partial_e X(\lambda)$ is equal to $p_{SF}^{-1}(\mathcal{F})$, where $\mathcal{F} \subset \partial\overline{S_0}$ is the Furstenberg boundary of X (the unique compact G-orbit of $\overline{S_0}$). If $z = p_c(\xi)$ is a generic point of $X(\infty)$, i.e., P_z is a minimal parabolic subgroup, then the condition $D \subset P_z$ shows that D contains a maximal unipotent subgroup of G as a cocompact subgroup. Hence, $\partial_e X(\lambda)$ is also the closure of the inverse image under p_c of the set of generic points in $X(\infty)$. To parametrize only $\partial_e X(\lambda)$ one needs both compactifications $\overline{X}^{SF}$ and $\overline{X}^{c}$. More precisely, each G-orbit in $\partial_e X(\lambda)$ has to be isomorphic to $\mathcal{F}$, due to the form of the minimal functions. In $X(\infty)$, the G-orbits of the generic points are isomorphic to $\mathcal{F}$, but the orbits of the non-generic points are isomorphic to the various factor spaces of $\mathcal{F}$: non-generic points exist as soon as the rank of X is at least two. Hence, in this case, in order to obtain $\partial_e X(\lambda)$ from $X(\infty)$ one needs to blow up $X(\infty)$ at the non-generic points, according to the Tits building structure. This process gives all minimal points of $\partial X(\lambda)$ but new directions are created that correspond to non-minimal eigenfunctions. In other words, one adds different boundary components for different non-generic points in $X(\infty)$. In fact, for any $z \in X(\infty)$, the fiber $p_c^{-1}(z)$ is the maximal Satake–Furstenberg compactification of the symmetric space X_z associated with the semisimple part of the parabolic subgroup P_z (see Propositions 3.45 and 8.25). It is the closure of the cell dual to the simplex defined by z.

If $b \in \mathcal{F} \subset \partial\overline{S_0}$, the fiber $p_{SF}^{-1}(b)$ is identified with the closure of a Weyl chamber at infinity (a closed simplex). More generally, if $D \in \partial\overline{S_o}$, the parabolic subgroups which contain D form a finite family. Namely, those that contain the minimum one $P(D)$ (see Corollary 14.30). Hence, the points $z \in X(\infty)$ such that $P_z \supset D$ form the closure of the face at infinity defined by $P(D)$; this face is the simplex dual to the cell of $\partial\overline{S_0}$ defined by D. Consequently, the fiber $p_{SF}^{-1}(b)$ can be identified with this closed simplex.

Application to Brownian motion

As Dynkin emphasized in his ICM talk [D3], one of the motivations for studying the Martin compactification is to determine the asymptotic behavior of Brownian motion. Using the above result on the Martin compactification of X, it is shown that almost every Brownian path in X converges to a point in $X \cup \partial X(0)$ that projects under the map $\partial X(\lambda) \to X(\infty)$ to the barycenter of a simplex of maximal dimension in the spherical Tits building.[5] This proves a result of Malliavin–Malliavin[M2] (see (§ 8.30 and § 8.31) for more details).

[5] Here the underlying topological space of the spherical Tits building is identified with the sphere at infinity $X(\infty)$ as above.

Eigenfunctions and Martin's method

Furstenberg [F3] showed how to relate the differential equation $Lf = 0$ and the convolution equation $h * p = h$, where p is a K-bi-invariant probability measure on G, for bounded f and h. Using similar ideas one gives a short proof of the Poisson formula for a semisimple group and one obtains the precise form of the minimal λ-eigenfunctions of the Laplacian. This gives new and short proofs of results of Furstenberg and Karpelevič. Further, these results show the basic role played by the maximal Furstenberg boundary $\mathcal{F} = G/P$, where P is a minimal parabolic subgroup of G. The methods of proof extend naturally to other classes of locally compact groups, including the reductive algebraic groups defined over a local field and convolution equations relative to a K-bi-invariant probability measure p.

Another proof is given of the results of Chapter VII, from the point of view of Chapter XI, with the emphasis on ground state properties. This proof also applies to the corresponding Martin compactification for K-bi-invariant random walks on reductive Lie groups, as well as on algebraic semisimple groups defined over a local field (see Chapters XII to XV).

Chapter XI is devoted to a study of Martin's method and the positive spectrum for a convolution kernel given by a well-behaved probability measure p on a locally compact group. The emphasis is on the eigenfunctions of convolution operators. For a special class of locally compact groups, including the reductive algebraic groups defined over a local field, the bottom of the positive spectrum is calculated in terms of a "Laplace transform" of p. In addition, the relation is determined of this number to the asymptotic behavior of the convolution powers p^n and the ground state.

While the results for K-bi-invariant probability measures on a semisimple group are very close to the results for the Laplacian (the set of minimal functions when the eigenvalue varies is the same), this is no longer the case for more general probability measures on a larger class of groups, including the Lie groups. Nevertheless, the results expected in such a case should not depend too much upon the choice of p, but should take into account some qualitative features of p, as well as the structural properties of the group G. Such a possible extension is one of the reasons for the probabilistic point of view taken in Chapters XI to XV.

Methods of proof

In this book, the determination of the Martin compactifications associated with the Laplacian is done in two ways. For $\lambda = \lambda_0$, by making use of the subgroups under which the possible limit functions (i.e., the Martin kernels) are invariant, and the forms of the Laplacian acting on solutions that have a suitable invariance, the limit functions are computed in Chapter VII without making use of the asymptotic behavior of the Green function G^{λ_0} of $L + \lambda_0 Id$. This result is also obtained in Corollary 14.22 by using the realization of $\overline{X}^{SF}$ as a G-space of closed subgroups (see Theorem 9.18).

However, for $\lambda < \lambda_0$ some asymptotic information about the Green function G^λ of $L+\lambda Id$ is necessary. Here use was made of work (to appear) of Anker–Ji [A6] that gives sharp estimates, accurate up to order, for the behavior of the Green function, rather than the precise asymptotics used by Olshanetsky in [O2]. In addition, one needs to use the precise form of the minimal functions, due to Karpelevič [K3], that is given in Theorem 13.23. Note that the results of Olshanetsky depend entirely upon his results about the asymptotics of the Green function. Unfortunately, this proof is incomplete.[6]

In the study of positive eigenfunctions for random walks an essential use is made of the so-called fixed-line property [C5] that was introduced by Furstenberg [F4] and Margulis [M3]. This property can be viewed as a refinement of the concept of amenability, that is of essential use in the study of bounded harmonic functions. Of course a kind of Harnack inequality is necessary for the usual limiting procedures to apply and for the study of group actions on spaces of measures. Specific tools from harmonic analysis, namely, intertwining operators, are very useful to determine which eigenfunctions are minimal. This is not surprising, since the study of positive eigenfunctions of convolution operators can be viewed as a part of representation theory, namely, that part dealing with representations that preserve a convex cone with a compact base [F4].

Open problems

The results obtained in this book suggest that many natural questions can be posed concerning Martin-type compactifications in other geometrical contexts: for example, random walks on Bruhat–Tits buildings; diffusions on Lie groups; and non-symmetric Hadamard manifolds, in particular

[6]The proof of the asymptotic behavior of the Green function off the walls of the Weyl chamber in [O2, Theorem 3.2] is incomplete. More precisely, using the notation of [O2], the proof on pp. 194–195 in [O2] only works under the assumption that $\lim_{|\tilde{r}|\to\infty} \tilde{r}/|\tilde{r}| = \tilde{v}$ exists and the direction $\tilde{v}$ belongs to the interior of the corresponding positive Weyl chamber face: in particular, the point $\tilde{v}$ is not a singular point of the corresponding c-function c_z. This is particularly obvious in [O2, Equation 3.47]. More importantly, this assumption is crucial for the interchange of the limit and the integral on p. 195. In particular, the proof does not give the asymptotics if the direction $\tilde{v}$ belongs to a face and the distance of x to that face goes to infinity. Hence, the region of validity of the asymptotics of the Green function is much smaller than the one claimed in [O2, Theorem 3.2]. Besides this problem, there are further problems in identifying the Martin compactification using [O2, Theorem 3.2]. The reason is that the asymptotics of the Green function along the walls of the Weyl chamber is not known in [O2, Theorem 3.2], and the trick on [O2, p. 200] does not overcome this problem. Since the Harish-Chandra convergent expansion of the spherical function is valid only inside the positive Weyl chamber, the assumption $x' \neq x'_0$ in [O2, Theorem 3.2] is crucial, as only the Harish-Chandra expansion is used. In fact, to get the asymptotics of the Green function on the walls, one has to use the much more complicated Trombi–Varadarajan asymptotic expansion of the spherical function on the walls. See [A6] and the references cited there for more details.

the rank one Hadamard manifolds. A short survey of open problems relating to groups is given at the end of Chapter XI.

In what follows, a construction of a compactification of a rank one Hadamard manifold is given that could be the first step towards identifying the Martin compactification of such manifolds.

A non-symmetric Hadamard manifold M is called a rank one manifold if it is not a product and admits a cocompact quotient. The problem of determining the Martin compactifications of rank one manifolds is completely open (see Yau [Y, Problem 47]).

Ballmann–Ledrappier [B6] proved that the Poisson boundary of M can be identified with the sphere at infinity $M(\infty)$. Since the Poisson boundary is contained in the Martin boundary when $\lambda = 0$, this result suggests that the Martin compactification could be larger than the conic compactification $M \cup M(\infty)$.

For every point $x \in M$, let ω_x be the harmonic measure associated to x. Denote the set of probability measures on $M(\infty)$ by $\mathcal{M}_1(M(\infty))$. Then one gets a map $x \in M \to \omega_x \in \mathcal{M}_1(M(\infty))$. If M is a symmetric space of rank one, $M(\infty)$ is the Furstenberg boundary and this map is an embedding. In general, this map may not be an embedding. However, if it is, the closure of the image of M in $\mathcal{M}_1(M(\infty))$ defines a compactification of M. To overcome this, let $M \cup \{\infty\}$ be the one point compactification. Define a diagonal map

$$\pi : M \to \mathcal{M}_1(M(\infty)) \times M \cup \{\infty\} : \ x \to (\omega_x, x).$$

Then π is an embedding and the closure of $\pi(M)$ in $\mathcal{M}_1(M(\infty)) \times M \cup \{\infty\}$ defines a compactification of M. It is a generalization of the Satake–Furstenberg compactification for symmetric spaces. For this reason, it seems reasonable to denote it by $\overline{M}^{SF}$.

Since the Dirichlet problem on $M(\infty)$ is solvable with respect to the conic compactification $M \cup M(\infty)$ (see Ballmann [B3]), it can be shown that the compactification $\overline{M}^{SF}$ dominates $M \cup M(\infty)$. If the sectional curvature is negatively pinched, it can be shown that $\overline{M}^{SF}$ is homeomorphic to $M \cup M(\infty)$, using the boundary Harnack inequality in [A4], [A2], and, in particular, bounds on the Poisson kernels. This implies that if M is a rank one manifold and contains flats of dimension greater than 1, $\overline{M}^{SF}$ could be strictly greater than $M \cup M(\infty)$.

The above analysis and the results in the book suggest that for a rank one manifold M, $\overline{M}^{SF}$ is homeomorphic to its Martin compactification (below the bottom of the spectrum). To show this it will be necessary to show that there are Martin kernels for all points on the boundary $\partial\overline{M}^{SF}$ and, in addition, the fibers of the projection map from $\overline{M}^{SF}$ to $M \cup M(\infty)$ will need to be determined. Based on the example in (Ballmann [B5]), it seems that these fibers are non-trivial only when there are many flats of dimension greater than 1 that form some complicated configurations. Another interesting problem is to understand the relation between the boundary $\partial\overline{M}^{SF}$ (or Martin boundary) and the Tits metric on $M(\infty)$ defined by Gromov in [B4].

CONVENTIONS

To conclude the introduction, one recalls briefly the basic concepts and definitions involved in this work. An attempt has been made to make the exposition relatively self-contained and easy to connect to the existing literature. Helgason [H2] [H3] and Warner [W1] are the main references used for information relating to symmetric spaces and parabolic groups.

In this book, G denotes a connected semisimple Lie group with finite center, K a maximal compact subgroup of G. In addition, in Chapter IV and Appendix B, as well as in those parts of Chapter IX referring to Satake's compactification, the group G is assumed to have no compact normal subgroups.

It is well-known that every symmetric space X of non-compact type can be written as $X = G/K$. Let $\mathfrak{g} = \mathfrak{k} + \mathfrak{p}$ be the Cartan decomposition of $\mathfrak{g}$ associated with the Cartan involution determined by K. Then T_oX can be identified with $\mathfrak{p}$, where $o = K \in X$. The Killing form B on the Lie algebra $\mathfrak{g} : B(X,Y) = \mathrm{tr}(adX \circ adY)$ is positive definite on $\mathfrak{p}$ and defines an inner product on T_oX. (The Killing form $B(X,Y)$ is also denoted by $\langle X,Y\rangle$ in this book.) Under left translation by G, this defines a G-invariant Riemannian metric on X.

The Laplace–Beltrami operator L of X is assumed to be negative and so the integral kernel $G^\lambda(x,y)$ of $(-L-\lambda Id)^{-1}$, the Green function of $L+\lambda Id$, is positive. As there is no standard notation in the literature for the Martin compactification of a space X in this book it will be denoted by $X \cup \partial X(\lambda)$. The boundary $\partial\overline{X}^{SF}$ is referred to as the maximal Furstenberg boundary instead of the maximal Satake–Furstenberg boundary in order to be consistent with the existing literature.

STUDY GUIDE

This monograph approaches the study of compactification of symmetric spaces from three distinct, closely interwoven, points of view: geometric, analytic, and probabilistic. All three make essential use of the basic information about parabolic groups that is summarized in Chapter II. Chapters VI to VII are devoted to the Martin compactifications of X associated with the Laplace–Beltrami operator. Together with Chapter X, they give the analytic point of view on compactification and form the essential core of the book, the basic motivation of the whole project. The geometric point of view is presented in Chapters III to V and brings out the rich geometric structure of the Martin compactification.

The probabilistic aspect has to do with random walks on G and on X and is to be found in Chapters XI to XV. While some of this material, especially results in Chapter XIII (proved originally by Karpelevič [K3]), is used to obtain the analytic results concerning the Martin compactifications associated with the Laplacian in Chapter VIII, it is largely independent of the earlier material. The remaining chapter, Chapter IX, is devoted to the group theoretical approach to the maximal Satake–Furstenberg compactification. It sets the stage, in part, for the results on random walks and plays an essential role in some of the proofs of Chapters III and IV.

While Chapter II is essential, some understanding of the conical compactification (see the early parts of Chapter III) and of the Satake compactification is helpful for the study of Chapters VII and VIII. In view of Remark 7.25, given this background, Chapter VII is essentially self-contained and the reader may prefer to begin with this chapter.

A more thorough understanding of the situation, especially the various modes of convergence to the boundary $\partial X(\lambda)$ requires some understanding of the Tits building structure on $X(\infty)$, as reflected in the dual cell boundary $\Delta^*(X)$, and the mixed cellular structures on $\partial(\overline{X}^{SF} \vee \overline{X}^{c})$.

Finally, Chapters XI–XV provide another perspective on the subject, thus opening the way for the investigation of compactifications of a larger class of groups than the semisimple ones defined over $\mathbb{R}$.

CHAPTER II

SUBALGEBRAS AND PARABOLIC SUBGROUPS

The key to understanding the geometrical structure of the compactifications of symmetric spaces of non-compact type is given by the family of parabolic subgroups of G. In this chapter the relation between these subgroups and sets of simple roots is discussed. Additional details for matters treated in this chapter may be found in Helgason [H2] or Warner [W1]. This chapter begins by introducing the two basic decompositions of G.

THE IWASAWA AND CARTAN DECOMPOSITIONS

§ **2.1.** Let G be a semisimple connected Lie group with finite center and let K denote a maximal compact subgroup. Let $\mathfrak{g} = \mathfrak{k} \oplus \mathfrak{p}$ denote the corresponding Cartan decomposition and θ the associated Cartan involution. There are two standard decompositions of G, both of which involve the choice of a maximal abelian subalgebra $\mathfrak{a}$ of $\mathfrak{p}$ and of a positive Weyl chamber $\mathfrak{a}^+$. To describe the first one, the so-called Iwasawa decomposition of G, fix a maximal abelian subalgebra $\mathfrak{a}$ of $\mathfrak{p}$. Let Σ denote the set of roots α of $\mathfrak{g}$ with respect to $\mathfrak{a}$ (the so-called restricted roots). The Lie algebra $\mathfrak{g} = \mathfrak{g}_0 + \sum_{\alpha \in \Sigma} \mathfrak{g}_\alpha$, where $\mathfrak{g}_\alpha \stackrel{\text{def}}{=} \{X \in \mathfrak{g} \mid [H, X] = \alpha(H)X \text{ for all } H \in \mathfrak{a}\}$ and $\mathfrak{g}_0 = \mathfrak{m} + \mathfrak{a}$, with $\mathfrak{m}$ the centralizer in $\mathfrak{k}$ of $\mathfrak{a}$.

The connected components of $\mathfrak{a} \backslash \cup_{\alpha \in \Sigma} \ker(\alpha)$ are called the **Weyl chambers** of $\mathfrak{a}$. Choose one of them to be the **positive Weyl chamber** and denote it by $\mathfrak{a}^+$. A root α is then said to be positive, denoted by writing $\alpha > 0$, if and only if it is positive on $\mathfrak{a}^+$. Let Σ^+ denote the set of positive roots.

Given this choice of $\mathfrak{a}$ and $\mathfrak{a}^+$, every element g of the group G has a unique decomposition as $g = kan, k \in K, a \in A \stackrel{\text{def}}{=} \exp \mathfrak{a}$ and $n \in N$, where N is the nilpotent Lie group with Lie algebra $\mathfrak{n} = \sum_{\alpha \in \Sigma^+} \mathfrak{g}_\alpha$. The decomposition $G = KAN$ is called an **Iwasawa decomposition** of G. If $g \in G$, let $k(g)$ denote its K-component and $H = H(g)$ denote the **logarithm** of its A-component, i.e., $g = k(g) \exp H(g)\, n$. Note that $(g, k) \to k(gk)$ defines an action of G on K, the **Iwasawa action**. Furthermore, AN is a solvable subgroup that will be denoted by S.

The roots also determine decompositions of $\mathfrak{k}$ and $\mathfrak{p}$. If $X \in \mathfrak{g}_\alpha$, then $X = \frac{1}{2}\{X + \theta(X)\} + \frac{1}{2}\{X - \theta(X)\}$. Clearly, $X + \theta(X) \in \mathfrak{k}$ and $X - \theta(X) \in \mathfrak{p}$. Let $\mathfrak{k}_\alpha = \{X + \theta(X) \mid X \in \mathfrak{g}_\alpha\}$ and $\mathfrak{p}_\alpha = \{X - \theta(X) \mid X \in \mathfrak{g}_\alpha\}$. It follows that $\mathfrak{k} = \mathfrak{m} \oplus \sum_{\alpha \in \Sigma} \mathfrak{k}_\alpha$ and $\mathfrak{p} = \mathfrak{a} \oplus \sum_{\alpha \in \Sigma} \mathfrak{p}_\alpha$.

The other Iwasawa decompositions correspond to different choices of a positive Weyl chamber in $\mathfrak{a}$, and to the other maximal abelian subalgebras of $\mathfrak{p}$. These are given by the adjoint action of K on $\mathfrak{p}$: recall that if $X \in \mathfrak{g}$, then $Ad(g)X$ is the derivative of $t \to g\exp tXg^{-1}$ at $t = 0$. For a fixed maximal abelian subalgebra $\mathfrak{a}$, the different chambers are given by the action of the **Weyl group** $W \stackrel{\text{def}}{=} M'/M$, where M' is the normalizer of $\mathfrak{a}$ in K and M is the centralizer of $\mathfrak{a}$ in K: $k \in M'$ if and only if $Ad(k)\mathfrak{a} = \mathfrak{a}$; $k \in M$ if and only if $Ad(k)H = H$ for all $H \in \mathfrak{a}$. The other maximal abelian subalgebras are obtained from $\mathfrak{a}$ by the action of K as stated in the following proposition.

2.2. Proposition. (See Helgason [H2, Lemma 6.3, p. 247], Ballman–Gromov–Schroeder [B4, p. 241]) *The maximal abelian subalgebras of* $\mathfrak{p}$ *are of the form* $Ad(k)\mathfrak{a}, k \in K$. *Hence,* $\mathfrak{p} = \cup_{k\in K} Ad(k)\mathfrak{a}$.

§ **2.3.** The second decomposition of G is the **Cartan decomposition**. It states that every element g of the group G can be written as $g = k_1 a k_2$, with $k_i \in K$ and a a unique element of $\overline{A^+ \cdot o} = \overline{\exp \mathfrak{a}^+ \cdot o}$ (see Helgason [H2, Theorem 1.1, p. 402]). Since by [H2, Theorem 1.1, p. 252] the symmetric space $X = \exp \mathfrak{p} \cdot o$, it follows from Proposition 2.2 that $X = \cup_{k\in K} k \cdot A \cdot o$. The Cartan decomposition sharpens this to show that $X = \cup_{k\in K} k \cdot \overline{A^+} \cdot o$. It implies that every point $x \in X$ may be expressed in **polar coordinates** or in **polar form** as $x = ka \cdot o$, where $a \in \overline{A^+}$ is unique.

Parabolic subgroups

Since M, the centralizer of $\mathfrak{a}$ in K, and A both normalize N, one has the following result.

2.4. Proposition. (See Helgason [H2], Warner [W1, Proposition 1.2.3.4]) $P = MAN$ *is a closed subgroup of* G. *The Lie algebra* $\mathfrak{b}$ *of* P *is* $\mathfrak{g}_0 + \sum_{\alpha\in\Sigma^+} \mathfrak{g}_\alpha = \mathfrak{m} + \mathfrak{a} + \mathfrak{n}$.

Remark. Since $\mathfrak{p}$ is used to denote the direct summand of $\mathfrak{k}$ in a Cartan decomposition, it is not appropriate to use it to denote the Lie algebra of P. It seems appropriate to use $\mathfrak{b}$ as, in the earlier literature, the group MAN is often denoted by B.

2.5. Definition. A closed subgroup P' of G is said to be **parabolic** if, for some $g \in G$, $gPg^{-1} \subset P'$. It is said to be **standard parabolic** if $P \subset P'$.

§ **2.6.** A positive root α is said to be **simple** if it cannot be written as the sum of two positive roots. Let $\Delta = \Delta(\mathfrak{g}, \mathfrak{a}^+)$ denote the set of simple roots. Then the cardinality $|\Delta|$, which equals $\dim \mathfrak{a}$, is defined to be the **rank of** G. Furthermore, if $\Delta = \{\alpha_1, \alpha_2, \ldots, \alpha_r\}$, every root $\alpha = \sum_{i=1}^n n_i\alpha_i$, with the n_i integers all having the same sign. Hence, $\alpha > 0$ if and only if all these integers are positive with at least one strictly positive.

The standard parabolic subgroups are in one-to-one correspondence with the subsets I of Δ as stated in Theorem 2.8. To explain this, some technical details about the decompositions of $\mathfrak{a}$ and $\mathfrak{n}$ are first needed.

SUBSETS OF Δ AND LIE SUBALGEBRAS

§ **2.7.** The roots α may be identified with their so-called **root vectors** $H_\alpha \in \mathfrak{a}$ by means of the Killing form $B : \alpha(H) = B(H, H_\alpha)$ for all $H \in \mathfrak{a}$.

Recall that the Killing form is positive definite on $\mathfrak{p}$ and so defines an inner product on $\mathfrak{p}$. By using the Cartan involution θ this inner product can be extended to $\mathfrak{g}$: set $B_\theta(X,Y) = -B(X,\theta(Y))$; this is an inner product that agrees with B on $\mathfrak{p}$. In what follows all notions of orthogonality will be with respect to B_θ.

Each subset I of Δ determines an orthogonal decomposition $\mathfrak{a}^I \oplus \mathfrak{a}_I$ of $\mathfrak{a}$. Define $\mathfrak{a}^I$ to be the linear span of the root vectors $H_{\alpha_i}, \alpha_i \in I$, and let $\mathfrak{a}_I$ be its orthogonal complement. As a result, $H \in \mathfrak{a}_I$ if and only if $\alpha_i(H) = 0$ for all $\alpha_i \in I$. Note that $A = A^I \times A_I$, where $A^I \overset{\text{def}}{=} \exp \mathfrak{a}^I$ and A_I is defined similarly. Further, since $\mathfrak{a}^\Delta = \mathfrak{a}_\emptyset = \mathfrak{a}$, it follows that $A = A^\Delta = A_\emptyset$.

This decomposition splits the set Σ of roots into two disjoint subsets Σ^I and Σ_I, where $\alpha \in \Sigma^I$ if and only if α vanishes on $\mathfrak{a}_I$. Let $\Sigma^{I,+}$ and Σ_I^+ denote the corresponding subsets of the set Σ^+ of positive roots. Note that a positive root that vanishes on $\mathfrak{a}_I \cap \overline{\mathfrak{a}^+}$ is necessarily in $\Sigma^{I,+}$ (see Lemma 3.5). These subsets of Σ^+ determine two subalgebras of $\mathfrak{n}$: $\mathfrak{n}^I = \sum_{\alpha\in\Sigma^{I,+}} \mathfrak{g}_\alpha$ and $\mathfrak{n}_I = \sum_{\alpha\in\Sigma_I^+} \mathfrak{g}_\alpha$. Since $[\mathfrak{g}_\alpha, \mathfrak{g}_\beta] \subset \mathfrak{g}_{\alpha+\beta}$ or equals $\{0\}$ according to whether or not $\alpha+\beta$ is a root, it follows that (i) $\mathfrak{n}^I$ and $\mathfrak{n}_I$ are subalgebras of $\mathfrak{n}$, and (ii) $[\mathfrak{n}^I, \mathfrak{n}_I] \subset \mathfrak{n}_I$.[1] Since $\mathfrak{n} = \mathfrak{n}^I \oplus \mathfrak{n}_I$, it follows that N_I is normal in N, where $N_I = \exp \mathfrak{n}_I$. Therefore, N equals the semidirect product $N^I \ltimes N_I$, where $N^I = \exp \mathfrak{n}^I$. Furthermore, $S = NA$ is also the semidirect product $S_1 \ltimes S_2$, where $S_1 = N^I A^I$ and $S_2 = N_I A_I$.

2.8. Theorem. (Standard parabolic subgroups: Moore [M8]) *If I is a subset of Δ, let P^I denote the normalizer in G of $\mathfrak{n}_I$. Then P^I is a standard parabolic subgroup. Conversely, every standard parabolic subgroup P' is of this form.*

The Lie algebra of P^I is $\mathfrak{b}_I = \mathfrak{b} + \sum_{\alpha\in\Sigma^I} \mathfrak{g}_\alpha = \mathfrak{b} + \overline{\mathfrak{n}}^I$, where $\mathfrak{b}$ denotes the Lie algebra of $P = MAN$ and $\overline{\mathfrak{n}}^I = \theta(\mathfrak{n}^I)$. Every Lie algebra containing $\mathfrak{b}$ is of this form.

The standard parabolic group P^I is the unique closed subgroup of G containing $P = MAN$ with Lie algebra $\mathfrak{b}_I$. In particular, $P^{I_1} \subset P^{I_2}$ if and only if $I_1 \subset I_2$.

Comments. The first statement is part of Theorem 3 in Moore [M8]. The second is Theorem 2 in [M8]. The final assertion follows from a consequence of Lemma 3 on p. 208 of [M8], which states that there is a fixed finite subgroup Z of M such that, for all subsets I of Δ, one has $P^I = ZP_0^I$, where P_0^I is the closed connected group with Lie algebra $\mathfrak{b}_I$.

---

[1] In the case of $\mathfrak{n}_I$ it is also necessary to use the observation that a positive root is in $\Sigma^{I,+}$ if it vanishes on $\mathfrak{a}_I \cap \overline{\mathfrak{a}^+}$.

2.9. Remark. A proof — using only the restricted roots — of the fact that the subalgebras of $\mathfrak{g}$ containing $\mathfrak{b} = \mathfrak{m} + \mathfrak{a} + \mathfrak{n}$ are of the form $\mathfrak{b}_I$ is to be found in Koranyi [K4, Lemma 2.2]. As Koranyi remarks, Theorem 2.8 can be deduced from the theory of Tits systems (see Brown [B15]). One may verify that (G, P, M', R), where R (denoted by S in [B15]) denotes the set of reflections s_{α_i} corresponding to the simple roots α_i, is in fact a so-called BN–pair (see Brown [B15, p. 107], Warner [W1, pp. 66–68]).

THE LANGLANDS DECOMPOSITION OF P^I AND THE SYMMETRIC SPACE X^I

Each parabolic group has an associated Langlands decomposition (see Corollary 2.16 and Warner [W1, p. 81]). What follows is a simplified version of the discussion in [W1] that concludes with a description of the Langlands decomposition of a standard parabolic subgroup. In the process, one defines the symmetric space X^I associated with a proper subset I of the set of simple roots. This space is one of the key ingredients in the description of the boundaries of Martin, Satake–Furstenberg and Karpelevič. The Langlands decomposition follows easily once the centralizer in G of $\mathfrak{a}_I$ has been determined. It turns out that the essential object to determine is the centralizer in K of $\mathfrak{a}_I$. Note that when $I = \emptyset$, this is exactly M. First however, one considers the centralizer $\mathfrak{z}(I)$ of $\mathfrak{a}_I$ in $\mathfrak{g}$, which is shown later to be the Lie algebra of the centralizer in G of $\mathfrak{a}_I$ (see Proposition 2.15(3)).

2.10. Proposition. *$\mathfrak{z}(I)$ is a reductive subalgebra of $\mathfrak{g}$, i.e., it is the direct sum of its center and a semisimple algebra.* (see Warner [W1, p. 42]).

Proof. Since $\mathfrak{z}(I) = \mathfrak{z}$ is stable under the Cartan involution θ given by the Cartan decomposition of $\mathfrak{g}$, $\mathfrak{z} = \mathfrak{z} \cap \mathfrak{k} \oplus \mathfrak{z} \cap \mathfrak{p}$. Further, since θ is an automorphism of $\mathfrak{g}$, it follows that the orthogonal complement $\mathfrak{z}^\perp$ of $\mathfrak{z}$ relative to B_θ is also θ invariant. The identity $B(U, [X, Y]) = B(X, [Y, U]) = B(Y, [U, X])$ (see Helgason [H2, p. 131]) implies that if $U, X \in \mathfrak{z}$ and $Y \in \mathfrak{z}^\perp$ then $B_\theta(U, [\theta(X), Y]) = -B([U, X], \theta(Y)) = 0$. Hence, if $X \in \mathfrak{z}$ and $Y \in \mathfrak{z}^\perp$ then $[X, Y] \in \mathfrak{z}^\perp$.

If $X \in \mathfrak{z} \cap \mathfrak{k}$, then $ad_{\mathfrak{g}} X$ is represented by a skew symmetric matrix (see Helgason [H2, p. 253]). Making use of a basis that is subordinate to the decomposition of $\mathfrak{g}$ as $\mathfrak{z} \oplus \mathfrak{z}^\perp$,[2] the above observation about $\mathfrak{z}^\perp$ implies that this matrix is in block-diagonal form. Hence, $(ad_{\mathfrak{g}} X)^2_{|\mathfrak{z}} = 0$ if and only if $(ad_{\mathfrak{g}} X)_{|\mathfrak{z}} = 0$, as $(ad_{\mathfrak{g}} X)_{|\mathfrak{z}}$ is skew symmetric. Similarly, for $X \in \mathfrak{z} \cap \mathfrak{p}$, $(ad_{\mathfrak{g}} X)^2_{|\mathfrak{z}} = 0$ if and only if $(ad_{\mathfrak{g}} X)_{|\mathfrak{z}} = 0$, since $(ad_{\mathfrak{g}} X)$ is represented by a symmetric matrix.

Hence, if $B_{\mathfrak{z}}$ denotes the Killing form of $\mathfrak{z}$, and X is orthogonal to the center $\mathfrak{c} = \mathfrak{c}(I)$ of $\mathfrak{z} = \mathfrak{z}(I)$ under B_θ, then $B_{\mathfrak{z}}(X, X) < 0$ if $X \in \mathfrak{z} \cap \mathfrak{k}$ and $B_{\mathfrak{z}}(X, X) > 0$ if $X \in \mathfrak{z} \cap \mathfrak{p}$.

If $X, Y \in \mathfrak{z}$, then $[\theta X, \theta Y]$ is orthogonal to $\mathfrak{c}$ since $B(Z, [X, Y]) = B(X, [Y, Z]) = 0$ if $Z \in \mathfrak{c}$. As a result, $X, Y \in \mathfrak{z}$ implies that $[X, Y]$ is

[2]In other words, a basis that is the union of a basis of $\mathfrak{z}$ and a basis of $\mathfrak{z}^\perp$.

orthogonal to $\mathfrak{c}$, from which it follows that $\mathfrak{z}' = \mathfrak{z} \cap \mathfrak{c}^\perp$ is a Lie algebra. It is semisimple in view of what has been proved.

In fact, $\mathfrak{z}'$ is the derived algebra $[\mathfrak{z}, \mathfrak{z}]$: by semisimplicity, $\mathfrak{z}' = [\mathfrak{z}', \mathfrak{z}'] \subset [\mathfrak{z}, \mathfrak{z}]$; and, by the above, $[\mathfrak{z}, \mathfrak{z}] \subset \mathfrak{z}'$. □

Remark. Note that $\mathfrak{z}'$ contains $\mathfrak{a}^I$ and $\mathfrak{n}^I$.

§ **2.11.** Let $\mathfrak{g}^I$ denote the derived algebra $\mathfrak{z}' = [\mathfrak{z}, \mathfrak{z}]$ of $\mathfrak{z}(I)$. It is semisimple by Proposition 2.10. Furthermore, it inherits a Cartan decomposition from that of $\mathfrak{g}$. Namely, if $\mathfrak{k}^I = \mathfrak{g}^I \cap \mathfrak{k}$ and $\mathfrak{p}^I = \mathfrak{g}^I \cap \mathfrak{p}$, then $\mathfrak{g}^I = \mathfrak{k}^I \oplus \mathfrak{p}^I$. Note that $\mathfrak{a}^I$ is a maximal abelian subalgebra of $\mathfrak{p}^I$ since $[\mathfrak{p}^I, \mathfrak{a}_I] = 0$, and that the (restricted) roots are the restrictions to $\mathfrak{a}^I$ of the roots in Σ^I. As a result, the orthogonal projections onto $\mathfrak{a}^I$ of the Weyl chambers in $\mathfrak{a}$ are the Weyl chambers of $\mathfrak{a}^I$. In particular, the projection onto $\mathfrak{a}^I$ of $\mathfrak{a}^+$ will be taken to be the positive Weyl chamber $\mathfrak{a}^{I,+}$ for $\mathfrak{a}^I$. The positive root spaces of $\mathfrak{g}^I$ are the $\mathfrak{g}_\alpha, \alpha \in \Sigma^{I,+}$ and the negative root spaces are the $\theta(\mathfrak{g}_\alpha), \alpha \in \Sigma^{I,+}$. The sum of the positive root spaces is $\mathfrak{n}^I$ and the sum of the negative root spaces is $\overline{\mathfrak{n}}^I$. In addition, the centralizer $\mathfrak{m}^I$ in $\mathfrak{k}^I$ of $\mathfrak{a}^I$ equals $\mathfrak{m} \cap \mathfrak{k}^I$ and $\mathfrak{g}^I = \mathfrak{m}^I \oplus \mathfrak{a}^I \oplus \mathfrak{n}^I \oplus \overline{\mathfrak{n}}^I$.

2.12. Proposition. (See Warner [W1, Lemma 1.2.4.4]) *The centralizer* $\mathfrak{m}$ *of* $\mathfrak{a}$ *in* $\mathfrak{k}$ *splits as* $\mathfrak{m} = \mathfrak{m}^I \oplus \{\mathfrak{c}(I) \cap \mathfrak{k}\}$, *where* $\mathfrak{c}(I)$ *is the center of* $\mathfrak{z}(I)$. *In addition, if* $\mathfrak{m}_I = \mathfrak{m} \oplus \mathfrak{n}^I \oplus \overline{\mathfrak{n}}^I \oplus \mathfrak{a}^I$, *then*

(1) $\mathfrak{m}_I \subset \mathfrak{z}(I)$;
(2) $\mathfrak{m}_I$ *is reductive;*
(3) *the derived algebra* $[\mathfrak{m}_I, \mathfrak{m}_I] = \mathfrak{g}^I$; *and*
(4) $\mathfrak{m}_I = \mathfrak{m} + \mathfrak{g}^I$.

Proof. The decomposition $\mathfrak{m} = \mathfrak{m}^I \oplus \{\mathfrak{c}(I) \cap \mathfrak{k}\}$ follows from the fact that $\mathfrak{m} \subset \mathfrak{z}(I) = \mathfrak{c}(I) \oplus \mathfrak{g}^I$ (see the proof of Proposition 2.10).

(1) follows from the definition of $\mathfrak{m}_I$. If X is in the center $\mathfrak{c}(\mathfrak{m}_I)$ of $\mathfrak{m}_I$, then $X \in \mathfrak{c}(I)$: if $X = X_1 + X_2, X_1 \in \mathfrak{c}(I), X_2 \in \mathfrak{g}^I$, then $X_2 \in \mathfrak{c}(\mathfrak{g}^I)$ and so $X_2 = 0$. It follows that $\mathfrak{m}_I = \mathfrak{c}(\mathfrak{m}_I) \oplus \mathfrak{g}^I$ and, hence, is reductive as $\mathfrak{g}^I$ is semisimple. (3) follows from the fact that the derived algebra of a reductive algebra is its semisimple part (see Warner [W1, p. 42]). The last statement is obvious. □

§ **2.13.** Let the connected Lie groups corresponding to the Lie algebras $\mathfrak{g}^I$ and $\mathfrak{k}^I$ be denoted by G^I and K^I. Since the Lie algebra of $K \cap G^I$ is $\mathfrak{k} \cap \mathfrak{g}^I$, it follows that $K \cap G^I = K^I$. In addition, the group K^I is maximal compact in G^I (see Helgason [H2, Theorem 1.1, p. 252]). The coset space $G^I/K^I \stackrel{\text{def}}{=} X^I$ is a symmetric space of non-compact type. This symmetric space is a subspace of $X = G/K$ since $K \cap G^I = K^I$: it is the orbit of o under G^I, i.e., $X^I = G^I \cdot o$. Note that $X = X^\Delta$.

The above choices of maximal abelian subalgebra $\mathfrak{a}^I \subset \mathfrak{p}^I$ and positive Weyl chamber $\mathfrak{a}^{I,+}$ determine the Iwasawa decomposition $G^I = K^I A^I N^I$ of G^I, where A^I and N^I are defined in § 2.7. Note that M normalizes $\mathfrak{z}(I)$

and $\mathfrak{c}(I)$. Hence, M normalizes G^I and so acts on X^I. Since it normalizes $A^I N^I$, it also normalizes K^I. As a result, $K^I M$ is a group.

2.14. Remarks. In general $\mathfrak{g}^I$ is larger than the Lie algebra $\mathfrak{g}(I)$ generated by $\mathfrak{n}_I$ and $\bar{\mathfrak{n}}_I$. This Lie algebra is defined in Warner [W1, p. 66] and is semisimple ([W1, Lemma 1.2.3.14]). If $X \in \mathfrak{g}_\alpha$ and $B_\theta(X,X) = -B(X,\theta(X)) = 1$, then $H_\alpha = [\theta(X), X]$ — it follows from the fact that $B(H,[\theta(X),X]) = B(\theta(X),[X,H])$. As a result, $\mathfrak{g}(I) \supset \mathfrak{a}^I \oplus \mathfrak{n}^I \oplus \bar{\mathfrak{n}}^I$. Hence, $\mathfrak{g}^I \cap \mathfrak{g}(I)^\perp \subset \mathfrak{m}^I$ and so $\mathfrak{m}^I + \mathfrak{g}(I) = \mathfrak{g}^I$. Consequently, $\mathfrak{m} + \mathfrak{g}(I) = \mathfrak{m} + \mathfrak{g}^I = \mathfrak{m}_I$. The Lie algebra $\mathfrak{g}(I)$ can be seen to be the algebra $\mathfrak{g}_n^I$ defined by Koranyi [K7] as the sum of the irreducible direct summands of $\mathfrak{g}^I$ that are of non-compact type. Also, if $G(I)$ is the closed connected Lie group with Lie algebra $\mathfrak{g}(I)$, it follows that $G^I \supset G(I) \supset A^I N^I$. Hence, $G(I) \cdot o = G^I \cdot o = X^I$.

2.15. Proposition. *Let $Z(I)$ denote the centralizer in G of $\mathfrak{a}_I$. Then*

(1) *$K^I M$ is the centralizer $M(I)$ of $\mathfrak{a}_I$ in K;*
(2) *$Z(I) = G^I M A_I$; and*
(3) *$\mathfrak{z}(I)$ is the Lie algebra of $Z(I)$.*

Proof. The first statement is proved later as Lemma 3.10. (Another proof is implicit in an argument of Moore [M8, Lemma 3, p. 208 and p. 216] involving the use of Furstenberg's ideas concerning the limiting behavior of sequences of measures on the Furstenberg boundary K/M, see Proposition 9.8.)

If $g = kan \in Z(I)$, then, for any $a_2 \in A_I$, $ga_2 = a_2 g$. Hence, $a_2 k a_2^{-1} \cdot a \cdot a_2 n a_2^{-1} = kan$. If $k'a'n'$ is the Iwasawa decomposition of $a_2 k a_2^{-1}$, this implies that $a' = e$ and $k' = k$. Hence, $n'a \cdot a_2 n a_2^{-1} = an$ and so

$$(*) \qquad a_2 n a_2^{-1} = a^{-1}(n')^{-1} a n.$$

By § 2.7, $n = n_1 n_2$ with $n_1 \in N^I$ and $n_2 \in N_I$. It follows from (*) that $n_2 = e$ as otherwise the left hand side of (*) varies with $a_2 \in A_I$. Hence, $n \in N^I \subset Z(I)$. As a result, $k \in Z(I)$ and so (2) follows from (1).

The last statement follows from (2), the fact that $\mathfrak{m} = \mathfrak{m}^I \oplus \{\mathfrak{c}(I) \cap \mathfrak{k}\}$ (see the proof of Proposition 2.12), and also that $\mathfrak{c}(I) = \{\mathfrak{c}(I) \cap \mathfrak{k}\} \oplus \{\mathfrak{c}(I) \cap \mathfrak{p}\} = \{\mathfrak{c}(I) \cap \mathfrak{k}\} \oplus \mathfrak{a}_I$. Observe that $\mathfrak{c}(I) \cap \mathfrak{p} = \mathfrak{a}_I$: if $X \in \mathfrak{c}(I) \cap \mathfrak{p}$, then $X \in \mathfrak{a}$ as $\mathfrak{a} \subset \mathfrak{z}(I)$; and $\mathfrak{c}(I) \cap \mathfrak{a} = \mathfrak{a}_I$. □

2.16. Corollary. (The Langlands decomposition) *Let P^I be the standard parabolic subgroup corresponding to $I \subset \Delta$, i.e., P^I is the normalizer of $\mathfrak{n}_I$. Then*

$$Z(I) \subset P^I \text{ and } P^I = M_I A_I N_I,$$

where $M_I \overset{\text{def}}{=} G^I M$.

Hence, $X = A_I N_I \cdot X^I$, i.e., for each $x \in X$, $x = a_I n_I \cdot x^I$, where a_I, n_I, and x^I (the generalized horocyclic coordinates of x, see § 7.6) are unique.

Proof. The Lie algebra $\mathfrak{b}_I$ of P^I contains $\mathfrak{m}_I$ (Proposition 2.12). Since $\mathfrak{g}^I \subset \mathfrak{m}_I$ it follows that $G^I \subset P^I$. Hence, by Proposition 2.15(2), $Z(I) \subset P^I$.

Moore proves in Theorem 3 of [M8] that $P^I = M(I)AN$. Since $M(I) = K^I M$, the Langlands decomposition follows. The observation that $X = A_I N_I \cdot X^I$ is immediate as $X^I = M_I \cdot o$. □

Remark. Moore's result that $P^I = M(I)AN$ also implies that $P^{I_1} \subset P^{I_2}$ if and only if $I_1 \subset I_2$ (see Theorem 2.8). This is a consequence of Proposition 2.15(1), since it is clear that $M(I_1) \subset M(I_2)$ if and only if $I_1 \subset I_2$.

Bruhat decompositions

2.17. Theorem. (The Bruhat decomposition) (Harish-Chandra [H1], also Warner [W1] and Helgason [H2]). *Each double coset $PgP, g \in G$ is of the form $PwP, w \in W$ and the map $w \to PwP$ is a bijection of the Weyl group W with the set of all double cosets $PgP, g \in G$.*

As usual w is realized by a representative $k_i \in M'$ (see § 3.2) and PwP denotes Pk_iP.

Warner [W1, p. 69] remarks, as is well known, that the Bruhat decomposition is a consequence of the axioms of a Tits system (see also Brown [B15]). However, for semisimple real Lie groups it is used to establish that (G, P, M', R) is a Tits system (see Warner [W1, pp. 67–68]). One standard consequence of the axioms of a Tits system is the following result.

2.18. Proposition. (See Warner [W1, Lemma 1.2.1.11]) *If $gP^{I_1}g^{-1} = P^{I_2}$, then $I_1 = I_2 = I$ and $g \in P^I$.*

Proof. The first part is exactly Warner's Lemma 1.2.1.11. The fact that $g \in P^I$ follows from a theorem of Tits ([W1, Theorem 1.2.1.1]), which states, among other things, that each parabolic subgroup is its own normalizer. □

2.19. Remark. One may avoid the use of the Tits system to prove this by first using Harish-Chandra's proof of the Bruhat decomposition and then making the following use of Lemma 3.5. If $gP^{I_1}g^{-1} = P^{I_2}$ and $g \in PkP, k \in M'$, then $kP^{I_1}k^{-1} = P^{I_2}$. It follows that $Ad(k)\mathfrak{b}_{I_1} = \mathfrak{b}_{I_2}$. For any I, the set Σ^I is the set of roots α such that $\mathfrak{g}_\alpha + \mathfrak{g}_{-\alpha} \subset \mathfrak{b}_I$. Consequently, $Ad(k)\mathfrak{a}_{I_1} = \mathfrak{a}_{I_2}$. Since $\Sigma^+_{I_1} \circ Ad(k) = \Sigma^+_{I_2}$, it follows that $Ad(k)C_{I_1} = C_{I_2}$. Hence, by Lemma 3.5, $I_1 = I_2 = I$ and $k \in (M^I)'M \subset K^I M \subset P^I$, where the last inclusion follows by Corollary 2.16.

If $I \subset \Delta$, the manifold G/P^I is called a **flag manifold**. If $g \in G$, the image gP^I of g under the canonical map $G \to G/P^I$ will be denoted by $\dot{g} = gP^I$. If $\overline{N}_I$ is the connected subgroup of G with Lie algebra $\theta(\overline{\mathfrak{n}}_I) \subset \overline{\mathfrak{n}} = \theta(\overline{\mathfrak{n}})$, the submanifold $\overline{N}_I \cdot \dot{e}$ is called the large cell of G/P^I. It can be identified with N_I since $\overline{n}_2^{-1}\overline{n}_1 \in P^I$, if $\overline{n}_i \in P^I$, if and only if $\overline{n}_1 = \overline{n}_2$. If $a \in A$ and $\eta \in \overline{N}_I$, then, under this identification, $a\eta a^{-1} \in \overline{N}_I$ is identified with $a\eta \cdot \dot{e}$, since $a\eta a^{-1} \cdot e = a\eta \cdot e$.

It will be shown that the large cell $\overline{N}_I \cdot \dot{e}$ is open and dense in G/P^I and that the map $\eta \to \eta \cdot \dot{e}$ of $\overline{N}_I$ onto $\overline{N}_I \cdot \dot{e} \subset G/P^I$ is a topological embedding.

Under the map $gP \to gP^I$, each flag manifold is the image of G/P. This flag manifold, often denoted by $\mathcal{F}$, is of central importance in this book and is called the Furstenberg boundary. It has a cellular decomposition, corresponding to the cellular decomposition of G (see [G1, pp. 76–81]), that can often be can be used to reduce calculations in flag manifolds to calculations of affine type.

2.20. Proposition. (The cellular Bruhat decomposition) *If $w \in W$, let $\overline{N}^w \overset{\text{def}}{=} \overline{N} \cap w\overline{N}w^{-1}$. Then, the map $\eta \to \eta \cdot \dot{w}$ of $\overline{N}^w$ into G/P is injective and G/P is the disjoint union of the cells $\overline{N}^w \cdot \dot{w}, w \in W$. Furthermore, if $a \in A$ and $\eta \in \overline{N}^w$, the map $\eta \to \eta \cdot \dot{w}$ identifies $a\eta a^{-1} \in \overline{N}^w$ with $a \cdot \eta\dot{w} \in G/P$.*

Proof. (See [G1] for additional details.) Since the Weyl group W is (even simply) transitive on the Weyl chambers of $\mathfrak{a}$, there is a unique element σ of the Weyl group such that $\sigma \cdot \mathfrak{a}^+ = -\mathfrak{a}^+$, equivalently, $\sigma N \sigma^{-1} = \overline{N}$. It follows from the Bruhat decomposition, Theorem 2.17, that $G = \cup_{w \in W} \overline{N} wP$. More explicitly, since $G = \sigma G = \cup_{w \in W} \sigma P w P = \cup_{w \in W} \sigma NAM \sigma^{-1} w P$, this decomposition follows since $\sigma NAM\sigma^{-1}wP = \overline{N}wP$.

This union is disjoint and may be used, as follows, to construct a decomposition of G/P into cells, each one given by a subgroup of $\overline{N}$. Since the stabilizer of wP in G is wPw^{-1}, the stabilizer of $\dot{w} = wP \in G/P$ in $\overline{N}$ is $\overline{N} \cap wPw^{-1} = \overline{N} \cap wNw^{-1}$. From this it follows that $\overline{N} \cdot wP$ can be identified with $\overline{N}/(\overline{N} \cap wNw^{-1})$.

On the other hand, since $\overline{N} = (\overline{N} \cap w\overline{N}w^{-1})(\overline{N} \cap wNw^{-1})$, it follows that $\overline{N} \cdot wP = \overline{N}^w \cdot wP$ can be identified with $\overline{N}^w$. Hence, G/P is the disjoint union of the cells $\overline{N}^w \cdot \dot{w}$. In fact, each cell is embedded topologically into G/P, as shown in [G1].

The dimension of $\overline{N} \cdot \dot{e}$ equals the dimension of G/P, while the cells $\overline{N}^w \cdot \dot{w}(w \neq e)$ have codimension at least one. It follows that $\overline{N} \cdot \dot{e}$ is open and dense in G/P.

The group A acts on $\overline{N}^w$ by conjugation since, if $a \in A$ and $w \in W$, $aw = wa'$ with $a' \in A$. Hence, A acts, as a subgroup of G acting on G/P, on each of the cells $\overline{N}^w \cdot \dot{w}$. Namely, $a \cdot \eta wP = (a\eta a^{-1}) \cdot (awP) = (a\eta a^{-1}) \cdot wP$ since $awP = wP$ as w conjugates A. □

2.21. Corollary. *The large cell $\overline{N}_I \cdot \dot{e}$ is open and dense in G/P^I.*

Proof. Since $P^I = G^I P$ and since $\overline{N} \cap G^I = \overline{N}^I = \theta(N^I)$ and $\overline{N} = \overline{N}^I \ltimes \overline{N}_I$ (see § 2.7), it follows that the projection of $\overline{N} \cdot \dot{e}$ into G/P^I equals the large cell $\overline{N}_I \cdot \dot{e}$. As $\overline{N} \cdot \dot{e}$ is open and dense in G/P, it follows that $\overline{N}_I \cdot \dot{e}$ is open and dense in G/P^I. □

CHAPTER III

GEOMETRICAL CONSTRUCTIONS OF COMPACTIFICATIONS

In this chapter several geometrical compactifications are described that are relevant to the rest of this book. The first one is the conic compactification (see § 3.1). When X is identified with $\mathfrak{p}$, it amounts to adjoining a sphere of codimension 1 at infinity to a Euclidean space in the usual way. It turns out that this sphere $X(\infty)$ at infinity may be given the structure of a simplicial complex $\Delta(X)$ (see Theorem 3.15 and Proposition 3.18) with respect to which it is a spherical Tits building (Definition 3.14). This is accomplished by identifying the sets of points in $X(\infty)$ stabilized by the various parabolic subgroups (see Proposition 3.9).

To explain this it is useful to discuss the structure of a maximal flat subspace of X, a so-called flat, and its relation to the Weyl group (see § 3.2 and Proposition 3.4). The Weyl group determines a triangulation of the unit sphere of the flat and as the flats vary (with a common base point) these triangulations are compatible and so define a triangulation $\Delta(X)$ of $X(\infty)$.

In addition, the triangulation of the unit sphere of the flat determines a dual cell complex and a corresponding compactification of the flat, the so-called polyhedral compactification. This construction of the dual cell complex, when extended to the Tits building $\Delta(X)$ so as to respect the structure of the symmetric space X, yields the dual cell complex $\Delta^*(X)$. This geometrical object is added at infinity to X to give the dual cell compactification $X \cup \Delta^*(X)$ (see Definition 3.40).

In this compactification a sequence converges to an ideal boundary point so as to respect the structure of $\Delta^*(X)$. It is important to note that this dual cell complex is not determined solely by the triangulation $\Delta(X)$. The structure of the symmetric space is involved: specifically, the relation between K, $\mathfrak{a}$, and $\mathfrak{p}$ (see Proposition 2.2 and § 2.3). This new compactification is characterized by three properties: (i) K acts continuously on it; (ii) the closure of a flat in it gives the polyhedral compactification of the flat; and (iii) the closure of the intersection of two flats is the intersection of their polyhedral compactifications (Theorem 3.39). In addition, a class of sequences is given, having explicit limits in $X \cup \Delta^*(X)$, that determines this compactification (Theorem 3.38).

A final compactification is described in which sequences converge at infinity so as to respect the structure of $\Delta^*(X)$ and the limiting direction. This compactification can be described as the least upper bound of $\overline{X}^c$ and $X \cup \Delta^*(X)$ in the partially ordered set of compactifications of X. It is denoted by $\overline{X}^c \vee (X \cup \Delta^*(X))$.

THE CONIC COMPACTIFICATION $\overline{X}^c$

§ 3.1. Let $X = G/K$ be a symmetric space of non-compact type. Any two directed unit speed geodesics $\gamma_1(t), \gamma_2(t), t \in \mathbb{R}$ are defined to be **equivalent** if

$$\lim_{t \to +\infty} d(\gamma_1(t), \gamma_2(t)) < +\infty.$$

Denote by $[\gamma]$ the equivalence class of the geodesic γ, and let $X(\infty)$ be the set of the equivalence classes of geodesics in X. One puts a metrizable topology on $X \cup X(\infty)$ as follows. First, choose a base point o. Then, a sequence $\{x_n\}$ in X, converging to infinity, converges to $[\gamma] \in X(\infty)$ if and only if the directed geodesic from o to x_n converges to a geodesic that is equivalent to γ. Secondly, in the topology on $X(\infty)$, a sequence of equivalence classes $[\gamma_n] \in X(\infty)$ converges to $[\gamma] \in X(\infty)$ if and only if there are representatives $\gamma_n(t)$ and $\gamma(t)$ such that for any $t \in \mathbb{R}$, $\gamma_n(t) \to \gamma(t)$ as $n \to +\infty$. Finally, the induced subset topology on X coincides with its original topology.

This is a compact topology, which will be called the **conic topology**, and the compactification $X \cup X(\infty)$ is called the **conic compactification** of X (see Ballman–Gromov–Schroeder [B4] and Eberlein–O'Neill [E1]).[1] It will also be denoted by $\overline{X}^c$. Since the Lie group G acts on X isometrically, the G-action preserves the equivalence classes and, hence, extends to a continuous action on the conic compactification $\overline{X}^c$. More specifically, if γ is a geodesic and $g \in G$, let $(g \cdot \gamma)(t) \stackrel{\text{def}}{=} g \cdot \gamma(t)$. This G-action preserves equivalence classes. Define $g \cdot [\gamma]$ to be $[g \cdot \gamma]$. This defines an action of G on $X(\infty)$ that is a continuous extension of the G-action on X. This also shows that the conic compactification is independent of the base point o.

The boundary $X(\infty)$ is called the sphere at infinity, and can be given another more concrete characterization that is useful for later purposes. Take $o = K$ as the base point in $X = G/K$. For any $[\gamma] \in X(\infty)$ there is a unique unit vector L in the tangent space T_oX such that the geodesic from o with initial tangent vector L is equivalent to γ. Therefore, there is a bijective correspondence between $X(\infty)$ and the unit sphere in T_oX with its center at the origin, and this map is a homeomorphism: if $L \in T_oX$ is a unit vector (a direction), let $L(\infty)$ denote the corresponding point of $X(\infty)$, i.e., $L(\infty) = [\gamma]$, where γ is the directed geodesic from o with tangent vector L at o.

The tangent space T_oX is isomorphic to $\mathfrak{p}$: each $Y \in \mathfrak{p}$ defines the curve $t \to \exp tY \cdot o$ on X and its derivative at $t = 0$ can be identified with Y. With this identification, the exponential map $Y \to \exp Y \cdot o$ is a diffeomorphism of $\mathfrak{p}$ with X (see Helgason [H2, Theorem 1.1, p. 252]) and the conic compactification of X corresponds to the compactification of $\mathfrak{p}$

[1] Since a basis of neighborhoods of the boundary points in $X(\infty)$ can be given by truncated cones based at any given point in X, the topology on $X \cup X(\infty)$ is called the conic topology (see Anderson–Schoen [A4]) and, for this reason, the compactification $X \cup X(\infty) = \overline{X}^c$ is called the conic compactification.

obtained by adding the unit sphere at infinity in the usual way. From now on, $X(\infty)$ will often be identified with the unit sphere in T_oX and, thereby, with the unit sphere in $\mathfrak{p}$.

Using this description, the topology on $X \cup X(\infty)$ is characterized by the fact that a sequence (x_n) in X, converging to infinity, converges to $L(\infty) \in X(\infty) = T_oX$ if and only if the directed geodesic from o to x_n converges to the directed geodesic from o whose tangent vector at o is equal to $L \in T_o(X)$.

Remarks. The **Busemann function** $d_\gamma(x) \stackrel{\text{def}}{=} \lim_{t\to\infty} d(x,\gamma(t)) - t$, where γ is a unit speed geodesic with $\gamma(0) = o$, depends only on $[\gamma]$. Let $d_z(x) \stackrel{\text{def}}{=} d_\gamma(x)$ if $z = [\gamma]$. The set of Busemann functions d_z is thereby identified with the sphere $X(\infty)$ at infinity, since $\gamma(0) = o$ selects a unique representative of $[\gamma]$. The conic compactification $X \cup X(\infty)$ can be identified with a compactification defined in terms of the Busemann functions (see [B4, p. 27]). These functions play a key role in relating the minimal functions for the Martin compactifications $X \cup \partial X(\lambda_0)$ and $X \cup \partial X(\lambda)$ (see the remark following Proposition 8.15, and also Theorem 13.28).

When $\dot\gamma(0) = L \in \mathfrak{a}^+$, then $d_\gamma(x) = -B(L, \log a) = B(L, \log a^{-1}) = B(L, H((na)^{-1}))$ if $x = na\cdot o$. To see this, let $a_t = \exp tL$ and $n_t = a_t^{-1} n a_t$. Then (i) $n_t^{-1} \to e$ as $t \to \infty$ and (ii) $d(na\cdot o, a_t\cdot o) - t \to -B(L,H)$ if $a = \exp H$. The first statement follows by a standard argument (used, for example, in the proof of Proposition 7.20).

The following calculation verifies the second statement: $d(na\cdot o, a_t\cdot o) - t = d(a_t^{-1}a\cdot o, n_t^{-1}\cdot o) - t = d(a\cdot o, a_t\cdot o) - t + \{d(a_t^{-1}a\cdot o, n_t^{-1}\cdot o) - d(a_t^{-1}a\cdot o, o)\} = \sqrt{B(H-tL, H-tL)} - t + \{d(a_t^{-1}a\cdot o, n_t^{-1}\cdot o) - d(a_t^{-1}a\cdot o, o)\}$. The term in the brackets goes to zero and $\sqrt{B(H-tL,H-tL)} - t \to -B(L,H)$ by the usual calculation in Euclidean space. If $x = g\cdot o$, then $d_\gamma(x) = B(L, H(g^{-1}))$: let $g^{-1} = k'a'n'$; then $g\cdot o = na\cdot o$ with $n = (n')^{-1}$ and $a = (a')^{-1}$.

By continuity of the normalized Busemann function $d_z(x)$ in z, the same formula holds for d_z when $\dot\gamma(0) \in \overline{\mathfrak{a}^+}$. (For a proof see Proposition 13.26.) Note that $d_{k\cdot\gamma}(x) = d_\gamma(k^{-1}\cdot x)$. As a result, if $z = k\cdot L(\infty) \in X(\infty)$, with L a unit vector in $\overline{\mathfrak{a}^+}$, then $d_z(g\cdot o) = B(L, H(g^{-1}k))$.

THE CONICAL DECOMPOSITION OF $\mathfrak{a}$ AND THE WEYL GROUP

The information in this section is discussed in full generality by Brown in Chapter I of [B15]. However, some of the more specific information (e.g., Proposition 3.4) is important for later purposes.

§ **3.2.** As stated in § 2.1 the Weyl chambers of $\mathfrak{a}$ are the connected components of $\mathfrak{a} \setminus \bigcup_{\alpha\in\Sigma} \ker(\alpha)$. One of them was chosen to be the positive Weyl chamber $\mathfrak{a}^+$. The other Weyl chambers are obtained by the adjoint action of the normalizer M' of $\mathfrak{a}$ in K, see § 2.1. Recall that the group M'/M is called the **Weyl group of the pair** (G,K) and is denoted by

W. If $k \in M'$ and α is a root, then $\beta = \alpha \circ Ad(k^{-1})$ is also a root. If $kM = w \in W$, one denotes β by $w \cdot \alpha$.

It is known that W is a finite group that acts simply transitively on the set of Weyl chambers (see Helgason [H2, Theorem 2.12, p. 288], and use the duality between semisimple Lie algebras of compact and non-compact type).

Under the pairing of $\mathfrak{a}$ and $\mathfrak{a}^*$ given by the Killing form, the group M'/M is isomorphic to the finite reflection group that is generated by the reflections s_{α_i} with respect to the simple roots α_i (see Warner [W1, p. 13]) — the Weyl group of the root system Σ in $\mathfrak{a}^*$. In other words, W can be realized on $\mathfrak{a}$ as the finite reflection group whose generators are the reflections with respect to the hyperplanes orthogonal to the root vectors H_{α_i}.

Let $|W|$ be the cardinality of W and let $\{k_1, k_2, \dots, k_{|W|}\}$ be a complete set of representatives of the Weyl group $W = M'/M$. Then the convex cones $Ad(k_i)\mathfrak{a}^+, 1 \le i \le |W|$, are the (pairwise disjoint) Weyl chambers. Their union is dense and if $H \in \mathfrak{a}$ is not in a chamber it is in the boundary of a chamber. The boundary of the positive Weyl chamber is described by convex cones corresponding to the non-void proper subsets I of the set of simple roots as follows.

3.3. Definition. Let I be a proper subset of $\Delta = \Delta(\mathfrak{g}, \mathfrak{a}^+)$. Define the **(Weyl) chamber face** C_I to be $\{H \in \overline{\mathfrak{a}^+} \mid \alpha_i(H) > 0$ if and only if $\alpha_i \notin I\}$.

Remarks. (1) $C_I = \{H \in \overline{\mathfrak{a}^+} \mid \alpha_i(H) = 0$ if and only if $\alpha_i \in I\}$.
(2) Since $C_I = \left(\cap_{\alpha_i \notin I}\{\alpha_i > 0\}\right) \cap \mathfrak{a}_I$, it is an open subset of $\mathfrak{a}_I$. Hence, C_I generates $\mathfrak{a}_I$.
(3) Note that $C_\emptyset$ equals $\mathfrak{a}^+$ and $C_\Delta = \{0\}$.
(4) If $H \in \mathfrak{a}$ and $\subset \Delta$, let H^I denote the projection of H on $\mathfrak{a}^I$ and let H_I denote its projection on $\mathfrak{a}_I$. Then $\mathfrak{a}^{I,+} \overset{\text{def}}{=} \{H^I \mid H \in \mathfrak{a}^+\}$ (see earlier in § 2.11) equals $\{H \in \mathfrak{a}^I \mid \alpha_i(H) > 0$ for all $\alpha_i \in I\}$: if $H \in \mathfrak{a}^+$ and $\alpha_i \in I$ then $\alpha(H_I) = 0$ and so $\alpha_i(H^I) > 0$; conversely, if $H \in \mathfrak{a}^{I,+}$ and $H_0 \in C_I$, it follows that, for large n, $H + nH_0 \in \mathfrak{a}^+$ and $(H + nH_0)^I = H$.

It follows that the boundary of the positive Weyl chamber $\mathfrak{a}^+$ is the disjoint union of the chamber faces C_I as I runs over the non-void subsets of the set Δ of simple roots. Similarly, the boundary of a Weyl chamber $Ad(k_i)\mathfrak{a}^+$ is the disjoint union of the chamber faces $Ad(k_i)C_I$.

Furthermore, if $0 \neq H \in \mathfrak{a}$, then H either lies in a chamber $Ad(k_i)\mathfrak{a}^+$ or in a chamber face $Ad(k_i)C_I, I \neq \Delta, \emptyset$. The chamber faces and the chambers are all convex cones that are open in the linear subspaces they generate.

It follows from the next result that these convex cones constitute a polyhedral cone decomposition of $\mathfrak{a}$ in the sense of Gerardin [G2] (also Taylor [T4]) — see Definition 3.23.

3.4. Proposition. *Two chamber faces $Ad(k_i)C_{I_1}$ and $Ad(k_j)C_{I_2}$ are either disjoint or coincide. They coincide if and only if $I_1 = I_2 = I$ and*

$k = k_j^{-1} k_i \in (M^I)'/M^I$, *where* $(M^I)'/M^I$ *is the* **Weyl group** W^I **of** $\mathfrak{a}^I$.

This result is an easy consequence of Propositions 2.18 and 3.9, whose proofs do not depend upon Proposition 3.4. It may also be proved directly, without using parabolic subgroups. Before beginning this direct argument, it will be useful to consider two lemmas. Recall that Σ^I is the set of roots α that vanish on $\mathfrak{a}_I$ (see § 2.7).

3.5. Lemma. (See Warner [W1, Lemma 1.2.4.1(i)]) *Let* $H \in C_I$ *and* $\alpha \in \Sigma$. *Then*

(1) $\alpha(H) = 0$ *if and only if* $\alpha \in \Sigma^I$; *and*
(2) $\alpha(H) > 0$ *if and only if* $\alpha \in \Sigma_I$ *and* $\alpha > 0$.

Furthermore, (1) *and* (2) *are equivalent statements.*

Given any finite set of non-trivial linear functionals on a real vector space, the connected components of the complement of the union of their kernels are convex cones that will be called **chambers**.

In this way, the roots $\beta \in \Sigma_I$, restricted to $\mathfrak{a}_I$, determine chambers in $\mathfrak{a}_I$ since none of them vanish on this subspace (see the proof of Theorem 3 in Moore [M8]). Clearly, each chamber face C_I is a chamber in $\mathfrak{a}_I$.

3.6. Lemma. *If* $k \in M'$ *and* $Ad(k)\mathfrak{a}_{I_1} = \mathfrak{a}_{I_2}$, *where* I_1, I_2 *are two proper non-void subsets of the set of simple roots, then* $Ad(k)$ *maps the chambers of* $\mathfrak{a}_{I_1}$ *onto the chambers of* $\mathfrak{a}_{I_2}$.

Proof. $\beta \in \Sigma_{I_2}$ if and only if $\beta \circ Ad(k) \in \Sigma_{I_1}$. □

Proof of Proposition 3.4.

If the conditions are satisfied, then, clearly, $Ad(k_1)C_{I_1} = Ad(k_2)C_{I_2}$. Now suppose $H_2 \in Ad(k)C_{I_1} \cap C_{I_2}, k \in M'$. Let $H_1 = Ad(k^{-1})H_2$. Since $\mathfrak{a}^I$ is generated by the root vectors $H_{\alpha_i}, \alpha_i \in I$, it follows from Lemma 3.5, applied to H_1 and H_2, that $Ad(k)\mathfrak{a}^{I_1} = \mathfrak{a}^{I_2}$. Consequently, $Ad(k)\mathfrak{a}_{I_1} = \mathfrak{a}_{I_2}$. Hence, if $Ad(k)C_{I_1} \cap C_{I_2} \neq \emptyset$, Lemma 3.6 implies that $Ad(k)C_{I_1} = C_{I_2}$.

Let $w \in W$ be such that $w.\alpha = \alpha \circ Ad(k)$. It follows from Lemma 3.5 that the only simple roots α_i for which the sign of $w.\alpha_i$ is negative belong to I_2. Let w_2 be the element of the Weyl group W^{I_2} that changes the signs of the simple roots in I_2 in exactly the same way — recall that W^I denotes the Weyl group of the pair (G^I, K^I). Since the Weyl group W is simply transitive on the Weyl chambers and $W^{I_2} \subset W$, $w = w_2$. This implies that $k \in (M^{I_2})'M^{I_2}$. Consequently, $Ad(k^{-1})\mathfrak{a}_{I_2} = \mathfrak{a}_{I_2}$. Hence, $I_1 = I_2$. □

It turns out that the chambers and their faces determine a triangulation of the unit sphere in $\mathfrak{a}$. This result (proved below) is a special case of the fact that the chambers and faces associated to any finite reflection group determine a triangulation of the unit sphere (see Brown [B15, Chapter I, §5]). This is not true of polyhedral cone decompositions in general.

3.7. Proposition. *The Weyl chambers and their chamber faces induce a triangulation of the unit sphere in* $\mathfrak{a}$. *More explicitly, define the simplex* $\sigma(k_i, I)$ *to be the intersection of* $Ad(k_i)C_I$ *with the unit sphere. Then* *these simplices form a simplicial complex with $\sigma(k_i, I)$ a face of $\sigma(k_j, J)$ if $\sigma(k_i, I) \subset \overline{\sigma(k_j, J)}$, equivalently, $Ad(k_i)C_I \subset \overline{Ad(k_j)C_J}$.*

Proof. Recall that $C_\emptyset = \mathfrak{a}^+$. Assume $Ad(k_i)C_I \subset \overline{Ad(k_j)C_J}$. Then, if $k = k_j^{-1}k_i, Ad(k)C_I \subset \overline{C_J}$. As a result, either (i) $Ad(k)C_I = C_J$ or (ii) $Ad(k)C_I = C_{I_1}, I_1 \supset J$. It follows from Proposition 3.4 that, in case (i), $I = J$ and $k = k_j^{-1}k_i \in K^I M$ and, in case (ii), $I = I_1$ and $k = k_j^{-1}k_i \in K^I M$. Hence, $Ad(k_i)C_I \subset \overline{Ad(k_j)C_J}$ implies $I \supset J$ and $k_j^{-1}k_i \in K^I M$.

Consequently, the set of all faces of a simplex $\sigma(k_j, J)$ is isomorphic with the partially ordered set of subsets of $\Delta \backslash J$.

It remains to show (see Brown [B15, p. 27]) that any two simplices have a largest common face (one may obviously exclude the cases corresponding to $Ad(k_i)C_I \subset \overline{Ad(k_j)C_J}$ or vice versa). If their closures are disjoint, this is so trivially. Assume $\overline{Ad(k_i)C_I} \cap \overline{Ad(k_j)C_J} \neq \emptyset$. Then, by Proposition 3.4, if $k = k_j^{-1}k_i$ either $Ad(k)C_I = C_J$ or they are disjoint. In the first instance the two simplices are the same. Assume they are distinct. Then $Ad(k)C_I \cap C_J = \emptyset$, and $\overline{Ad(k)C_I} \cap \overline{C_J} \subset \partial(Ad(k)C_I) \cap \partial(C_J)$.

Applying Proposition 3.4 again, one sees that, if $H \in \overline{Ad(k)C_I} \cap \overline{C_J}$, the chamber face containing H also lies in the intersection. Consequently, the intersection $\overline{Ad(k)C_I} \cap \overline{C_J}$ is a union of chamber faces. Choose one of maximal dimension, say C_{I_1}. Since the dimension of C_{I_1} is maximal and the intersection $\overline{Ad(k)C_I} \cap \overline{C_J}$ is convex, $\overline{C_{I_1}} = \overline{Ad(k)C_I} \cap \overline{C_J}$. Hence, if $k_1 \in M$, the simplex $\sigma(k_1, I_1)$ is the largest common face of $\sigma(k, I)$ and $\sigma(k_1, J)$. □

Parabolic subgroups and stabilizers of the points in $X(\infty)$

The group action on $X(\infty)$ relates the boundary points of X to parabolic subgroups by the following fact.

3.8. Proposition. (See [B4, pp. 248–249]) *A closed subgroup of G is parabolic if and only if it is the isotropy group P_z of a point $z \in X(\infty)$.*

The proof of this result for a standard parabolic group P^I makes use of information in the previous section about the geometrical structure determined on a flat $A \cdot o$ (see Definition 3.11) — equivalently on $\mathfrak{a}$ — by the Weyl chambers and their chamber faces. While the proof for the general parabolic group gP^Ig^{-1} may be obtained formally by moving the base point o to $g \cdot o = gK$ and using the Iwasawa decomposition given by conjugating $G = KAN$ with g, it follows as an immediate corollary of the result for a standard parabolic group (see later) since $gP^Ig^{-1} = k(g)P^Ik(g)^{-1}$.

Proposition 3.8 can be sharpened in the case of a standard parabolic group to include the identification of the class of geodesics stabilized by P^I.

3.9. Proposition. *The standard parabolic subgroup P^I is the stabilizer of $[\gamma]$ if $\gamma(t) = \exp tH \cdot o$, with $H \in C_I$ of unit length.*

Proof. In [B4, p. 245] it is shown that, for any geodesic γ from o directed by a unit vector H in $\overline{\mathfrak{a}^+}$, $[an \cdot \gamma] = [\gamma]$ if $an \in AN$. Consequently, for such a geodesic, if $[g \cdot \gamma] = [\gamma]$ and $g = kan$ then $[k \cdot \gamma] = [\gamma]$.

Since $P^I = K^I MAN$ (see the proof of Corollary 2.16), it is clear that P^I stabilizes γ if $\gamma(t) = \exp tH \cdot o, H \in C_I$. The stabilizer of $[\gamma]$ is therefore a standard parabolic subgroup P' that contains P^I.

If $[k \cdot \gamma] = [\gamma]$, then $d(\exp tAd(k)H \cdot o, \exp tH \cdot o), t > 0$ is bounded. In view of the non-positivity of the curvature of X this is only possible if $Ad(k)H = H$.

The following lemma shows that the Lie algebra of $P' \cap K$ is contained in $\mathfrak{b}_I$. Given this, it follows from Theorem 2.8 that $P' \subset P^I$. □

3.10. Lemma. *Let $H \in C_I$ and $Ad(k)H = H$. Then $k \in K^I M$. Hence, the centralizer $M(I)$ in K of $\mathfrak{a}_I$ is $K^I M$.*

Proof. (See [B4, pp. 248–249].) The exponential map $\mathfrak{k} \to K$ is surjective, see [H2, Proposition 6.10, p. 135] or [H5, Theorem 3.2, p. 150]. Hence, it suffices to show that, if $k = \exp U$ then $U \in \mathfrak{b}_I \cap \mathfrak{k} = \mathfrak{k}^I + \mathfrak{m}$.

Now it is clear that $Ad(k)H = H$ if and only if $(\exp -H)k \exp H = k$, i.e., if and only if $e^{-adH}U = U$. Let $U = \sum_{\alpha \in \Sigma} U_\alpha + U_0$ with $U_\alpha \in \mathfrak{k}_\alpha$ (see § 2.1). Since $H \in C_I$, it follows from Lemma 3.5 that $e^{-adH}U = U$ if and only if $U_\alpha \neq 0$ implies $\alpha \in \Sigma^I$. Hence, $U \in \mathfrak{b}_I \cap \mathfrak{k} = \mathfrak{k}^I + \mathfrak{m}$. □

Remark. Another proof, using an argument due to Moore [M8], is given later (see Proposition 9.8). It involves the use of Furstenberg's boundary theory.

FLATS THROUGH THE BASE POINT AND PROPOSITION 3.8

3.11. Definition. A **flat** F in X is a complete, totally geodesic flat submanifold of maximal dimension. This dimension is independent of the flat, and is called the **rank of** X, denoted by $r(X)$. The rank of G, as defined earlier, equals the rank of X, and is denoted by $r(G)$.

If the rank of X is equal to one, then the flats are geodesics, and the structure of X is simple.

The main interest here is in the higher rank case. Fix the basepoint in X, as before, to be $o = K$. Any geodesic through o is contained in a flat. If it is contained in only one flat, the geodesic is called **regular**, otherwise it is called **singular**. Similarly, any point $x \in X$ different from o is called regular (or singular) if the geodesic determined by o and x is regular (or singular, respectively). The structure of the sets of regular and of singular points can be described explicitly in terms of maximal abelian subalgebras of $\mathfrak{p}$ and their associated roots, and plays an important role in the construction of the compactification $X \cup \Delta^*(X)$ that will be described later (see also the proof of Theorem A.5 in Appendix A). This structure for a fixed flat was described in the discussion of the conical decomposition of $\mathfrak{a}$ (see § 3.2, Definition 3.3, and Proposition 3.4).

The maximal abelian subalgebras of $\mathfrak{p}$ are involved because every flat through o is the image under the exponential map of such a subalgebra. Hence, $A \cdot o$ is a flat through o and, in view of Proposition 2.2, the flats through o are the translates $kA \cdot o$ by $k \in K$ of the basic flat $A \cdot o$. In other words, they are the images of the linear subspaces $Ad(k)\mathfrak{a}$ of $\mathfrak{p}$ under the exponential map, for some $k \in K$. The regular points in a flat are those that correspond to points in the Weyl chambers and the singular ones to points in the chamber faces.

Proof of Proposition 3.8.

It follows from the Cartan decomposition that every point $z \in X(\infty)$ is of the form $k \cdot u$, with u given by a unit vector L in $\overline{\mathfrak{a}^+}$. Consequently, there is a unique subset I of Δ such that $L \in C_I$. Since, by Proposition 3.9, the stabilizer of u is P^I, it follows that kP^Ik^{-1} is the stabilizer of z. On the other hand, if P' is a parabolic group then, for some subset I of Δ, it is of the form $kP^Ik^{-1}, k \in K$ in view of the Iwasawa decomposition for an arbitrary group element. Hence, it fixes all the points of the form $k \cdot u$, where u is given by a unit vector in C_I. □

§ **3.12.** It follows from this proof and Proposition 3.9 that a parabolic subgroup is the stabilizer of every point in a specified subset of $X(\infty)$; in the case of P^I this consists of all points corresponding to a unit vector in C_I and in the case of kP^Ik^{-1} of all points corresponding to unit vectors in $Ad(k)C_I$. For any convex cone $J \subset \mathfrak{p}$ that contains no line, let $J(\infty)$ denote the points in $X(\infty)$ corresponding to the unit vectors in J. Using this notation, the stabilizer of every point of $C_I(\infty)$ (respectively, of $Ad(k)C_I(\infty)$) is P^I (respectively, kP^Ik^{-1}). Note that, in view of the action of G on $X(\infty)$, see § 3.1, $Ad(k)C_I(\infty) = k \cdot C_I(\infty)$.

These subsets of $X(\infty)$ in fact form a simplicial complex of subsets as shown in Proposition 3.18. It is the so-called **Tits building** $\Delta(X)$ of X.

3.13. Definition. The subset $k \cdot C_I(\infty) = Ad(k)C_I(\infty)$ of $X(\infty)$, where $I \subset \Delta$ and $k \in K$, is said to be a **Weyl chamber face at infinity** if $I \neq \Delta$ or $\emptyset$ and to be a **Weyl chamber at infinity** if $I = \Delta$.

In other words, the Weyl chamber faces at infinity form the Tits building $\Delta(X)$.

THE TITS BUILDING $\Delta(G)$ OF G AND ITS GEOMETRICAL REALIZATION $\Delta(X)$

3.14. Definition. (See Brown [B15, p. 76], **Tits** [T6, p. 38]) A **spherical Tits building** is a simplicial complex Δ that can be expressed as the union of a collection of subcomplexes Σ, which are called apartments, such that:

(1) each apartment Σ is a finite Coxeter complex determined by a finite reflection group and, hence, is a triangulation of a sphere;

(2) for any two simplexes $A, B \in \Delta$, there is an apartment Σ containing both of them; and
(3) if Σ and Σ' are two apartments containing A and B, then there is an isomorphism between Σ and Σ' that fixes A and B pointwise.

Since (G, P, M', R) is a Tits system, the set $\Delta(G)$ of all parabolic subgroups of G ordered by the opposite of the inclusion relation, where G is included as a special parabolic subgroup, is a simplicial complex. Define an apartment to be the set $\Sigma(g), g \in G$, of parabolic subgroups that contain gAg^{-1}. Note that, for a parabolic subgroup P', one has $P' \supset A$ if and only if $P' = \ell P^I \ell^{-1}$ for some $\ell \in M'$ and $I \subset \Delta$. In other words, the apartment corresponding to the group A, which consists of the parabolic groups that contain A, is exactly $\{\ell P^I \ell^{-1} \mid I \subset \Delta, \ell \in M'\}$.

To prove this, first observe that if $P' = kP^I k^{-1}, k \in K$, and $P' \supset A$, then $\mathfrak{b}_I = \mathfrak{m}_I \oplus \mathfrak{a}_I \oplus \mathfrak{n}_I \supset Ad(k^{-1})\mathfrak{a}$. Since $\mathfrak{b}_I$ can also be written as $\mathfrak{b}_I = \mathfrak{m}(I) \oplus \mathfrak{a} \oplus \mathfrak{n}$, where $\mathfrak{m}(I)$ is the Lie algebra of $K^I M = M(I)$, it follows that if $X \in \mathfrak{b}_I \cap \mathfrak{p}$, then $X \in \mathfrak{a}$. To see this observe that if $X = U + H + Y, U \in \mathfrak{m}(I), H \in \mathfrak{a}_I$, and $Y \in \mathfrak{n}_I$, then $U + Y \in \mathfrak{p}$ and so $\theta(U) + \theta(Y) = U - Y$. As a result, $Y = -\theta(Y) \in \bar{\mathfrak{n}}$ and so $Y = 0$. Consequently, $X = U + H$, which implies, by the Cartan decomposition, that $X = H$. As a result, if $P^I \supset kAk^{-1}$, it follows that $\mathfrak{a} \supset Ad(k)\mathfrak{a}$. This implies that $k \in M'$.

Another proof of this fact is to be found in Brown [B15, pp. 112–113] as part of the proof the next result.

3.15. Theorem. (See Tits [T6, Theorem 5.2] and Brown [B15, p.112]) *The abstract simplicial complex $\Delta(G)$, with apartments taken to be the sets $\Sigma(g), g \in G$, is a spherical Tits building.*

There is an intimate relation between the parabolic subgroups of G and the Weyl chamber faces at infinity that is stated as follows.

3.16. Proposition. *For any two points $z_1, z_2 \in X(\infty)$ the following are equivalent:*

(1) *their stabilizers in G are equal;*
(2) *they lie in the same Weyl chamber face at infinity. In other words, there exist $k \in K$ and $I \subsetneq \Delta$ such that $z_1, z_2 \in k \cdot C_I(\infty)$.*

Proof. Proposition 3.9 shows that (2) implies (1). Let $z_i \in k_i \cdot C_{I_i}(\infty)$. Then $k_1 P^{I_1} k_1^{-1} = k_2 P^{I_2} k_2^{-1}$. Hence, by Proposition 2.18, $I_1 = I_2 = I$ and $k_1^{-1} k_2 \in K^I M$. Therefore, $k_1 \cdot C_I(\infty) = k_2 \cdot C_I(\infty)$. □

From this it follows that the map $kP^I k^{-1} \to k \cdot C_I(\infty)$ is a bijection of $\Delta(G)$ with the set $\Delta(X)$ of Weyl chamber faces at infinity.

3.17. Definition. Given two Weyl chamber faces $F_1 = k_1 \cdot C_{I_1}(\infty)$ and $F_2 = k_2 \cdot C_{I_2}(\infty)$, F_1 is said to be a **face** of F_2 if $F_1 \subset \overline{F_2}$.

The next result implies that, under the partial order "face of", the set $\Delta(X)$ of Weyl chamber faces is a simplicial complex that is a spherical building isomorphic to the building $\Delta(G)$. The family of apartments in $\Delta(X)$ corresponding to the flats $F = k \cdot A \cdot o, k \in K$, determines the simplicial complex $\Delta(X)$. Each of these apartments is isomorphic to the triangulation of the unit sphere in $Ad(k)\mathfrak{a}$, $k \in K$, induced by the Weyl chambers and their chamber faces.

3.18. Proposition. *For any two Weyl chamber faces at infinity* $F_1 = k_1 \cdot C_{I_1}(\infty)$ *and* $F_2 = k_2 \cdot C_{I_2}(\infty)$, F_1 *is a face of* F_2 *if and only if* $P_1 = k_1 P^{I_1} k_1^{-1} \supset k_2 P^{I_2} k_2^{-1} = P_2$. *In particular, the Weyl chambers at infinity correspond to the minimal parabolic subgroups.*

Proof. If $z_1 \in F_1$ and $F_1 \subset \overline{F}_2$, then, by continuity, the group P_2 stabilizes z_1. Hence, $P_1 \supset P_2$. Conversely, if $P_1 \supset P_2$, the face F_1 is stabilized by P_2. This implies that $F_1 \subset \overline{F}_2$ in view of the following lemma. □

3.19. Lemma. *The standard parabolic group* P^I *stabilizes* $z \in X(\infty)$ *if and only if* $z \in \overline{C_I(\infty)}$.

Proof. If $z \in k_0 \cdot C_{I_0}(\infty)$ is stabilized by P^I, then $k_0 P^{I_0} k_0^{-1} \supset P^I$. It follows that $k_0 P^{I_0} k_0^{-1}$ is a standard parabolic group and so equals P^{I_1} for some I_1. It follows from Proposition 2.18 that $I_0 = I_1$ and $k_0 \in K^{I_1} M$. Theorem 2.8 implies that $P^{I_1} \supset P^I$ if and only if $I_1 \supset I$. This implies that $z \in C_{I_1}(\infty) \subset \overline{C_I}(\infty)$. The converse follows by continuity. □

The above results prove the following proposition.

3.20. Proposition. (A geometrical realization of $\Delta(G)$.) *Let* $\Delta(X)$ *be the simplicial complex of Weyl chamber faces at infinity. For any flat* F *in* X, *define the corresponding apartment to be the subcomplex of* $\Delta(X)$ *contained in the sphere at infinity* $F(\infty)$ *of the flat* F. *Then* $\Delta(X)$ *is a spherical Tits building, and it is isomorphic to* $\Delta(G)$ *under the map* $k \cdot C_I(\infty) \to k P^I k^{-1}$.

Finally, one observes that, since the sphere at infinity $X(\infty)$ does not depend on the choice of basepoint o, Proposition 3.18 implies the following important result.

3.21. Proposition. *The definition of the Weyl chamber faces at infinity and, hence, the disjoint decomposition of* $X(\infty)$ *into Weyl chamber faces does not depend on the choice of the basepoint* o.

The polyhedral compactification of a flat

§ **3.22.** In this section it will be shown how the Weyl group induces a natural compactification of $\mathfrak{a}$ (and, hence, of any flat), the so-called polyhedral compactification (see Gerardin [G2], Taylor [T4]) which is an elaboration of ideas in Ash–Mumford–Rapaport–Tai [A7, pp. 1–10].[2] The boundary

[2] The polyhedral compactification of a Euclidean space corresponds to the closure of the non-compact part of the torus in the torus embeddings.

$\Delta^*(\mathfrak{a})$ of this compactification is constructed by duality from the triangulation of the unit sphere in $\mathfrak{a}$ determined by the Weyl chambers and their walls (Proposition 3.7).

For example, in the case of $SL(3, \mathbb{R})$, the flat $\mathfrak{a}$ is of dimension 2 and the Weyl chambers and their walls are shown in Figure 1.

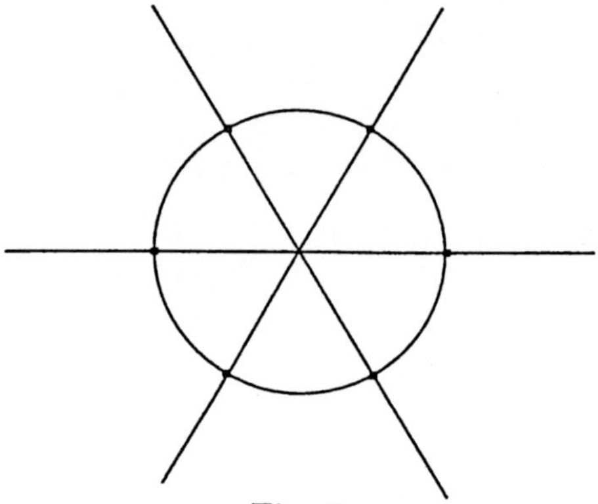

Fig. 1

The resulting triangulation of the unit circle involves six simplices of dimension 1 and six vertices. Replacing the 1-simplices by points and the 0-simplices by 1-dimensional cells yields the boundary of a hexagon as shown in Figure 2.

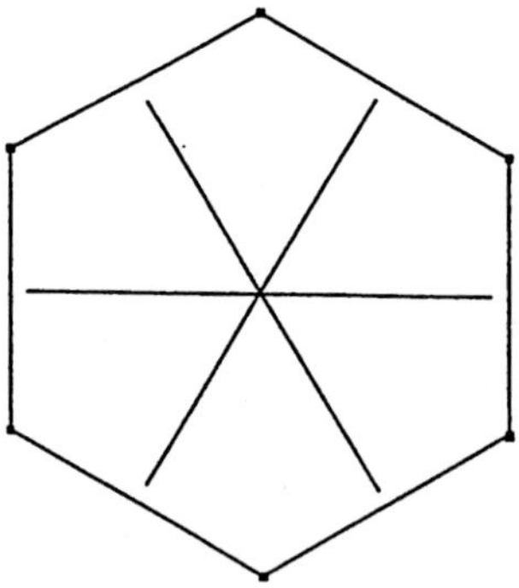

Fig. 2

It is to be viewed as the boundary of a hexagon rather than yet another triangulation of the circle because there is a built-in linear structure in the cells.

The boundary of the hexagon is viewed as being at infinity and the resulting compactification is such that sequences converge at ∞ so as to respect the hexagonal structure. More explicitly, sequences that go to ∞ through a fixed chamber and whose distance from the walls also goes to infinity all converge to one and the same point regardless of their limiting direction, and a sequence that goes to ∞ along a wall or parallel to a wall converges to an ideal boundary point on the straight line at ∞ joining the two vertices that correspond to the chambers in whose boundaries the wall lies. (Note that the parallel possibilities are parametrized by the linear space, viewed at ∞, that is orthogonal to the wall.)

This is indicated schematically in Figure 3.

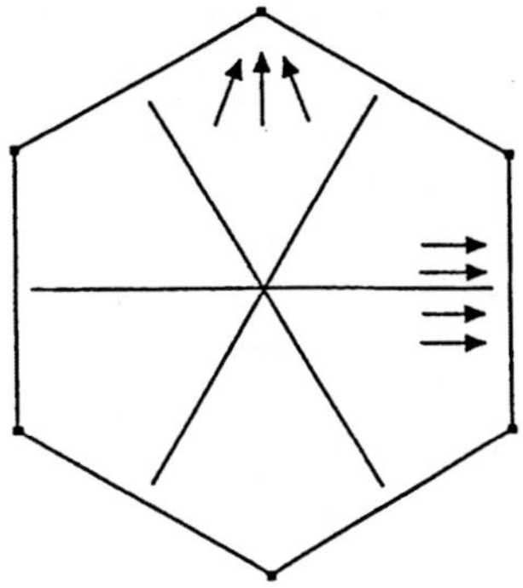

Fig. 3

The cellular nature of the dual object becomes clearer when looking at the ℓ-polydisc, $\ell = 3$. Here the flat $\mathfrak{a}$ is $\mathbb{R}^3$ with eight Weyl chambers corresponding to the eight octants of $\mathbb{R}^3$ and with the walls lying in the coordinate planes. The dual object here is the boundary of a cube with, for example, the top face corresponding to the positive z-axis.

In order to discuss the general case a new concept is useful, which has nothing per se to do with Weyl chambers and flats.

3.23. Definition. (see Taylor [T4]) Given any d-dimensional Euclidean space V, and an inner product on V, a **polyhedral cone decomposition** (or **conical decomposition**) Π is a partition of $V \backslash \{0\}$ into convex cones C, of the form $(\cap_{i=1}^{m}\{\ell_i > 0\}) \cap (\cap_{i=m+1}^{m+n}\{\ell_i = 0\})$ — where the ℓ_i are linear functionals — such that for any two distinct cones C_1 and C_2, if $C_1 \cap \overline{C_2} \neq \emptyset$ then $C_1 \subset \overline{C_2}$. If $C \in \Pi$, let $V(C)$ denote the linear subspace generated by C.

A polyhedral cone decomposition is determined by a finite number of supporting hyperplanes through the origin. In general, the cell decomposition determined by these hyperplanes (see Brown [B15, p. 7]) contains more cones than the polyhedral cone decomposition (e.g., the decomposition of $\mathbb{R}^2$ consisting of the open right half plane, the open quadrants II and III, plus three half rays). Note that while the inner product can be dispensed with, as in Gerardin [G2], it allows the identification of the quotient space V/W with the orthogonal complement $W^{\perp}$. Also, it is clear that a 1-dimensional space has a unique non-trivial polyhedral cone decomposition. It consists of two cones, each one a half-line.

A conical decomposition Π of $V\backslash\{0\}$ determines a cell complex $\Delta(V)$ on the unit sphere. Note that if Π is determined by an essential finite reflection group (see Brown [B15, p. 2]), then, as pointed out earlier, this cell complex is a simplicial complex (see [B15, Proposition 1, p. 24]). The cell complex has a corresponding dual object $\Delta^*(V) = \Delta^*(V, \Pi)$ and an associated compactification of the vector space V, the so-called polyhedral compactification of V (see Definition 3.30). The duality is given by associating to each cell of $\Delta(V)$ of dimension $(d-1)-k$, the k-dimensional linear space $V(C)^\perp$ that is orthogonal to the cone $C \in \Pi$ whose intersection with the unit sphere is the given cell. Set theoretically, the dual object is the disjoint union of the linear subspaces $V(C)^\perp, C \in \Pi$. These linear spaces are then assembled in a disjoint manner so as to produce a compact cell complex. In other words, $\Delta^*(V) = \cup_{C\in\Pi} V(C)^\perp$, equipped with a certain topology. The topology may be defined inductively, by assembling $\Delta^*(V)$ using its skeletons and the polyhedral compactification for lower dimensional subspaces to attach the k-dimensional linear space $V(C)^\perp$ to the $(k-1)$ skeleton (see Taylor [T4]).

The idea of the argument is as follows. First, the 0-skeleton consists of the points that correspond to the cones of dimension d. Then, the 1-skeleton is obtained by attaching to the 0-skeleton the spaces $V(C)^\perp$, where $C \in \Pi$ has codimension 1. This is done by observing that such a cone C is the face of exactly two cones $C^+, C^- \in \Pi$ of dimension d. Projecting these two cones onto $V(C)^\perp$ determines two half-lines: a polyhedral cone decomposition of $V(C)^\perp$. The polyhedral compactification of the one-dimensional space $V(C)^\perp$ is defined to be the two-point compactification $V(C)^\perp \cup \{+\infty\} \cup \{-\infty\}$. One attaches the compactification of $V(C)^\perp$ to the 0-skeleton by identifying the points corresponding to the cones $C^\pm$ with the points $\{\pm\infty\}$.

Assuming the $(k-1)$-skeleton to have been assembled, consider a cone C of codimension k. Project onto $V(C)^\perp$ all the cones C_1 of the original polyhedral cone decomposition Π for which C is a face. This gives a polyhedral cone decomposition Π' of $V(C)^\perp$. Each cell of the cell complex $\Delta(V(C)^\perp)$ of $V(C)^\perp$ can be identified with an appropriate cone C_1 in Π. The dual object corresponding to the cone C_1 is already incorporated into the $(k-1)$-skeleton of $\Delta^*(V)$. Since the dimension of $V(C)^\perp$ is less than the dimension of V one may assume, as part of the inductive hypothesis, that the dual cell complex and the polyhedral compactification of $V(C)^\perp$ relative to Π' exist. Then it is evident how to attach (as for codimension 1) the boundary $\Delta^*(V(C)^\perp, \Pi')$ of the polyhedral compactification $V(C)^\perp \cup \Delta^*(V(C)^\perp, \Pi')$ of $V(C)^\perp$ to the $(k-1)$-skeleton of $\Delta^*(V)$.

In this way the k-skeleton is assembled and so, eventually, one obtains the compact cell-complex $\Delta^*(V)$. Finally, to complete the inductive step of the argument, $\Delta^*(V)$ also needs to be added at infinity to form the polyhedral compactification $V \cup \Delta^*(V)$.

In order to do this it is useful to have some way, in terms of V, to describe an ideal boundary point. In Taylor [T4], the traces on V of the filter of neighborhoods of an ideal boundary point are used. Alternatively, one may view the ideal boundary point as the class of sequences that are to converge to it or one may define it formally.

Proceeding formally, the ideal boundary points are taken to be of the form $(C(\infty), y^C)$ where $y^C \in V(C)^{\perp}$. This boundary point may be thought of as the point $y^C \in V(C)^{\perp}$ located at the "end" of the cone C. It will also be denoted by $y^C(\infty)$.

3.24. Definition. Given a fixed cone C of the cone decomposition Π, any point $y \in V$ can be written as $y = y_C + y^C$, with $y_C \in V(C)$ and $y^C \in V(C)^{\perp}$. A sequence (y_n) will be said to **converge formally to** $y^C(\infty)$ if $y_{n,C} \xrightarrow{C} \infty$, and $y_n^C \to y^C$ — where $y_{n,C} \xrightarrow{C} \infty$ means that, for any $c \in C$, the sequence $(y_{n,C})$ is eventually in $c+C$. Such a sequence will be called a **C-fundamental sequence** and $y^C(\infty) = (C(\infty), y^C)$ will be called its **formal limit**. A sequence will be said to be **fundamental** if it is **C-fundamental** for some cone $C \in \Pi$.

3.25. Proposition. *The class of fundamental sequences has the following properties:*

(1) *given a compact subset of V, every fundamental sequence is eventually in its complement;*
(2) *every sequence in V, converging to infinity, has a fundamental subsequence; and*
(3) *a subsequence of a fundamental sequence is fundamental.*

Proof. Statements (1) and (3) are evident. The proof of (2) will be given later as it involves induction on the dimension of V.

3.26. Remark. Let $\overline{X}$ be a metric space that contains a locally compact metric space X as a dense subspace. If there is a class of sequences on X satisfying conditions (1), (2), and (3) of Proposition 3.25, each of which converges in $\overline{X}$, then $\overline{X}$ is compact.

3.27. Definition. A **compactification** of a locally compact space X is a pair (K, i) consisting of a compact space K and an embedding $i : X \to K$ with dense image, which is necessarily open. (When X is viewed as a subset of K no explicit reference will be made to the embedding i.) If H is a group that acts continuously on X and K, the compactification is called an **H-compactification** if i is H-equivariant, i.e., $i(g \cdot x) = g \cdot i(x)$ for all $x \in X$ and $g \in H$. Two compactifications (K_1, i_1) and (K_2, i_2) are said to be **isomorphic** if there is a homeomorphism $\phi : K_1 \to K_2$ such that $\phi \circ i_1 = i_2$. If, in addition, they are H-compactifications, they are said to be **H-isomorphic** if ϕ is H-equivariant. A compactification (K_1, i_1) **dominates** a compactification (K_2, i_2) if there is a continuous map $\phi : K_1 \to K_2$ such that $\phi \circ i_1 = i_2$.

An immediate corollary of the following lemma is the fact that there is at most one metrizable compactification of V such that (a) every fundamental sequence converges in it and (b) two fundamental sequences have the same limit point if and only if their formal limits agree.

3.28. Lemma. *Let (K_1, i_1) and (K_2, i_2) be two metrizable compactifications of a locally compact topological space X. Assume a class of sequences σ in X is given with the following properties:*

(a) *for any compact subset of X, each σ in the class is eventually in its complement (i.e., σ converges to infinity);*
(b) *for each σ in the class, $i_1(\sigma)$ converges in K_1 — where $i_1(\sigma) = (i_1(y_n))$ if $\sigma = (y_n)$;*
(c) *every sequence of points of X that converges to infinity has a subsequence in the class.*

Assume that

(1) *for each σ in the class, $i_2(\sigma)$ converges in K_2; and*
(2) *if $i_1(\sigma)$ and $i_1(\sigma')$ have the same limit in K_1, then $i_2(\sigma)$ and $i_2(\sigma')$ have the same limit in K_2.*

Then, (K_1, i_1) dominates (K_2, i_2).

Proof. Define $\varphi(i_1(y)) = i_2(y)$ if $y \in X$, and if $b(1) = \lim_{n\to\infty} i_1(y_n) \in K_1 \backslash X \stackrel{\text{def}}{=} \partial K_1$ where (y_n) is a sequence in X that converges to infinity, define $\varphi(b(1))$ to be the common limit in K_2 of any subsequence σ of (y_n) that belongs to the class. To show that φ is continuous, it suffices to show that it is continuous at each point $b(1)$ of ∂K_1. Assume that $(y_n) \subset X$ and $i_1(y_n) \to b(1)$. Then $i_2(y_n) = \varphi(i_1(y_n)) \to b(2) \stackrel{\text{def}}{=} \varphi(b(1))$. Otherwise, there will be a subsequence (y_{n_k}) and an $\epsilon > 0$ with $d_2(i_2(y_{n_k}), b(2)) \geq \epsilon$ where d_2 is a metric on K_2. This leads to a contradiction since it follows from (2) that $i_2(\sigma)$ converges in K_2 to $b(2)$ if σ is a fundamental subsequence of (y_{n_k}).

Now let $(b_n(1))$ be a sequence on ∂K_1 that converges to $b(1)$ and set $b_n(2) = \varphi(b_n(1))$ and $b(2) = \varphi(b(1))$. One may determine a sequence $(y_n) \subset X$ such that $d_1(i_1(y_n), b_n(1)) < 2^{-n}$ and $d_2(i_2(y_n), b_n(2)) < 2^{-n}$, where d_1 is a metric on K_1. Since $i_1(y_n) \to b(1)$, it follows that $\varphi(i_1(y_n)) \to b(2)$. Hence, $\varphi(b_n(1)) \to b(2)$. □

The next result states that the class of fundamental sequences determines a unique compactification of V that will be called its polyhedral compactification.

3.29. Theorem. (See Taylor [T4]) *There is a unique metrizable compactification $\tilde{V}$ of V such that*

(1) *every fundamental sequence converges in $\tilde{V}$; and*
(2) *the limits in $\tilde{V}$ of two fundamental sequences agree if and only if their formal limits agree.*

Proof. The uniqueness follows from Lemma 3.28. If $O \subset V$, let $O^* = O \cup \{y^C(\infty) \mid \text{ for some } \varepsilon > 0 \text{ and } c \in C,\ c + C + B^{\perp}(y^C; \varepsilon) \subset O\}$, where

$B^{\perp}(y^C;\varepsilon)$ is the open ball in $V(C)^{\perp}$ about y^C of radius ε. The existence is proved by showing that the sets O^*, where O is open in V, are a basis for a compact topology $\mathcal{T}$ on $V \cup \Delta^*(V)$. For details see Proposition A.1 in Appendix A. For another proof see [T4]. □

3.30. Definition. The above compactification will be called the **polyhedral compactification of** V associated with the polyhedral cone decomposition Π. It will be denoted by $V \cup \Delta^*(V)$ or $V \cup \Delta^*(V, \Pi)$.

3.31. Remark. It is also shown in Appendix A that the topology $\mathcal{T}$, restricted to $\Delta^*(V)$, coincides with the topology obtained by assembling this space using the skeletons and the polyhedral compactifications of the lower-dimensional subspaces as outlined following Definition 3.23.

Proof of Proposition 3.25(2).

This requires an inductive hypothesis. Assume that, for $1 \leq n < d$, if W is an n-dimensional inner product space equipped with a polyhedral cone decomposition Π', Theorem 3.29 is valid. This amounts to saying that Proposition 3.25 holds for $W = (W, \Pi')$ if $n < d$. Assume that V has dimension d.

Given any sequence (y_n) in V, converging to infinity, there is a subsequence (y_{n_k}) with a limiting direction that lies in, say, the cone C. If the subsequence $(y^C_{n_k})$ is bounded, then it has a subsequence that converges to $y \in V(C)^{\perp}$. This implies that a subsequence of (y_n) is C-fundamental.

On the other hand, if the subsequence $(y^C_{n_k})$ is converging to infinity, one makes use of the inductive hypothesis. Let $W = V(C)^{\perp}$ equipped with the polyhedral cone decomposition that is given by projecting onto $V(C)^{\perp}$ all the cones C_1 of Π that have C as a face. Since $(y^C_{n_k})$ is converging to infinity, some subsequence $(y^C_{n_{k_j}})$ of it converges in the polyhedral compactification of $V(C)^{\perp}$. This implies that there is a cone $C_1 \in \Pi$ with C as a face, for which the sequence (y_n) has a C_1-fundamental subsequence: if $x_j = y_{n_{k_j}}$ and C_1' is the projection of C_1 onto $V(C)^{\perp}$ then $x_j = x_{j,C} + x^C_j = x_{j,C} + (x^C_j)_{C_1'} + (x^C_j)^{C_1'}$; since $(x^C_j)^{C_1'} = x^{C_1}_j$, it converges in $V(C_1)^{\perp}$ and $x_{j,C} + (x^C_j)_{C_1'} = x_{j,C_1} \overset{C_1}{\to} \infty$ (see Lemma A.3 in Appendix A). This completes the proof of Proposition 3.25. □

3.32. Definition. The **polyhedral compactification of** $\mathfrak{a}$ is the polyhedral compactification $\mathfrak{a} \cup \Delta^*(\mathfrak{a})$ associated with the polyhedral cone decomposition of $\mathfrak{a}$ given by the Weyl chambers and their faces, where the inner product is given by the Killing form.

The **polyhedral compactification of a flat** $A \cdot o \subset X$ is the compactification $(A \cdot o) \cup \Delta^*(A \cdot o)$ for which a sequence $(a_n \cdot o)$, converging to infinity, converges if and only if the sequence $\log a_n$ converges in $\mathfrak{a} \cup \Delta^*(\mathfrak{a})$. In other words, the diffeomorphism $\mathfrak{a} \to A \cdot o$ given by $H \to \exp H \cdot o$ extends as an isomorphism $\mathfrak{a} \cup \Delta^*(\mathfrak{a}) \to (A \cdot o) \cup \Delta^*(A \cdot o)$.

3.33. Remarks. (1) If $k \in K$, then $Ad(k)$ transports the polyhedral cone decomposition given by the Weyl chambers and the chamber faces of $\mathfrak{a}$ isomorphically to the corresponding structure on $Ad(k)\mathfrak{a}$. Therefore, $Ad(k)$ extends to a homeomorphism, which will also be denoted by $Ad(k)$, of $\mathfrak{a} \cup \Delta^*(\mathfrak{a})$ with $Ad(k)\mathfrak{a} \cup \Delta^*(Ad(k)\mathfrak{a})$. If $x \in \mathfrak{a} \cup \Delta^*(\mathfrak{a})$, let $Ad(k)(x)$ denote its image under the extended map. Note that this implies $Ad(k)(C(\infty), y^C) = ((Ad(k)C)(\infty), Ad(k)y^C) = (k \cdot (C(\infty), k \cdot y^C)$.

(2) Under the isomorphism $\mathfrak{a} \to A \cdot o$ given by $H \to \exp H \cdot o$, all the notions connected with the polyhedral compactification may be carried over to $kA \cdot o$. For example, the diffeomorphism $A \cdot o \to kA \cdot o$ given by $a \cdot o \to ka \cdot o$ defines an isomorphism of the polyhedral compactification $(A \cdot o) \cup \Delta^*(A \cdot o)$ with the polyhedral compactification $(kA \cdot o) \cup \Delta^*(kA \cdot o)$. If $x \in (A \cdot o) \cup \Delta^*(A \cdot o)$, let $k \cdot x$ denote its image in $(kA \cdot o) \cup \Delta^*(kA \cdot o)$ under this isomorphism.

(3) Similarly, for a flat $kA \cdot o \subset X$, a sequence $(ka_n \cdot o)$ is taken to be fundamental if the sequence $(\log a_n)$ is fundamental in $\mathfrak{a}$.

(4) If C_I is a Weyl chamber face, the closure of the open cell $(C_I(\infty), \mathfrak{a}^I)$ $= \mathfrak{a}^I(\infty)$ in $\Delta^*(\mathfrak{a})$ is isomorphic to the polyhedral compactification of $\mathfrak{a}^I$. This follows from Lemma 3.4 of [T4]. It is evident when one assembles the cell complex $\Delta^*(\mathfrak{a})$ from its skeletons.

(5) The observation in (4) can be verified by determining convergence on the boundary $\Delta^*(\mathfrak{a})$ in terms of the description of the boundary points. It turns out, from the description of the topology given in the proof of Theorem 3.29, that a sequence of points $(C_I(\infty), H_n^I)$ converges to $(C_J(\infty), H^J)$ if and only if (i) $I \supset J$ and (ii) if $J = I$, $H_n^I \to H^J = H^I$, and if $I \neq J$, (H_n^I) is C_J^I-fundamental with limit $(C_J^I(\infty), H^J)$. (When $I \neq J$, if $H_n^I = H_{n,J}^I + H_n^J$ is the decomposition given by $\mathfrak{a}^I = (\mathfrak{a}^I \cap \mathfrak{a}_J) \oplus \mathfrak{a}^J$, then $H_{n,J}^I \overset{C_J^I}{\to} \infty$ and $H_n^J \to H^J$. (See Lemma 3.37 for the definition of C_J^I.) As a result, the set of limit points in the closure of $(C_I(\infty), \mathfrak{a}^I) = \mathfrak{a}^I(\infty)$ is $\{k_j \cdot (C_J(\infty), H^J) \mid k_j \in M' \cap K^I M, J \subsetneq I, H^J \in \overline{\mathfrak{a}^{J,+}}\}$. This set is naturally identified with $\{k_j \cdot (C_J^I(\infty), H^J) \mid k_j \in (M^I)', J \subsetneq I, H^J \in \overline{\mathfrak{a}^{J,+}}\}$, which is $\Delta^*(\mathfrak{a}^I)$. This outlines a proof of (4).

THE DUAL CELL COMPLEX $\Delta^*(X)$

§ **3.34.** It is important to note that the construction of the dual cell complex associated with the polyhedral cone decomposition of a flat through o does not automatically extend to the spherical Tits building $\Delta(X)$ even if one identifies X with $\mathfrak{p}$. The Tits building defines a simplicial complex on the sphere at infinity centered at o (see Proposition 3.20). The natural linear space to put at infinity in place of a Weyl chamber face $Ad(k)C_I(\infty)$ is not the orthogonal complement in $\mathfrak{p}$ of the vector space generated by the cone $Ad(k)C_I$. Rather it is $Ad(k)\mathfrak{p}^I$, which corresponds to the symmetric space $k \cdot X^I$. This is because of the relation between $K, \mathfrak{a}$, and $\mathfrak{p}$ (see Proposition 2.2). What results is an object $\Delta^*(X) = \Delta^*(\mathfrak{p})$ on which K

acts and that is completely determined by the dual cell complex associated with polyhedral compactifications of pairs of flats $Ad(k_1)\mathfrak{a}$ and $Ad(k_2)\mathfrak{a}$ (see Theorem 3.39). It is a union of cells, with each cell a symmetric space of non-compact type — each cell is, therefore, homeomorphic to a Euclidean space.

The discussion that follows is given in terms of the symmetric space X. Equivalently, it may be formulated in terms of $\mathfrak{p}$ using the identification of $\mathfrak{p}$ with X given by the map $Y \to \exp Y \cdot o$. From time to time the equivalent formulation will be mentioned for clarity. Set theoretically, $\Delta^*(X)$ is the disjoint union of the distinct symmetric spaces $k \cdot X^I$ as k varies over K and I runs over the proper subsets of the set $\Delta = \Delta(\mathfrak{g}, \mathfrak{a}^+)$ of simple roots (equivalently, $\Delta^*(\mathfrak{p})$ is the union of the distinct linear subspaces $Ad(k)\mathfrak{p}^I$).

A topology will be defined on $\Delta^*(X)$ via a compactification procedure for X with ideal boundary equal to $\Delta^*(X)$, and its relation to attaching cells to lower-dimensional skeletons will be explained. It will now be indicated why it is not so obvious that an inductive procedure similar to the one used for the polyhedral compactification of a flat is applicable.

Assume that, for any symmetric space Y with rank less than the rank r of X, the dual cell complex $\Delta^*(Y)$ and the compactification $Y \cup \Delta^*(Y)$ are defined.

When the rank of Y is one, the Tits building of a rank one symmetric space Y is trivial in the sense that all the points of the sphere at infinity are 0-simplices and there are no simplices of dimension larger than zero. Also, for any flat in a rank one space Y, the symmetric space added at infinity, to correspond to one of the two half-lines in a flat, is a point. Set theoretically, the dual cell complex $\Delta^*(Y)$ is the union of these points and so coincides with $\Delta(Y)$. It is also the union of the dual cell complexes for each of the flats through a fixed basepoint, which in each case consist of the two points at infinity of the two-point compactification of the flat. This is also the polyhedral compactification of the flat, since the flat has dimension one.

Now the union of the two-point compactification of a flat, taken over all the flats through a fixed basepoint, agrees set theoretically with the conic compactification of Y. One, therefore, defines the compactification $Y \cup \Delta^*(Y)$ for a rank one symmetric space Y to be its conic compactification.

The 0-skeleton of the dual cell complex $\Delta^*(X)$ consists of the set of Weyl chambers $k \cdot \mathfrak{a}^+(\infty)$, one point for each chamber. This set is isomorphic to the so-called Furstenberg boundary $\mathcal{F} = K/M$, a compact homogeneous space. The chamber faces of dimension $(r-1)$ are the convex cones $k \cdot C_I(\infty)$, where $|I| = 1$.

By Proposition 3.9 and Corollary 2.16, the stabilizer in K of $C_I(\infty)$ is $K^I M$. The subspace $\exp \mathfrak{a}^I \cdot o$ is a flat in X^I and its orbit under $K^I M$ is exactly X^I. This is the symmetric space corresponding to $C_I(\infty)$ that is to be added to the 0-skeleton. Similarly, one adds $k \cdot X^I$ to the 0-skeleton as the symmetric space corresponding to $k \cdot C_I(\infty)$ (in the equivalent formulation, $Ad(k)\mathfrak{p}^I$ is the linear space corresponding to the cone $Ad(k)C_I$ that is added to the 0-skeleton). Since $|I| = 1$, the symmetric space $k \cdot X^I$ has rank one. Hence, the compactification $k \cdot X^I \cup \Delta^*(k \cdot X^I)$ is the conic compactification of $k \cdot X^I$. This compactification is then attached to the 0-skeleton K/M by using the fact that the ideal boundary is $kK^I k^{-1}/kM^I k^{-1}$. This space can be attached naturally to K/M by first identifying K^I/M^I with the orbit $K^I \cdot \dot{e}$ of the base point $\dot{e} = M$ in K/M under K^I and then left translating this orbit by k.

To construct the skeleton of level ℓ from the skeleton of level $(\ell - 1)$, one considers chamber faces $k \cdot C_I(\infty)$ with $|I| = \ell$, i.e., of dimension $(r-1)-\ell$. These chamber faces correspond to the symmetric spaces in $\Delta^*(X)$ of rank ℓ. The subspace $k \cdot \exp \mathfrak{a}^I \cdot o$ is a flat in $k \cdot X^I$ and its orbit under $kK^I M k^{-1}$ is exactly $k \cdot X^I$. One attaches the compactification $k \cdot X^I \cup \Delta^*(k \cdot X^I)$ to the skeleton of level $(\ell - 1)$ as in the case of codimension 1. Namely, one identifies the symmetric spaces in the ideal boundary $\Delta^*(k \cdot X^I)$ of $k \cdot X^I$ with symmetric spaces in the skeleton of level $(\ell - 1)$.

One may do this by observing that the Tits building $\Delta(k \cdot X^I)$ is a subbuilding of $\Delta(X)$. As already mentioned, the flats through the origin o are of the form $k \cdot \exp \mathfrak{a}^I \cdot o, k \in K^I M$ and their cone decompositions determine $\Delta(k \cdot X^I)$. Since the cone decomposition of $\mathfrak{a}^I$ given by the Weyl group W^I is the projection onto $\mathfrak{a}^I$ of the cone decomposition of $\mathfrak{a}$ given by the Weyl group W, it follows that to each simplex of dimension $j \leq \ell - 1$ in $\Delta(k \cdot X^I)$ there corresponds a unique simplex in $\Delta(X)$ of dimension $(r - \ell) - j$, where $\ell = |I|$ is the rank of X^I. Each of these simplices is of the form $kk_1 C_J(\infty), k_1 \in K^I$, with $|J| = j + 1$. The corresponding symmetric space $kk_1 \cdot X^J$ is a symmetric space contained in $k \cdot X^I$. This space is already embedded in the skeleton of level $(\ell - 1)$. As a result, one may identify $\Delta^*(k \cdot X^I)$ with a subspace of the skeleton of level $(\ell - 1)$, and thus attach $k \cdot X^I \cup \Delta^*(k \cdot X^I)$ to the skeleton of level $(\ell - 1)$.

In the equivalent formulation, the Tits building $\Delta(Ad(k)\mathfrak{p}^I)$ is obtained by projecting orthogonally onto $Ad(k)\mathfrak{p}^I$ the cones that are the Weyl chambers and walls with $Ad(k)C_I$ as a face. The Weyl chamber faces at infinity corresponding to these cones form the star $St(Ad(k)C_I(\infty))$ of the simplex $Ad(k)C_I(\infty)$. These simplices of lower codimension already have dual objects in the $(\ell - 1)$-skeleton and so it is clear how to construct the map attaching the compactification $\mathfrak{p}^I \cup \Delta^*(\mathfrak{p}^I)$ to the $(\ell - 1)$-skeleton.

Rather than try to carry out the details of the above program, it is easier first of all to define the class of fundamental sequences, i.e., those that tend to infinity so as to respect the structure of $\Delta(\mathfrak{p}) = \Delta(X)$, and to then show that, as for the polyhedral compactification, there is a unique compactification associated with this class of sequences.

By the Cartan decomposition, every point y in X can be written as $y = ka \cdot o$, where $a \in \overline{A^+}$ is unique. (Under the identification of X with $\mathfrak{p}$, this amounts to saying that every $Y \in \mathfrak{p}$ is of the form $Y = Ad(k)H$, where $H \in \overline{\mathfrak{a}^+}$ is unique.)

3.35. Definition. A sequence (y_n) in X where $y_n = k_n \cdot \exp H_n = k_n a_n \cdot o$ with $H_n \in \overline{\mathfrak{a}^+}$ will be said to be **C_I-fundamental in** X if

(1) (k_n) converges; and
(2) (H_n) is C_I-fundamental in $\mathfrak{a}$ (see Definition 3.24) and $I \subsetneq \Delta$.

A sequence is **fundamental in** X if it is C_I-fundamental for some $I \subsetneq \Delta$. The **formal limit** of a C_I-fundamental sequence is $(k \cdot C_I(\infty), ka^I \cdot o)$, where (i) k_n converges to k and (ii) the projection H_n^I of H_n onto $\mathfrak{a}^I$ converges to $H^I = \log a^I$. Note that $H^I \in \overline{\mathfrak{a}^{I,+}}$ as $\overline{\mathfrak{a}^{I,+}}$ is the projection of $\overline{\mathfrak{a}^+}$ onto $\mathfrak{a}^I$.

Remark. The formal limit $(k \cdot C_I(\infty), ka^I \cdot o)$ is to be viewed as the point $ka^I \cdot o \in k \cdot X^I$ at infinity. It will also be denoted by $k \cdot (C_I(\infty), a^I \cdot o)$ and $(ka^I \cdot o)(\infty)$. Using this notation, the dual cell complex $\Delta^*(X)$ can be viewed as the set of formal limits since a point y_1 in $k \cdot X^I$ can be written as $k'a^I \cdot o$ with $k' = k\ell, \ell \in K^I$ and $(k \cdot C_I(\infty), y_1) = (k' \cdot C_I(\infty), k'a^I \cdot o)$. One can also represent a point in $\Delta^*(X)$ as $(k \cdot C(\infty), k \cdot y)$ where C is a Weyl chamber face $Ad(k_j)C_I$ and y is a point in the symmetric space $k \cdot X^I$ over $C(\infty)$ if $C = Ad(k)C_I$. This symmetric space $k \cdot X^I$ is called a boundary symmetric space (see § 5.2).

One immediate problem with this definition is that it is not evident that the concept of formal limit is well-defined. The definition also raises the question of the relation between this concept of fundamental and the notion of fundamental sequence in a flat. These questions are addressed by the following lemma, in which it is also shown that every sequence converging to infinity has a fundamental subsequence.

3.36. Lemma.

(1) *Let (y_n) be a C_I-fundamental sequence, $y_n = k_n \exp H_n \cdot o = k'_n \exp H_n \cdot o$ with $k = \lim_n k_n, k' = \lim_n k'_n$. Then $Ad(k)C_I = Ad(k')C_I$ and $Ad(k)H^I = Ad(k')H^I$, where $H^I = \lim_n H_n^I$ and H_n^I is the projection onto $\mathfrak{a}^I$ of H_n. Hence, the formal limit of a fundamental sequence is well-defined.*
(2) *If $(k_n \exp H_n \cdot o)$ is a fundamental sequence contained in the flat $kA \cdot o$, then there is a fundamental sequence (y_n) in $A \cdot o$ with $k_n \exp H_n \cdot o = k \cdot y_n$. Furthermore, the formal limit of $(k_n \exp H_n \cdot o)$ equals $(k \cdot C(\infty), k \cdot y^C)$, where $(C(\infty), \log y^C)$ is the formal limit of $(\log y_n) \subset \mathfrak{a}$ and C is a Weyl chamber face in $\mathfrak{a}$.*
(3) *Two formal limits $(k_1 \cdot C_{I_1}(\infty), k_1 \cdot a_1)$ and $(k_2 \cdot C_{I_2}(\infty), k_2 \cdot a_2)$, $a_i \in \overline{A^{I_i,+}}$ agree if and only if $I_1 = I_2 = I$, $a_1 = a_2 = a$, and $k_1^{-1}k_2 \in (K^I \cap aK^I a^{-1})M$.*
(4) *Every sequence (y_n) in X that converges to infinity has a fundamental subsequence.*

Proof. (1) Since $H_n \in C_{I_n}$ for some subset $I_n \subset I$ of Δ and $Ad(k_n)H_n = Ad(k'_n)H_n$, it follows from Lemma 3.10 that $k_n^{-1}k'_n \in K^{I_n}M \subset K^I M$ and so $k^{-1}k' \in K^I M$. Hence, $Ad(k)C_I = Ad(k')C_I$. Also, for all n, $Ad(k_n)\mathfrak{a}_I =$

it follows that, for all n, $Ad(k_n)H_n^I = Ad(k_n')H_n^I$. Hence, $Ad(k)H^I = Ad(k')H^I$.

(2) If $(k_n \exp H_n \cdot o) \subset kA \cdot o$, then it follows from the Cartan decomposition that, for all n, $k_n \exp H_n \cdot o = kk_{j(n)} \exp H_n \cdot o$, where $j(n) \in \{1, 2, \ldots, |W|\}$ and $k_{j(n)}$ is one of the representatives of the Weyl group in M' (see § 3.2). Let j_0 and j_0' be two values that $j(n)$ takes infinitely often. It follows from (1) that $Ad(j_0)C_I = Ad(j_0')C_I$ is a well-determined cone C of the polyhedral cone decomposition of $\mathfrak{a}$. As a result, if $Y_n = Ad(k_{j(n)})H_n$, then (Y_n) is a fundamental sequence in $\mathfrak{a}$ relative to its polyhedral cone decomposition. Let its formal limit be $(C(\infty), Y^C)$. By considering a subsequence for which $j(n_k) = j_0$, where j_0 is one of the values assumed infinitely often, it follows from (1) that the formal limit of $(k_n \exp H_n \cdot o)$ is $(k \cdot C(\infty), k \cdot y^C)$, where $y^C = \exp Y^C$.

(3) The conditions are clearly sufficient for equality of the formal limits. If $k_1 \cdot C_{I_1}(\infty) = k_2 \cdot C_{I_2}(\infty)$, it follows from Propositions 2.18, 3.9, and 3.16 that $I_1 = I_2 = I$ and $k_1^{-1}k_2 \in K^I M$.

If the formal limits agree then, obviously, $k_1 a_1 \cdot o = k_2 a_2 \cdot o$. It follows from the uniqueness in the Cartan decomposition that $a_1 = a_2 = a$. Since $aK^I a^{-1}$ is the isotropy group of $a \cdot o \in X^I$, the result follows.

Finally, if (y_n) is a sequence that converges to infinity and $y_n = k_n a_n \cdot o$, with $\log a_n = H_n \in \overline{\mathfrak{a}^+}$, then there is a subsequence (y_{n_j}) such that (H_{nj}) is fundamental in $\mathfrak{a}$ and (k_{n_j}) converges. The subsequence (y_{n_j}) is therefore fundamental. □

The group $(K^I \cap aK^I a^{-1})M$, where $a \in \overline{A^{I,+}}$, can be identified with $K^J M$ if $\log a$ belongs to the projection onto $\mathfrak{a}^I$ of C_J, $J \subset I$. This is part of the following lemma.

3.37. Lemma. *If $I \supset J$, let C_J^I denote the projection of C_J onto $\mathfrak{a}^I$. Then C_J^I is the face of the Weyl chamber $\mathfrak{a}^{I,+}$ determined by the set of simple roots J. In other words,*

$$C_J^I = \{H \in \overline{\mathfrak{a}^{I,+}} \mid \alpha(H) > 0 \text{ if and only if } \alpha \in I \backslash J\}.$$

If $\log a \in C_J^I$, then $(K^I \cap aK^I a^{-1})M = K^J M$.

Proof. If $H \in C_J$, then $H = H_I + H^I$. Since $\mathfrak{a}_J \supset \mathfrak{a}_I$, one has $\mathfrak{a}_J = \mathfrak{a}_I \oplus (\mathfrak{a}^I \cap \mathfrak{a}_J)$. The projection of H onto $\mathfrak{a}^I$ is $H^I \in \mathfrak{a}^I \cap \mathfrak{a}_J$. It follows that, for any $\alpha \in I$, one has $\alpha(H^I) \geq 0$ since $\alpha(H_I) = 0$. Furthermore, $\alpha(H^I) > 0$ if and only if $\alpha \in I\backslash J$, and so $C_J^I \subset \{H \in \overline{\mathfrak{a}^{I,+}} \mid \alpha(H) > 0$ if and only if $\alpha \in I\backslash J\}$.

Conversely, if $H_1 \in \overline{\mathfrak{a}^{I,+}}$ and $\alpha(H_1) > 0$ if and only if $\alpha \in I\backslash J$ then, for $H_0 \in C_I$, there is an integer $n \geq 1$ such that $\alpha(nH_0 + H_1) > 0$ for all $\alpha \in \Delta\backslash I$. This implies that $nH_0 + H_1 \in C_J$. Its projection onto $\mathfrak{a}^I$ is H_1.

Let $\log a = H_1 \in C_J^I$, and let $a_2 = \exp nH_0$, where $H_0 \in C^I$ and $nH_0 + H_1 \in C_J$. If $k \in (K^I \cap aKa^{-1})M$, then $ka_2 a \cdot o = a_2 ka \cdot o = a_2 a \cdot o$. Proposition 3.9 implies that $k \in K^J M$. Conversely, if $k \in K^J M$ then

$k \in K^I M$ and so $ka_2a \cdot o = a_2a \cdot o$ and $ka_2a \cdot o = a_2ka \cdot o$. Hence, $ka \cdot o = a \cdot o$ and so $k \in aKa^{-1}$.

However, since $a \cdot o \in X^I$ and $k \in K^I M$, it follows that if $k = \ell m, \ell \in K^I, m \in M$, then $\ell a \cdot o = a \cdot o$ and so $\ell \in aK^I a^{-1}$. □

As in the case of a polyhedral cone decomposition, the fundamental sequences determine a unique compactification of X (equivalently, of $\mathfrak{p}$).

3.38. Theorem. *There is a unique compactification $\tilde{X}$ of X such that*

(1) *every fundamental sequence converges in $\tilde{X}$; and*
(2) *the limits of two fundamental sequences in $\tilde{X}$ agree if and only if their formal limits agree.*

Hence, a sequence $(y_n) \subset X$ converges if and only if the formal limits of all its fundamental subsequences coincide.

Proof. Uniqueness follows from Lemma 3.28, as in the proof of Theorem 3.29, since it is clear from Definition 3.35 and Lemma 3.36(4) that the set of fundamental sequences has the three properties stated in Proposition 3.25. The existence proof is given in Appendix A. It amounts to showing that the polyhedral compactifications of the flats $kA \cdot o = \exp\{Ad(k)\mathfrak{a}\} \cdot o$ fit together, which is to be expected, since the existence of the Tits building $\Delta(X)$ implies that the various triangulations of the unit spheres in the flats induced by their Weyl chambers and chamber faces are compatible.

The last statement follows since every sequence has a fundamental subsequence. □

3.39. Theorem. *The compactification $\tilde{X}$ of X determined by Theorem 3.38 has the following properties:*

(1) *it is a K-compactification (see Definition 3.27);*
(2) *the closure of a flat $kA \cdot o$ is the polyhedral compactification of that flat; and*
(3) *the closure of the intersection of two flats $k_1A \cdot o$ and $k_2A \cdot o$ is the intersection of their closures, i.e., of their polyhedral compactifications.*

These three properties characterize this compactification.

Proof. For a fixed $k \in K$, the sequences $(k\ell_n \exp H_n \cdot o)$, with $\ell_n \in M'$ that are fundamental in the sense of Definition 3.35 are exactly the fundamental sequences of the flat $kA \cdot o$ (see Remarks 3.33). It follows from Theorem 3.29 and Lemma 3.36(2) that the closure in the compactification $\tilde{X}$ of the flat $kA \cdot o$ is its polyhedral compactification.

To verify (3) for $\tilde{X}$, it suffices to show that if $(H_n(1))$ and $(H_n(2))$ are two fundamental sequences in $\overline{\mathfrak{a}^+}$ with formal limits $(C_{I_1}(\infty), H_1)$ and $(C_{I_2}(\infty), H_2)$, with $H_i \in \overline{\mathfrak{a}^{I_i,+}}$, for which the formal limits of the sequences $(k_1 \exp H_n(1) \cdot o)$ and $(k_2 \exp H_n(2) \cdot o)$ agree, then there is a fundamental sequence in $(k_1A \cdot o) \cap (k_2A \cdot o)$ (equivalently, in $Ad(k_1)\mathfrak{a} \cap Ad(k_2)\mathfrak{a}$) that has the same formal limit.

By Lemma 3.36(3), the coincidence of these formal limits implies that $I_1 = I_2 = I, k_1^{-1}k_2 \in K^I M$, $H_1 = H_2 = H \in \mathfrak{a}^I$, and $k_1 \exp H_1 \cdot o = k_2 \exp H_2 \cdot o$. Consequently, $Ad(k_1)C_I = Ad(k_2)C_I$. Let $H_n = H + H_{n,I}(1)$, where $H_{n,I}(1)$ is the projection of $H_n(1)$ onto $\mathfrak{a}_I$. Since $Ad(k_1^{-1}k_2)$ is the identity on $\mathfrak{a}_I$, $(k_1 \exp H_n \cdot o)$ is a fundamental sequence in $(k_1 A \cdot o) \cap (k_2 A \cdot o)$ — in other words, $(Ad(k_1)H_n)$ is a fundamental sequence in $Ad(k_1)\mathfrak{a} \cap Ad(k_2)\mathfrak{a}$ — whose formal limit is the common formal limit of $(k_1 \exp H_n(1) \cdot o)$ and of $(k_2 \exp H_n(2) \cdot o)$.

It is clear that the compactification $\tilde{X}$ is a K-compactification in view of the obvious action of K on the fundamental sequences and their formal limits: if $(k \cdot C_I(\infty), k \exp H^I \cdot o)$ is the formal limit of the fundamental sequence $(k_n \exp H_n \cdot o)$ and $k'_n \to k'$ then $(k'k \cdot \exp C_I(\infty), k'k \exp H^I \cdot o)$ is the formal limit of the fundamental sequence $(k'_n k_n \exp H_n \cdot o)$.

Consider a compactification of X that satisfies (1), (2) and (3). From (2), it follows that fundamental sequences of the form $(k\ell_n \exp H_n \cdot o)$, with $\ell_n \in M'$ all converge. Property (1) implies that if a sequence (y_n) converges and (k_n) converges, then $(k_n \cdot y_n)$ converges. Hence, all the fundamental sequences converge. It remains to verify that their limits are determined by their formal limits, i.e., that condition (2) of Theorem 3.38 is valid.

Suppose that $(k_n \exp H_n \cdot o)$ and $(k'_n \exp H'_n \cdot o)$ are two fundamental sequences in the sense of Definition 3.35. The sequences (H_n) and (H'_n) are also fundamental and, by (2) and Theorem 3.29, they converge to the same limit in the compactification if and only if their formal limits $(C_I(\infty), \exp H^I \cdot o)$ and $(C'_I(\infty), \exp H'^{I'} \cdot o)$ agree. Let $k = \lim_n k_n$ and $k' = \lim_n k'_n$.

Assume that $(k \exp H_n \cdot o)$ and $(k' \exp H'_n \cdot o)$ converge to the same point x_0 in the given compactification of X. It follows from (3) that there is a fundamental sequence in $(k_1 A \cdot o) \cap (k_2 A \cdot o)$ that also converges to x_0. From (2) and Lemma 3.36(2) it follows therefore, that $(k \exp H_n \cdot o)$ and $(k' \exp H'_n \cdot o)$ have the same formal limit.

Conversely, assume that two fundamental sequences $(k_n \exp H_n \cdot o)$ and $(k'_n \exp H'_n \cdot o)$ are such that their formal limits, $(k \cdot C_I(\infty), k \exp H^I \cdot o)$ and $(k' \cdot C_{I'}(\infty), k' \exp H'^{I'} \cdot o)$, are the same. If $k_n \to k$ and $k'_n \to k'$, then $(k \exp H_n \cdot o)$ and $(k' \exp H'_n \cdot o)$ are also fundamental with the same formal limit $(k \cdot C_I(\infty), k \exp H^I \cdot o) = (k' \cdot C_{I'}(\infty), k' \exp H'^{I'} \cdot o)$.

Lemma 3.36(3) implies that $I = I', k^{-1}k' \in K^I M$, $H^I = H'^{I'}$, and $k \exp H^I \cdot o = k' \exp H^I \cdot o$.

It follows from property (3), that there is a fundamental sequence in $(kA \cdot o) \cap (k'A \cdot o)$ which has the same common formal limit: e.g., (y_n) with $y_n = k \exp(H^I + nH_0) \cdot o$, where $H_0 \in C_I$. It follows from (2) and Lemma 3.36(2) that the sequences $(k \exp H_n \cdot o)$ and $(k' \exp H'_n \cdot o)$ converge to the same point in the compactification. □

THE DUAL CELL COMPACTIFICATION $X \cup \Delta^*(X)$

3.40. Definition. The **dual cell compactification of** X is the compactification of X that is characterized by Theorem 3.38, equivalently by Theorem 3.39. It will be denoted by $X \cup \Delta^*(X)$. Equivalently, it will also be referred to as the dual cell compactification of $\mathfrak{p}$ and denoted by $\mathfrak{p} \cup \Delta^*(\mathfrak{p})$. Observe that, as defined, it depends upon the base point o.

Remark. By identifying $X \cup \Delta^*(X)$ with the maximal Satake compactification (see Theorem 4.43) or applying the method directly in [J2], one can show that $X \cup \Delta^*(X)$ is a topological ball of dimension $\dim X$, in particular, the boundary $\Delta^*(X)$ is a topological sphere of dimension $\dim X - 1$.

§ **3.41.** In Theorem 3.39, it is stated that the dual cell compactification of X is a K-compactification, i.e., the group K acts continuously on this compactification (see Definition 3.27). In fact, it is a G-compactification of X. This action is closely related to the Iwasawa decomposition and has as a consequence the fact that the compactification does not depend upon the base point.

First note that there is a natural action of G on the Tits building $\Delta(X)$. Since this building is isomorphic to the Tits building $\Delta(G)$ of G, it suffices to transport the action of G on $\Delta(G)$ to $\Delta(X)$. The group G acts on the set $\Delta(G)$ of parabolic subgroups by conjugation: $P' \to gP'g^{-1}$. If $P' = g'P^I g'^{-1}$ and $gg' = kan$, then $gP'g^{-1} = kP^I k^{-1}$. By passing to the simplices stabilized by parabolic subgroups (Proposition 3.20), this action, combined with the Iwasawa action (see § 2.1), gives the following action of G on $\Delta(X)$: if $C(\infty) = g' \cdot C_I(\infty) = k' \cdot C_I(\infty)$, where $k' = k(g')$, define $g \cdot C(\infty)$ to be $k(gg') \cdot C_I(\infty) = k(gk') \cdot C_I(\infty)$.

To see how $g \in G$ acts on a point $(C(\infty), y) = (k' \cdot C_I(\infty), k'a_1 \cdot o) \in \Delta^*(X)$ where $a_1 \in \overline{A^{I,+}}$, one uses generalized horocyclic coordinates (see § 7.6) on X. Let $g \cdot k' = k(gk')$ denote the Iwasawa action of G on K, where $gk' = (g \cdot k')\, an = (g \cdot k')s_1 s_2$ with $s_1 \in S_1 = A^I N^I, s_2 \in S_2 = A_I N_I$. Define the G-action as follows:

$$(*) \quad g \cdot (C(\infty), y) = g \cdot (k' \cdot C_I(\infty), k'a_1 \cdot o) = (k(gk') \cdot C_I(\infty), k(gk')s_1 a_1 \cdot o).$$

Clearly, if g belongs to K, this is the previous action of K on $\Delta^*(X)$. In addition, it follows from the Iwasawa decomposition applied to G^I that, under this group action, the set $\{(k' \cdot C_I(\infty), k'a_1 \cdot o) \in \Delta^*(X) \mid k' \in K, a_1 \in \overline{A^{I,+}}\}$ is invariant.

The reason for this definition will become more evident once the connection is made between the dual cell compactification and the Martin compactification $X \cup \partial X(\lambda_0)$ in Theorem 7.33. In the meantime, the following result shows that this action of G on $\Delta^*(X)$ is in fact a group action.

3.42. Proposition. *Let R^I denote the group $K^I M A_I N_I$. The mapping $(k' \cdot C_I(\infty), k'a_1 \cdot o) \to k'a_1 R^I$ is a G-equivariant bijection of $\{(k' \cdot C_I(\infty), k'a_1 \cdot o) \in \Delta^*(X) \mid k' \in K, a_1 \in \overline{A^{I,+}}\}$ onto G/R^I.*

Proof. Note that R^I is a group since $K^I M$ normalizes $A_I N_I$. By Lemma 3.36(3), if $(k' \cdot C_I(\infty)$ then k' is defined up to an element of K^I that commutes with a_1. Hence, the coset $k' a_1 R^I$ is well-defined.

On the other hand, if $k' a_1' R^I = k a_1 R^I$ with $k' \in K, a_1' \in \overline{A^{I,+}}$, then $a_1^{-1} k^{-1} k' a_1' \in R^I \subset P^I$. Hence, $k_1^{-1} k' \in P^I \cap K = K^I M$. It follows that $a_1^{-1} k^{-1} k' a_1' = \ell \in G^I M \cap R^I = K^I M$. Since $k^{-1} k' a_1' = \ell a_1$, the Cartan decomposition implies that $a_1 = a_1'$. Hence, $k^{-1} k' \in (K^I \cap a_1 K^I a_1^{-1}) M$. It follows from Lemma 3.36(3) that $(k' \cdot C_I(\infty), k' a_1' \cdot o) = (k \cdot C_I(\infty), k a_1 \cdot o)$.

Every point of G/R^I can be written in the form $k' a_1 R^I$: if $g = kan = k s_1 s_2, s_i \in S_i$, and $s_1 = k_1 a_1 k_1'$ with $k_1, k_1' \in K^I$ and $a_1 \in \overline{A^{I,+}}$, then $g = k' a_1 k_1' s_2$ and $k_1' s_2 \in R^I$. It follows that the map $(k' \cdot C_I(\infty), k' a_1 \cdot o) \to k' a_1 R^I$ is a bijection. Since $g k' a_1 R^I = k(gk') s_1 a_1 R^I$ if $gk' = k(gk') s_1 s_2$, with $s_i \in S_i$, the map is G-equivariant. □

This shows that G acts on the boundary $\Delta^*(X)$ of the dual cell compactification. In fact, this compactification is a G-compactification as stated in the following result.

3.43. Theorem. *The dual cell compactification $X \cup \Delta^*(X)$ is a G-compactification of X and the action of G on $\Delta^*(X)$ is the action defined by eq. (*) above.*

As explained earlier, the boundary $\Delta^*(X)$ of X is the union of cells corresponding to the symmetric spaces $k \cdot X^I$. For example, one associates with the simplex $C_I(\infty)$ in $X(\infty)$ the symmetric space X^I. In terms of the formal limits, this symmetric space at infinity is $\{(C_I(\infty), k_1 \exp H^I \cdot o) \mid k_1 \in K^I, H^I \in \overline{\mathfrak{a}^{I,+}}\}$. If one could assemble the boundary $\Delta^*(X)$ by adding symmetric spaces to lower-dimensional skeletons, it would be clear, as in Remark 3.33(4), that the closure of X^I in $\Delta^*(X)$ could be identified with $X^I \cup \Delta^*(X^I)$. The discussion in Remark 3.33(5) of the topology in the boundary $\Delta^*(\mathfrak{a})$, and the continuity of the K-action, shows that a sequence of points $(C_I(\infty), k_n a_n \cdot o) = k_n \cdot (C_I(\infty), a_n \cdot o)$, with $k_n \in K^I M, a_n \in \overline{A^{I,+}}$, converges to $k \cdot (C_J(\infty), a \cdot o)$ if and only if $(C_I(\infty), H_n^I) \to (C_J(\infty), H^J)$, where $H_n^I = \log a_n$ and $H^J = \log a$. Further, the limit point k' of any convergent subsequence of the sequence (k_n) equals k modulo $K^I M$ where $H^J \in C_I^J$. This indicates the main part of the proof of the next result.

3.44. Proposition. *Let $X^I(C_I(\infty))$ denote the set of points in $\Delta^*(X)$ corresponding to the set $\{(C_I(\infty), k_1 \exp H^I \cdot o) \mid k_1 \in K^I, H^I \in \overline{\mathfrak{a}^{I,+}}\}$ of formal limits. The boundary in $\Delta^*(X)$ of $X^I(C_I(\infty))$ is the set of points corresponding to the formal limits in the set $\{k_1 \cdot (C_J(\infty), \exp H^J \cdot o) \mid k_1 \in K^I, H^J \in \overline{\mathfrak{a}^{J,+}}, J \subsetneq I\}$.*

As a result, there is a natural embedding ϕ of $X^I \cup \Delta^*(X^I)$ into $\Delta^*(X)$ whose image is the closure of $X^I(C_I(\infty))$.

Proof. It remains to define ϕ. If $x = k_1 a_1 \cdot o \in X^I$ with $a_1 \in \overline{A^{I,+}}$, let $\phi(x) = (C_I(\infty), k_1 a_1 \cdot o)$. If $x = k_1 \cdot (C_J^I(\infty), a \cdot o)$ with $k_1 \in K^I M$, and $a \in \overline{A^{J,+}}$, let $\phi(x) = k_1 \cdot (C_J(\infty), a \cdot o)$. □

Finally, observe that one may define another geometrical compactification by using the Tits building $\Delta(X)$ and the conical compactification $\overline{X}^c$. Given any two compactifications (K_1, i_1) and (K_2, i_2) of X, let $i_1(x)$ and $i_2(x)$ be denoted by x to simplify the notation. The map $x \to (x, x)$ embeds X into the diagonal in $K_1 \times K_2$, and the closure of the image of X is another compactification that will be denoted by $K_1 \vee K_2$. It dominates each of the compactifications K_i — project $(z_1, z_2) \in K_1 \vee K_2$ to $z_i \in K_i$ — and has the property that if K is a compactification that dominates each of the compactifications K_i, then K dominates $K_1 \vee K_2$.

3.45. Proposition. *Let* $\varphi_1 : \overline{X}^c \vee (X \cup \Delta^*(X)) \to \overline{X}^c$ *be the map* $\varphi_1(z_1, z_2) = z_1$. *If* $z \in X(\infty) = \partial \overline{X}^c$, *then* $\varphi_1^{-1}(\{z\}) = X_z \cup \Delta^*(X_z)$. *In particular,* $\overline{X}^c \vee (X \cup \Delta^*(X)) = X \cup (\cup_{z \in X(\infty)} \{X_z \cup \Delta^*(X_z)\})$.

Proof. Clearly, a sequence (y_n), converging to infinity, converges in $\overline{X}^c \vee (X \cup \Delta^*(X))$ if and only if it converges in $\overline{X}^c$ and in $(X \cup \Delta^*(X))$. The fiber $\varphi_1^{-1}(z)$ is then identifiable with the set of limit points in $\Delta^*(X)$ of all fundamental sequences whose limiting direction is z.

In view of the K-action, it suffices to consider the case when $z = L(\infty), L \in C_I$. In the flat $A \cdot o$, the set of limits of exponentials of fundamental sequences (H_n) with limiting direction L is the space $\mathfrak{a}^I(\infty)$, and its closure in $X \cup \Delta^*(X)$ is identified with $\mathfrak{a}^I \cup \Delta^*(\mathfrak{a}^I)$. Since fundamental sequences on X arise, with the same limiting direction, by applying $k \in K^I M$ to any of the sequences $(\exp H_n \cdot o)$, it follows that the fiber contains $X^I \cup \Delta^*(X^I)$.

Now assume that $(k_n \exp H_n \cdot o)$ is fundamental with limiting direction $z = L(\infty)$. Since by assumption (k_n) converges to k, it follows that the sequence (H_n) has a limiting direction $L' \in \overline{\mathfrak{a}^+}$ and that $Ad(k)L' = L$. This implies that $L = L'$ and $k \in K^I M$. Hence, the fiber $\varphi_1^{-1}(L(\infty)) = X^I \cup \Delta^*(X^I)$. □

It remains to examine the fibration given by the map of $\overline{X}^c \vee (X \cup \Delta^*(X))$ onto $X \cup \Delta^*(X)$.

3.46. Proposition. *Let* $\varphi_2 : \overline{X}^c \vee (X \cup \Delta^*(X)) \to X \cup \Delta^*(X)$ *be the map* $\varphi_2(z_1, z_2) = z_2$. *If* $k \cdot (C_I(\infty), a^I \cdot o) \in \Delta^*(X)$, *then* $\varphi_2^{-1}(\{z\}) = \overline{k \cdot C_I(\infty)}$.

Proof. It is clear that if a sequence in X is I-fundamental and converges to $k \cdot (C_I(\infty), a^I \cdot o)$ and has a limiting direction that direction lies in $\overline{k \cdot C_I}$. Conversely, for any direction L in $\overline{k \cdot C_I}$, there is an I-fundamental sequence with L as its limiting direction that converges to $k \cdot (C_I(\infty), a^I \cdot o)$. □

CHAPTER IV

THE SATAKE–FURSTENBERG COMPACTIFICATIONS

The compactifications of symmetric spaces defined by Satake [S1], referred to nowadays as Satake compactifications, were motivated by the theory of automorphic forms and of representations. Furstenberg [F3] considered boundary value problems at infinity for the Laplacian on symmetric spaces and was led to isomorphic compactifications, as was shown by Moore [M8]. While these two families of compactifications are isomorphic, they are defined by quite different methods.

The Satake compactifications have played an important role in the celebrated Mostow rigidity theorem (see Mostow [M10]) and subsequent developments in rigidity theory. In this theory, as well as in the superrigidity theorem of Margulis [M4], the study of boundaries plays a basic role. Furthermore, boundary theory as well as the related Satake compactifications have important applications in ergodic theory and probability theory in the context of limit sets of subgroups of $\mathrm{GL}(n, \mathbb{R})$ or of Lyapunov exponents (see [F5], [G7], [G8] [G15] and the references cited).

The Satake compactifications of symmetric spaces were also used by Satake [S2] to define compactifications of quotients of symmetric spaces by arithmetic subgroups, i.e., compactifications of locally symmetric spaces, which play an essential role in the modern theory of automorphic forms and the cohomology of arithmetic subgroups (see [B11] and [J3] and the references cited). The results in this book also show that the so-called maximal Satake compactification is closely related to the potential theory on a symmetric space of non-compact type.

One of the main purposes of this chapter is to show that the dual cell compactification $X \cup \Delta^*(X)$ defined in Chapter III is isomorphic to the maximal Satake compactification, a compactification that will also appear throughout the later chapters of this book in its various realizations. Under this isomorphism, the G-action on the dual cell compactification corresponds to a very natural and explicit G-action on the maximal Satake compactification.

This chapter is organized as follows. It begins with a description of the finite dimensional representations of (real) semisimple Lie algebras. Then these results are used to give another characterization of parabolic subgroups. Following Satake [S1], the Satake compactifications are defined via irreducible, faithful representations (Definition 4.39) and a characterization of the maximal Satake compactification is recalled (Proposition 4.42). Using this, one proves easily that the dual cell compactification is isomorphic to the maximal Satake compactification (Theorem 4.43). The chapter concludes with a brief discussion of Moore's work [M8] on Furstenberg boundaries and compactifications. While additional details on this topic are to be found in Chapter IX and Appendix B, the results that are presented motivate the basic result of Moore, namely, that the maximal compactifications of Satake and Furstenberg are G-isomorphic.

In this chapter, G is assumed to have no compact normal subgroups.

FINITE DIMENSIONAL REPRESENTATIONS

§ 4.1. The space $\mathrm{SL}(n,\mathbb{C})/\mathrm{SU}(n)$ of positive definite Hermitian matrices of determinant 1 can be compactified by taking its closure in the real projective space $P(\mathcal{H}_n)$ associated with the complex vector space $\mathcal{H}_n$ of $n \times n$ Hermitian symmetric matrices. Satake showed that every symmetric space $X = G/K$ of non-compact type could be embedded as a totally geodesic submanifold of $\mathrm{SL}(n,\mathbb{C})/\mathrm{SU}(n)$ and, hence, compactified. The various compactifications of X produced in this way are called the **Satake compactifications of X**.

Each of these embeddings of X into $\mathrm{SL}(n,\mathbb{C})/\mathrm{SU}(n)$ turns out to be equivalent to realizing the Lie algebra $\mathfrak{g}$ of G as a Lie subalgebra of $\mathfrak{sl}(n,\mathbb{C})$. Such a realization, in turn, is given by any faithful representation of the Lie algebra $\mathfrak{g}$.

§ 4.2. Some basic facts concerning linear representations of semisimple Lie groups and their Lie algebras are now reviewed. In particular, the relation between the well-known theory of linear representation for semisimple complex Lie algebras [H6] and that for semisimple real Lie algebras and connected semisimple real Lie groups is discussed.

As well as being important for the study of Satake compactifications, these facts also play an important role in another approach to parabolic subgroups that is given in this chapter.

4.3. Definition. A **representation** of a semisimple Lie algebra $\mathfrak{g}$ is a Lie algebra homomorphism $\tau : \mathfrak{g} \to \mathfrak{sl}(V)$, where V is a finite dimensional complex vector space. The representation is called **irreducible** if there is no non-trivial subspace of V invariant under $\mathfrak{g}$. It is **faithful** if the homomorphism τ is injective.

4.4. Definition. Let $\mathrm{PSL}(V)$ denote the quotient of $\mathrm{SL}(V)$ by the (finite) subgroup of scalar matrices, i.e., scalar multiples of the identity matrix. A **representation** (respectively, **projective representation**) of a semisimple group G is a differentiable homomorphism τ of G into $\mathrm{SL}(V)$ (respectively, $\mathrm{PSL}(V)$). The representation is **faithful** if τ is injective and **irreducible** if no non-trivial subspace of V (respectively, $P(V)$) is G-invariant. The representation of $\mathfrak{g}$ associated by differentiation with a representation τ of G is called the **tangent representation associated with τ**. It is also denoted by τ.

4.5. Remarks. (1) In the definition of a representation it is usual to consider the algebra $\mathfrak{gl}(V)$ rather than $\mathfrak{sl}(V)$. Since $\mathfrak{g}$ is semisimple, it follows that every Lie algebra homomorphism from $\mathfrak{g}$ to $\mathbb{R}$ is trivial. Hence, if $\tau(\mathfrak{g}) \subset \mathfrak{gl}(V)$, it follows that $\operatorname{tr}(\tau(X)) = 0$ for every X in $\mathfrak{g}$ and so $\tau(\mathfrak{g}) \subset \mathfrak{sl}(V)$.

(2) In Definition 4.4, τ is assumed to be differentiable. This implies that τ is analytic — in fact, the continuity of τ is sufficient. Furthermore, the complete list of such representations of G is known [O3].

4.6. Proposition. *There is a one-to-one correspondence between irreducible projective representations of G and irreducible representations of $\mathfrak{g}$ that is given by differentiation. If the center of G is trivial, then faithfulness is also preserved.*

Proof. Since $\mathrm{PSL}(V)$ is a quotient of $\mathrm{SL}(V)$ by a finite subgroup, the Lie algebra of $\mathrm{PSL}(V)$ is also $\mathfrak{sl}(V)$. Hence, by differentiation, any projective representation of G gives a representation of $\mathfrak{g}$.

Conversely, let $\tilde{G}$ be the universal covering group of G. Then, the Lie algebra of $\tilde{G}$ is equal to $\mathfrak{g}$. Also, there exists a discrete subgroup in the center of $\tilde{G}$ such that G is the quotient of $\tilde{G}$ by this subgroup. In addition, as $\tilde{G}$ is simply connected, any representation $\tau : \mathfrak{g} \to \mathfrak{sl}(V)$ lifts to a representation $\tilde{\tau}$ of $\tilde{G}$ in $\mathrm{SL}(V)$. Since, by assumption, the representation τ is irreducible, Schur's lemma implies that any element in the center of $\tilde{G}$ is mapped to a scalar matrix. Hence, the representation of $\tilde{G}$ induces a projective representation of G.

If a projective representation of G is faithful, then the representation of $\mathfrak{g}$ is clearly faithful. Conversely, assume that the representation of $\mathfrak{g}$ is faithful and irreducible. Let τ be the corresponding projective representation of G. If the kernel of τ is non-trivial, it is a normal subgroup of G. If it is discrete, it will be in the center of G, which contradicts the assumption that G has trivial center. Hence, the Lie algebra of $\ker(\tau)$ is non-trivial and the representation of the Lie algebra is not faithful. This contradicts the faithfulness of τ. Consequently, the projective representation is also faithful. □

4.7. Remark. The proof of this proposition shows that any representation of $\mathfrak{g}$ can be lifted to a representation of a semisimple group $\tilde{G}_1$ with Lie algebra $\mathfrak{g}$ and finite center. More explicitly, if π is the covering map $\tilde{G} \to G$, consider the covering of G given by $\tilde{G}_1 = \tilde{G}/(\ker \pi \cap \ker \tilde{\tau})$. Clearly, the diagonal map δ of $\tilde{G}_1$ into $\tilde{G}/\ker \pi \times \tilde{G}/\ker \tilde{\tau} = G \times \tilde{\tau}(\tilde{G})$ is injective. The image $\delta(\tilde{G}_1)$ projects onto G and $\tilde{\tau}(\tilde{G})$. It follows that the center of $\delta(\tilde{G}_1)$ is contained in the product of the centers of G and of $\tilde{\tau}(\tilde{G}) \subset \mathrm{SL}(V)$. Since $\tilde{\tau}(\tilde{G})$ is a semisimple subgroup of $\mathrm{SL}(V)$, its center is finite. Hence, the center of $\delta(\tilde{G}_1)$ is finite and the group $\tilde{G}_1$ is the required covering of G.

Consequently, results for representations of $\mathfrak{g}$ can either be proved directly or by using the corresponding representation of a finite cover of G.

Depending on the situation, either a representation of $\mathfrak{g}$ or of G will be used in what follows. In the case of projective representations, one can always assume that G has trivial center and, hence, that the corresponding representation is faithful. Given a symmetric space $X = G/K$ of non-compact type, one can always suppose that G is connected and without center since it can always be divided out. However, as soon as representations of $\mathfrak{g}$ are considered, one cannot assume that the corresponding representation of G given by Proposition 4.6 has a trivial center.

WEIGHTS AND HIGHEST WEIGHTS

§ **4.8.** Basic facts about weights and highest weights are useful in the study of Satake compactifications and parabolic subgroups (see [H6] and [O3]). The finite dimensional irreducible representations of $\mathfrak{g}$ are classified by the **highest weight** of a Cartan subalgebra $\hat{\mathfrak{c}}$ of $\hat{\mathfrak{g}} = \mathfrak{g}_{\mathbb{C}}$ (see [H6, Chap VI]). For this reason one considers first the case of an irreducible representation $\hat{\tau}$ of a complex Lie algebra $\hat{\mathfrak{g}}$ in a complex vector space V.

For $\mu \in \hat{\mathfrak{c}}^*$, let V_μ denote the subspace

$$V_\mu = \{v \in V \mid \hat{\tau}(X)v = \mu(X)v| \text{ for all } X \in \hat{\mathfrak{c}}\}.$$

If $V_\mu \neq 0, \mu$ is called a **weight** of $\hat{\tau}$ and V_μ is called a **weight subspace.** If $\hat{\Lambda}$ is the set of weights of $\hat{\tau}$ then $V = \oplus_{\mu\in\hat{\Lambda}} V_\mu$. If one chooses a basis $\hat{\Delta}$ of the set of roots $\hat{\Sigma}$ of $\hat{\mathfrak{g}}$, the positive roots are defined (as the non-negative integral combinations of the roots in Δ), as is the maximal solvable subalgebra $\hat{\mathfrak{b}} = \hat{\mathfrak{c}} \oplus \sum_{\mu>0} \hat{\mathfrak{g}}_\mu$, which is a Borel subalgebra.

There exists a unique one-dimensional subspace $\mathbb{C}v \subset V$ that is a weight space and is $\hat{\tau}(\hat{\mathfrak{b}})$-invariant. The corresponding weight is called the **highest weight** and is denoted by μ_τ. The vector v is denoted by v_τ, i.e., $V_{\mu_\tau} = \mathbb{C}v_\tau$. For every other weight μ of $\hat{\tau}$ there are integers $c_\beta \geq 0$ such that $\mu = \mu_{\hat{\tau}} - \sum_{\beta\in\hat{\Delta}} c_\beta\beta$. Moreover, for each $\beta \in \hat{\Delta}$, $2\frac{\langle\mu_{\hat{\tau}},\beta\rangle}{\langle\beta,\beta\rangle}$ is a non-negative integer.

The collection of weights, for all representations $\hat{\tau}$, forms a lattice $\hat{\mathfrak{c}}^*(\mathbb{Z})$ in $\hat{\mathfrak{c}}^*$ called the **weight lattice.** The **dominant weights** are the weights $\mu \in \hat{\mathfrak{c}}^*(\mathbb{Z})$ such that $\langle\mu,\beta\rangle \geq 0$ for every $\beta \in \hat{\Delta}$. Every dominant weight is the highest weight of a unique irreducible representation of $\hat{\mathfrak{g}}$ [H6].

If $\mathfrak{g}$ is not complex and $\hat{\mathfrak{g}} = \mathfrak{g}_{\mathbb{C}}$, consider a Cartan subalgebra $\mathfrak{c} = \mathfrak{a} \oplus \mathfrak{a}'$ of $\mathfrak{g}$, where $\mathfrak{a}'$ is a maximal abelian subalgebra of $\mathfrak{m}$. Then $\hat{\mathfrak{c}} = (\mathfrak{a} \oplus \mathfrak{a}')_{\mathbb{C}}$ is a Cartan subalgebra of $\hat{\mathfrak{g}}$ and

$$\hat{\mathfrak{a}} = \mathfrak{a} \oplus i\mathfrak{a}' = \{x \in \hat{\mathfrak{c}} \mid \beta(x) \in \mathbb{R} \text{ for all } \beta \in \hat{\Sigma}\}$$

plays the role in $\hat{\mathfrak{g}}$ that $\mathfrak{a}$ plays in $\mathfrak{g}$.

The restriction $\bar{\beta}$ of $\beta \in \hat{\Sigma}$ to $\mathfrak{a}$ is 0 or a (restricted) root in $\Sigma \subset \mathfrak{a}^*$. Let $\hat{\Sigma}_0 \stackrel{\text{def}}{=} \{\beta \in \hat{\Sigma} \mid \bar{\beta} = 0\}, \hat{\Delta}_0 \stackrel{\text{def}}{=} \hat{\Delta} \cap \hat{\Sigma}_0$, and $\hat{\Delta}_1 \stackrel{\text{def}}{=} \hat{\Delta}/\hat{\Delta}_0$. Then

$\mathfrak{m}_{\mathbb{C}} = \mathfrak{a}'_{\mathbb{C}} \oplus_{\beta \in \hat{\Sigma}_0} \hat{\mathfrak{g}}_\beta$ and $(\mathfrak{g}_\lambda)_{\mathbb{C}} = \oplus_{\bar{\beta}=\lambda} \hat{\mathfrak{g}}_\beta$. Given $\mathfrak{a}, \mathfrak{a}'$, and Δ one can choose a basis of $(\mathfrak{a} + i\mathfrak{a}')^*$ adapted to Δ and $(i\mathfrak{a}')^*$. The set of simple roots $\hat{\Delta}$ associated with the corresponding lexicographic order has the property that the projection of $\hat{\Delta}_1$ onto $\mathfrak{a}^*$ is equal to Δ (see [O3 pp. 272–273] and [H2, p. 259–260]. In particular, the projection of $\hat{\Sigma}^+ \cup \{0\}$ is $\Sigma^+ \cup \{0\}$ and the Borel subalgebra $\hat{\mathfrak{b}}$ of $\hat{\mathfrak{g}}$ is contained in $(\mathfrak{m} + \mathfrak{a} + \mathfrak{n})_{\mathbb{C}}$. Furthermore, if $\alpha \in \Delta, \beta \in \hat{\Delta}_1$ and $\bar{\beta} = \alpha$, then β takes one or two values depending on α (this special information will not be needed here).

It is well known that the restriction of the Killing form of $\hat{\mathfrak{g}}$ to $\mathfrak{g}$ is the Killing form of $\mathfrak{g}$ (cf. [H2, p. 180]). As a result, the restriction of the scalar product of $\mathfrak{a} + i\mathfrak{a}'$ to $\mathfrak{a}$ is the scalar product in $\mathfrak{a}$. Furthermore, as $\mathfrak{a}$ and $i\mathfrak{a}'$ are orthogonal, $\mathfrak{a}^*$ can be identified with the orthogonal complement of $i\mathfrak{a}'$ in $(\mathfrak{a} + i\mathfrak{a}')^*$ and $(i\mathfrak{a}')^*$ with the orthogonal complement of $\mathfrak{a}$ in $(\mathfrak{a} + i\mathfrak{a}')^*$. With these identifications, the restriction of the scalar product of $(\mathfrak{a}+i\mathfrak{a}')^*$ to $\mathfrak{a}^*$ is the scalar product in $\mathfrak{a}^*$. Furthermore, $\mathfrak{a}^*$ and $(i\mathfrak{a}')^*$ are orthogonal in $(\mathfrak{a} + i\mathfrak{a}')^*$.

It follows from these remarks that, if $\gamma \in \hat{\Delta}$, the projection of $H_\gamma, \gamma \in \hat{\Delta}$, on $\mathfrak{a}$ is either zero or the vector H_α where $\bar{\gamma} = \alpha \in \Delta$.

If τ is an irreducible representation of $\mathfrak{g}$ in a complex vector space V, let $\hat{\tau}$ denote its extension to $\mathfrak{g}_{\mathbb{C}}$. This is an irreducible representation of $\hat{\mathfrak{g}} = \mathfrak{g}_{\mathbb{C}}$. Choose $\hat{\Delta}$ as above and, for $\lambda \in \mathfrak{a}^*$, set $V_\lambda = \{v \in V \mid \tau(X)v = \lambda(X) \text{ for all } X \in \mathfrak{a}\}$. If $V_\lambda \neq 0$, then λ is said to be a **weight** of τ and V_λ is called the associated **weight subspace**. Clearly, V_λ is the sum $\oplus_{\bar{\mu}=\lambda} V_\mu$ of the weight spaces of $\hat{\tau}$ for those weights μ of $\hat{\tau}$ for which the restriction $\bar{\mu}$ of μ to $\mathfrak{a}$ equals λ. If Λ is the set of weights of τ, it is clear that $V = \oplus_{\lambda \in \Lambda} V_\lambda$. The restriction μ_τ of $\mu_{\hat{\tau}}$ to $\mathfrak{a}$ is called the **highest weight** of τ. In general, the dimension $\dim V_{\mu_\tau}$ of the weight space V_{μ_τ} is greater than 1. From the complex case it follows that every weight λ of τ can be written as

$$\lambda = \mu_\tau - \sum_{\alpha \in \Delta} c_\alpha \alpha \tag{4.9}$$

where the c_α are non-negative integers.

If τ is a representation of a semisimple Lie group G, the **weights** of τ are defined to be the weights of the tangent representation of the Lie algebra $\mathfrak{g}$. Moreover, the properties of $\mathfrak{m}_{\mathbb{C}}, (\mathfrak{g}_\lambda)_{\mathbb{C}}, \hat{\Delta}$, and Δ show that the Borel subalgebra $\hat{\mathfrak{b}}$ of $\mathfrak{g}_{\mathbb{C}}$ is contained in $(\mathfrak{m} + \mathfrak{a} + \mathfrak{n})_{\mathbb{C}}$. If $\dim V_{\mu_\tau} > 1$, this algebra is used to define the unique $\hat{\mathfrak{b}}$-invariant one dimensional subspace $\mathbb{C}v_\tau \subset V (v_\tau \in V_{\mu_\tau})$. Clearly, the weight system of τ is invariant under the action of the (restricted) Weyl group of $\mathfrak{g}$. The following lemma shows how the space V is generated by v_τ and the action of $\mathfrak{g}$.

4.10. Lemma. *Let γ and $\lambda \in \mathfrak{a}^*$. Assume that $X \in \mathfrak{g}_\gamma$ and $v \in V_\lambda$. Then $\tau(X)v \in V_{\lambda+\gamma}$, i.e., $\tau(\mathfrak{g}_\gamma)V_\lambda \subset V_{\lambda+\gamma}$.*

Proof. Since $\tau(H)\tau(X)v = [\tau[H, X] + \tau(X)\tau(H)]v = (\gamma + \lambda)(H)\tau(X)v$, the result is obvious. □

4.11. Definition. An element $\lambda \in \mathfrak{a}^*$ is said to be **integral** if $\frac{2\langle\lambda,\alpha\rangle}{\langle\alpha,\alpha\rangle} = 2\frac{\lambda(H_\alpha)}{\langle\alpha,\alpha\rangle}$ is an integer for any $\alpha \in \Sigma$. It is said to be **dominant** if $\frac{2\langle\lambda,\alpha\rangle}{\langle\alpha,\alpha\rangle} \geq 0$ for all $\alpha \in \Delta$, where $\langle\ ,\ \rangle$ denotes the Killing form on $\mathfrak{a}^*$, i.e., if $\mathfrak{a}^*_+ = \{\lambda \in \mathfrak{a}^* \mid \langle\lambda,\alpha\rangle > 0 \text{ for all } \alpha \in \Delta\}$, then

$$\lambda \in \overline{\mathfrak{a}}^*_+ = \{\lambda \in \mathfrak{a}^* \mid \langle\lambda,\alpha\rangle \geq 0 \text{ for all } \alpha \in \Delta\}.$$

The set of integral elements in $\mathfrak{a}^*$ forms a lattice $\mathfrak{a}^*(\mathbb{Z})$ called the **weight lattice** and its intersection with $\overline{\mathfrak{a}}^*_+$ is the set of **dominant integral weights.** Two elements α, β of $\mathfrak{a}^*(\mathbb{Z})$ are said to be **connected** if they are not orthogonal, i.e., $\langle\alpha,\beta\rangle \neq 0$. A subset $\Gamma \subset \mathfrak{a}^*(\mathbb{Z})$ is said to be **connected** if it is not the disjoint union of two orthogonal subsets.

Remark. The inner product $\langle\ ,\ \rangle$ on $\mathfrak{a}^*$ is induced from the Killing form on $\mathfrak{a}$ by duality. More precisely, if $\alpha \in \Delta \subset \mathfrak{a}^*(\mathbb{Z})$ is represented by $H_\alpha \in \mathfrak{a}$, then $\langle\alpha,\beta\rangle = \beta(H_\alpha) = B(H_\alpha, H_\beta)$.

4.12. Lemma. *The weights of τ are integral and, hence, belong to the weight lattice.*

Proof. When $\mathfrak{g}$ is complex this lemma is well-known. In general, the proof is similar to that of the complex case if one uses the existence of a subalgebra of $\mathfrak{g}$ isomorphic to $\mathfrak{sl}(2,\mathbb{R})$ and associated with the root $\alpha \in \Delta$ (see [O3, p. 271]). This algebra is generated by three elements, H_α, X_α, and Y_α which satisfy the relations $[X_\alpha, Y_\alpha] = 2H_\alpha, [H_\alpha, X_\alpha] = 2X_\alpha, [H_\alpha, Y_\alpha] = -2Y_\alpha$. Let $H'_\alpha = 2\frac{H_\alpha}{\langle\alpha,\alpha\rangle}$ and $X_\alpha \in \mathfrak{g}_\alpha$ and $Y_\alpha \in \mathfrak{g}_{-\alpha}$. Then, as a result, for every $\beta \in \Sigma$, $\beta(H'_\alpha) = \frac{2\langle\beta,\alpha\rangle}{\langle\alpha,\alpha\rangle} \in \mathbb{Z}$. From the representation theory of $\mathfrak{sl}(2,\mathbb{R})$ (see [H6]) one knows that all the eigenvalues of $\tau(H'_\alpha)$ are integers. Hence, $\mu_\tau(H'_\alpha) = \frac{2\langle\mu_\tau,\alpha\rangle}{\langle\alpha,\alpha\rangle}$ is an integer. □

4.13. Proposition. *The set of weights of finite dimensional representations of $\mathfrak{g}$ form a sublattice of the weight lattice.*

Proof. Let $\hat{\mathfrak{c}}$ be a Cartan subalgebra of $\hat{\mathfrak{g}} = \mathfrak{g}_\mathbb{C}$ containing $\mathfrak{a}$(see § 4.8). Then every integral element in $\hat{\mathfrak{a}}^*$ is a weight of some representation (see [H6, Chap. VI]). The weights of a representation τ in $\mathfrak{a}^*$ are obtained by restriction from $\hat{\mathfrak{a}}$ to $\mathfrak{a}$. By Lemma 4.12, the projection on $\mathfrak{a}^*$ of the weight lattice in $\hat{\mathfrak{a}}^*$ forms a sublattice of the weight lattice in $\mathfrak{a}^*$. □

4.14. Lemma. *Let $\mathfrak{c} = \mathfrak{a} \oplus \mathfrak{a}'$ denote a Cartan subalgebra of $\mathfrak{g}$ with $\mathfrak{a}' \subset \mathfrak{m}$. Let Δ and $\hat{\Delta}$ be the two sets of simple roots for $\mathfrak{g}$ and $\hat{\mathfrak{g}}$, respectively, that were defined in §4.8. If $\gamma \in \hat{\Delta}$ and its restriction $\bar{\gamma}$ to $\mathfrak{a}$ is not zero, then $||\bar{\gamma}||^2/||\gamma||^2$ is rational.*

Proof. Let $\mathfrak{k} + \mathfrak{p}$ denote the Cartan decomposition of $\mathfrak{g}$ and denote by θ the corresponding Cartan involution. Let θ also denote its linear extension to $\mathfrak{g}_\mathbb{C}$. Since θ is an involutive automorphism of $\hat{\mathfrak{g}}$, it preserves $\hat{\Sigma}$ and $\theta(\hat{\mathfrak{g}}_\gamma) =$

$\hat{\mathfrak{g}}_{\gamma\circ\theta}$ for any $\hat{\gamma} \in \hat{\Sigma}$. Let γ^θ denote $\gamma \circ \theta$ and observe that $\theta(X) = -X$ if $X \in \mathfrak{a}, \theta(Y) = Y$ if $Y \in i\mathfrak{a}'$. As indicated in § 4.8, one can decompose $\gamma \in (\mathfrak{a} + i\mathfrak{a}')^*$ into the sum of its restrictions to $\mathfrak{a}$ and $i\mathfrak{a}'$, i.e., $\gamma = \bar{\gamma} + \bar{\gamma}'$ with $\bar{\gamma} \in \mathfrak{a}^*$ and $\bar{\gamma}' \in (i\mathfrak{a})^*$. Then $\bar{\gamma}' \circ \theta + \bar{\gamma} \circ \theta = \gamma \circ \theta = \bar{\gamma}' - \bar{\gamma}$. Hence, $2\bar{\gamma} = \gamma - \gamma \circ \theta = \gamma - \gamma^\theta$. The theory of complex semisimple Lie algebras implies that, for any $\gamma, \delta \in \hat{\Sigma}$, the scalar product $\langle \gamma, \delta \rangle$ is rational [H6]. Since γ and γ^θ are roots if $\gamma \in \hat{\Sigma}$, it follows that $\|\gamma\|^2$ and $4\|\bar{\gamma}\|^2 = \|\gamma - \gamma^\theta\|^2$ are rational numbers. Consequently, $\frac{\|\bar{\gamma}\|^2}{\|\gamma\|^2}$ is rational. □

4.15. Proposition. *There exists an integer N such that, for every dominant weight λ in $\mathfrak{a}^*(\mathbb{Z})$, the multiple $N\lambda$ is the highest weight of an irreducible representation of $\mathfrak{g}$.*

Proof. Let $H_\alpha^* \in \mathfrak{a}^*, \alpha \in \Delta$, and $H_\gamma^* \in (\mathfrak{a} + i\mathfrak{a}')^*, \gamma \in \hat{\Delta}$, denote the respective dual bases of $\mathfrak{a}^*$ and $(\mathfrak{a}+i\mathfrak{a}')^*$ corresponding to the bases $H_\alpha, \alpha \in \Delta$, and $H_\gamma, \gamma \in \hat{\Delta}$, of $\mathfrak{a}$ and $\mathfrak{a} + i\mathfrak{a}'$. The restriction of $\mu \in (\mathfrak{a} + i\mathfrak{a}')^*$ to $\mathfrak{a}$ is denoted by $\bar{\mu}$ and the orthogonal projection of $H \in \mathfrak{a} + i\mathfrak{a}'$ onto $\mathfrak{a}$ by $\overline{H}$. As pointed out in 4.8, if $\gamma \in \hat{\Delta}$ either $\bar{\gamma} \in \Delta$ or $\bar{\gamma} = 0$ and $\overline{H_\mu} = H_{\bar{\mu}}$. Since $H = \sum_{\gamma \in \hat{\Delta}} H_\gamma^*(H) H_\gamma$ if $H \in \mathfrak{a} + i\mathfrak{a}'$, it follows, by projecting orthogonally onto $\mathfrak{a}$, that $\overline{H} = \sum_{\gamma \in \hat{\Delta}} H_\gamma^*(H) H_{\bar{\gamma}} = \sum_{\alpha \in \Delta} (\sum_{\bar{\gamma} = \alpha} H_\gamma^*(H)) H_\alpha$. As a result, $H_\alpha^* = \sum_{\bar{\gamma} = \alpha} \bar{H}_\gamma^*$.

Let μ_γ denote $\frac{\|\gamma\|^2}{2} H_\gamma^*$ (these are the fundamental weights) and μ_α denote $\frac{\|\alpha\|^2}{2} H_\alpha^*$. If $\mu \in (\mathfrak{a} + i\mathfrak{a}')^*$, then

$$\mu = \sum_{\gamma \in \hat{\Delta}} \mu(H_\gamma) H_\gamma^* = \sum_{\gamma \in \hat{\Delta}} \frac{2\langle \mu, \gamma \rangle}{\|\gamma\|^2} \mu_\gamma.$$

Hence, by the theory of representations of semisimple Lie algebras, it follows that μ is a weight if and only if $\mu = \sum_{\gamma \in \hat{\Delta}} u_\gamma \mu_\gamma$ with $u_\gamma \in \mathbb{Z}$. In the same way, for any $\lambda \in \mathfrak{a}^*$, one has

$$\lambda = \sum_{\alpha \in \Delta} \lambda(H_\alpha) H_\alpha^* = \sum_{\alpha \in \Delta} \frac{2\langle \lambda, \alpha \rangle}{\|\alpha\|^2} \mu_\alpha = \sum_{\alpha \in \Delta} u_\alpha \mu_\alpha.$$

In other words, $\lambda \in \mathfrak{a}^*(\mathbb{Z})$ if and only if each $u_\alpha \in \mathbb{Z}$. Furthermore, λ is dominant if and only if the coordinates u_α of λ in the basis μ_α are positive integers. Since $H_\alpha^* = \sum_{\bar{\gamma} = \alpha} \bar{H}_\gamma^*$, it follows that $\mu_\alpha = \sum_{\bar{\gamma} = \alpha} \frac{\|\alpha\|^2}{\|\gamma\|^2} \bar{\mu}_\gamma$.

Observe that if two irreducible representations τ, τ' are given, then $\mu_\tau + \mu_{\tau'}$ is the highest weight of an irreducible component of the tensor product representation $\tau \otimes \tau'$. Hence, if N is the common denominator of the rational numbers $\frac{\|\alpha\|^2}{\|\gamma\|^2}$, $N\mu_\alpha$ is a linear combination with positive integers of the fundamental weights μ_γ, $\gamma \in \hat{\Delta}$, for which $\bar{\gamma} = \alpha$. It follows that $N\mu_\alpha$ is the highest weight of an irreducible representation of $\mathfrak{g}$. As a result, $N\lambda$ is the highest weight of an irreducible representation of $\mathfrak{g}$ if λ is a dominant weight. □

4.16. Lemma. *Suppose $\mathfrak{g}$ is simple and non-compact. Then a representation τ is faithful if and only if its highest weight μ_τ is not equal to zero.*

Proof. The weights μ of τ are of the form $\mu = \mu_\tau - \sum_{\alpha\in\Delta} c_\alpha \alpha$, $c_\alpha \geq 0$. Since τ is faithful, for some weight μ and positive root α, it follows that $c_\alpha > 0$. Otherwise, all the weights μ equal μ_τ. This implies that $\mathfrak{a}$ acts on V by homotheties. Since the trace vanishes on $\mathfrak{a}$, the action of $\mathfrak{a}$ is trivial. As $\mathfrak{g}$ is simple and $\mathfrak{a} \neq \{0\}$, it follows that the action of $\mathfrak{g}$ is also trivial. The fact that τ is faithful implies that $\mathfrak{g} = \{0\}$, which is impossible.

Hence, if $d = \dim V, H \in \mathfrak{a}^+$, it follows that Tr $(\tau(H)) < d\mu_\tau(H)$. Since $H \in \mathfrak{a}^+$, the fact that Tr $(\tau(H)) = 0$ implies that $\mu_\tau(H) > 0$. As a result, $\mu_\tau \neq 0$.

On the other hand, suppose $\mu_\tau \neq 0$. Then, since $\mathfrak{g}$ is simple, τ is faithful, as the representation is non-trivial. □

4.17. Proposition. *Let $\mathfrak{g}$ be a semisimple Lie algebra, and let τ an irreducible representation of $\mathfrak{g}$. Let $I^\perp(\tau)$ denote the union of the components of Δ which are orthogonal to μ_τ. Then the kernel of τ is the semisimple ideal $\mathfrak{g}^{I^\perp(\tau)}$. In particular τ is faithful if and only if $\{\mu_\tau\} \cup \Delta$ is connected, i.e., every connected component of Δ contains a root that is not orthogonal to μ_τ.*

Proof. Let $\mathfrak{g} = \mathfrak{g}_1 + \mathfrak{g}_2 \cdots + \mathfrak{g}_k$ be the decomposition of $\mathfrak{g}$ into simple ideals, each one corresponding to a component of Δ. Since the kernel of τ is an ideal, it is the sum of a certain number of the ideals $\mathfrak{g}_i$. Hence, it is the semisimple ideal $\mathfrak{g}^I \subset \mathfrak{g}$, where I is a union of components of Δ.

Clearly, for $\alpha \in I, H_\alpha \in \mathfrak{a}^I$ and $\mu_\tau(H_\alpha) = \langle \mu_\tau, \alpha \rangle = 0$. Hence, $I \subset I^\perp(\tau)$. The restriction of τ to $\mathfrak{g}_i$ is a direct sum of p_i equivalent irreducible representations τ_i of $\mathfrak{g}_i$. Hence, for $X \in \mathfrak{a} \cap \mathfrak{g}_i$, it follows that $\mu_\tau(X) = p_i \mu_{\tau_i}(X)$. Lemma 4.16 implies that either τ_i is faithful or $\mu_{\tau_i} = 0$ and $\tau_i = 0$. It follows that $\mathfrak{g}^I$ is the sum of the simple ideals $\mathfrak{g}_i$ such that $\mu_{\tau_i} = 0$. If $X \in \mathfrak{a} \cap \mathfrak{g}_i$, it follows, from the formula $\mu_\tau(X) = p_i\mu_{\tau_i}(X)$, that $\mu_{\tau_i} = 0$ if and only if $\mu_\tau(\mathfrak{a} \cap \mathfrak{g}_i) = 0$. Hence, μ_τ is orthogonal to the component of Δ associated with $\mathfrak{g}_i$. Consequently, $I^\perp(\tau) \subset I$, and so $I = I^\perp(\tau)$. Finally, τ is faithful if and only if $I^\perp(\tau) = \emptyset$, equivalently, if and only if $\{\mu_\tau\} \cup \Delta$ is connected. □

Representation and parabolic subgroups

In this section a geometrical and dynamical interpretation of parabolic subgroups is given (see Propositions 4.19 and 4.27) using the above results on representations. In particular, another characterization is given of the parabolic subgroups of a connected semisimple group (see Theorem 4.29). It is used when discussing Satake compactifications, and also in Chapter IX and Appendix B.

4.18. Proposition. *Let τ be an irreducible representation of G in a vector space V. Denote by $V_\tau = V_{\mu_\tau}$ the weight subspace corresponding to the* *highest weight* μ_τ*. Let* P_τ *denote the stabilizer of* V_τ*. Then* P_τ *is a standard parabolic subgroup.*

Proof. If $\alpha \in \Sigma$, it follows from Lemma 4.10 that $\tau(\mathfrak{g}_\alpha)V_\tau \subset V_{\mu_\tau+\alpha}$. From formula (4.9), one sees that if $\alpha \in \Sigma^+$, then $\mu_\tau + \alpha$ is not a weight. Hence, $\tau(\mathfrak{g}_\alpha)(V_\tau) = \{0\}$. Since $\mathfrak{n} = \sum_{\alpha>0} \mathfrak{g}_\alpha$, it follows that $\tau(\mathfrak{n})V_\tau = \{0\}$. As a result, $\tau(AN)$ leaves V_τ invariant and acts on V_τ by homotheties of ratio e^{μ_τ}, i.e., $\tau(an)v = e^{\mu_\tau}(a)v$, where $e^{\mu_\tau}(a) = e^{\mu_\tau(\log a)}$. Since M centralizes $\mathfrak{a}$, M acts trivially on $\mathfrak{a}^*$ and so $\tau(M)V_\tau = V_\tau$. Consequently, $\tau(MAN)V_\tau = V_\tau$. This implies that $MAN = P \subset P_\tau$ and so P_τ is a standard parabolic subgroup. □

4.19. Proposition. *Let* μ_τ *be the highest weight of* τ*. Let*

$$I_\tau^\perp = \{\alpha \in \Delta \mid \langle \mu_\tau, \alpha \rangle = 0\}.$$

Then the stabilizer of V_τ *in* G *is the standard parabolic subgroup* $P^{I_\tau^\perp}$*.*

The following lemmas are used to prove this result.

4.20. Lemma. *If* $\alpha \in \Delta$ *and* $\mu_\tau - \alpha$ *is a weight of* τ*, then* $\langle \mu_\tau, \alpha \rangle \neq 0$.

Proof. It is clear that the Weyl group W of G acts on the set of weights of τ. Furthermore, to each $\alpha \in \Delta$ there corresponds a reflection s_α with respect to the hyperplane $\ker \alpha$ (see [W1, p. 13] and [O3, p. 272]), i.e., $s_\alpha(H) = H - 2\frac{\langle H, H_\alpha \rangle}{\langle H_\alpha, H_\alpha \rangle} H_\alpha$. Dualizing to $\mathfrak{a}^*$, it follows that $s_\alpha \cdot \lambda = \lambda - 2\frac{\langle \lambda, \alpha \rangle}{\langle \alpha, \alpha \rangle}\alpha$. Hence, if $\mu_\tau - \alpha$ is a weight, $s_\alpha \cdot (\mu_\tau - \alpha) = (\mu_\tau - \alpha) - 2\frac{\langle \mu_\tau - \alpha, \alpha \rangle}{\langle \alpha, \alpha \rangle}\alpha$. Consequently, if $\langle \mu_\tau, \alpha \rangle = 0$, it follows that $s_\alpha \cdot (\mu_\tau - \alpha) = \mu_\tau + \alpha$. Since, by formula (4.9), $\mu_\tau + \alpha$ is not a weight, it follows that $\langle \mu_\tau, \alpha \rangle \neq 0$. □

4.21. Lemma. *Let* $p_\tau = \dim V_\tau$*. The character* e^{μ_τ} *of* A*, defined by* $e^{\mu_\tau}(a) = e^{\mu_\tau(\log a)}$*, is the restriction to* A *of a homomorphism* γ_τ *of* P_τ *into* $\mathbb{R}_+^*$*.*

Proof. Let $\theta_\tau(g) \in \mathbb{C}^*$ denote the determinant of the restriction of $g \in P_\tau$ to V_τ. Define $\gamma_\tau(g)$ to be $|\theta_\tau(g)|^{\frac{1}{p_\tau}}$. Since μ_τ is real on $\mathfrak{a}$, if $a = \exp H$, where $H \in \mathfrak{a}$, $\gamma_\tau(a) = |\theta_\tau(a)|^{\frac{1}{p_\tau}} = e^{\mu_\tau}(a)$ □

4.22. Lemma. *The group* $P^{I_\tau^\perp}$ *preserves* V_τ*. For any* M*-invariant scalar product on* V_τ*, the subgroup* $G^{I_\tau^\perp}MAN$ *of* P_τ *acts on* V_τ *by homothety and isometry. More explicitly, if* $v \in V_\tau$ *and* $g = g_1 \ell a n$*, with* $g_1 \in G^{I_\tau^\perp}, \ell \in M, a \in A$*, and* $n \in N$*, then* $g \cdot v = e^{\mu_\tau(\log a)}\tau(\ell)v$ *and, hence,* $\|g \cdot v\| = \gamma_\tau(g)\|v\| = e^{\mu_\tau(\log a)}\|v\|$ *for* $v \in V_\tau$*.*

Proof. From the proof of Proposition 4.18 it follows that $\tau(AN)$ acts on V_τ by homotheties of ratio e^{μ_τ} and M preserves V_τ. If $\overline{\mathfrak{n}}^{I_\tau^\perp}$ is the sum of the root subspaces in $\overline{\mathfrak{n}}$ corresponding to the roots in $I_\tau^\perp$, then $\tau(\overline{\mathfrak{n}}^{I_\tau^\perp})V_\tau = \{0\}$.

To see this, note that from Lemma 4.10 it follows that, if $\beta \in \Sigma$, $\tau(\mathfrak{g}_\beta)V_\tau \subset V_{\mu_\tau+\beta}$. As a result, if $\beta = -\alpha \in I_\tau^\perp$, then $\tau(\mathfrak{g}_{-\alpha})V_\tau \subset V_{\mu_\tau-\alpha}$. If $\alpha \in I_\tau^\perp$, it follows from lemma 4.20 that $\mu_\tau - \alpha$ is not a weight of τ. Hence, $\tau(\mathfrak{g}_{-\alpha})V_\tau = 0$. This shows that $\tau(\overline{\mathfrak{n}}^{I_\tau^\perp})V_\tau = 0$.

Let $\mathfrak{g}(I_\tau^\perp)$ denote the Lie algebra generated by $\overline{\mathfrak{n}}^{I_\tau^\perp}$ and $\mathfrak{n}^{I_\tau^\perp}$ (see Remarks 2.14). Then $\tau(\mathfrak{g}(I_\tau^\perp)+\mathfrak{n})V_\tau = \{0\}$. Hence, the action of $\tau(Y+H+Z)$ on V_τ, for $Y \in \mathfrak{g}(I_\tau^\perp), H \in \mathfrak{a}$, and $Z \in \mathfrak{n}$, reduces to the action of $\tau(H)$, i.e., to the homothety of ratio $\mu_\tau(H)$. As a result, if $h = g_1 an$ with $g_1 \in G(I_\tau^\perp), a \in A$, and $n \in N$, then $\tau(h)$ acts on V_τ by the homothety of ratio $e^{\mu_\tau}(a)$.

Consequently, if $g = mh$ with $m \in M$, the action of $\tau(g)$ is the composition of the homothety of ratio e^{μ_τ} and the isometry $\tau(m)$. The result follows. □

Proof of proposition 4.19.

It follows from Proposition 4.18 that $P_\tau \supset P$ is a standard parabolic subgroup. Hence, for some $I \subset \Delta$, it follows that $P_\tau = P^I = G^I P$, where G^I is semisimple. Let γ_τ be the homomorphism of P_τ into $\mathbb{R}_+^*$, defined in Lemma 4.21, that extends e^{μ_τ}. Then, since G^I is semisimple and connected, $\gamma_\tau(G^I) = \gamma_\tau(A^I) = \{1\}$.

Hence, $\langle \mu_\tau, \alpha \rangle = 0$ for $\alpha \in I$ since $\langle \mu_\tau, \alpha \rangle = \mu_\tau(H_\alpha)$ and μ_τ is real on $\mathfrak{a}$. This proves that $I \subset I_\tau^\perp$ and $P_\tau \subset P^{I_\tau^\perp}$. From Lemma 4.22 it follows that $P^{I_\tau^\perp}$ preserves V_τ. Hence, $P^{I_\tau^\perp} \subset P_\tau$ and so $P^{I_\tau^\perp} = P_\tau$ □

4.23. Remark. The proof shows that $g \in G(I_\tau^\perp)AN$ acts on V_τ by a homothety of ratio $e^{\mu_\tau}(a)$ if $g = ha\eta$, where $h \in G(I_\tau^\perp), a \in A$, and $\eta \in N$.

4.24. Proposition. *Let τ be an irreducible representation of G in V. Denote by p_τ the dimension of V_τ. Let $v_\tau \in \wedge^{p_\tau} V$ denote a p_τ-vector corresponding to the subspace V_τ. Then $\wedge^{p_\tau}[\tau(G)]v_\tau$ generates a subspace $V^\wedge \subset \wedge^{p_\tau} V$. The corresponding representation $\wedge^{p_\tau}\tau = \tau^\wedge$ in $V^\wedge$ is irreducible with highest weight $p_\tau \mu_\tau$. If $\bar{v}_\tau \in P(V^\wedge)$ is the direction of v_τ, then $\tau^\wedge(G) \cdot \bar{v}_\tau$ is a compact analytic submanifold of $P(V^\wedge)$ and $\tau^\wedge(G)$ acts on it by projective transformations. Further, the stabilizer of $\bar{v}_\tau$ is P_τ. Hence, $G/P_\tau = \tau^\wedge(G) \cdot \bar{v}_\tau$.*

Proof. Since G is semisimple, the representation $\wedge^{p_\tau}\tau$ of G in $\wedge^{p_\tau} V = U$ decomposes as a sum of irreducible representations, i.e., $U = \oplus_{i=1}^r W_i$. If U_λ is a weight subspace of $\wedge^{p_\tau}\tau$, then $U_\lambda = \oplus_{i=1}^r U_\lambda \cap W_i$ and U_λ is a p-wedge product of weight subspaces of τ. Namely, if $V = \oplus_{\lambda \in \Lambda_\tau} V_\lambda$, then $\wedge^{p_\tau} V = \oplus_{\mu \in \Lambda'_\tau} U_\mu$, where $U_\mu = V_{\lambda_1} \wedge V_{\lambda_2} \wedge \cdots \wedge V_{\lambda_{p_\tau}}$, $\mu = \lambda_1 + \lambda_2 + \cdots \lambda_{p_\tau}$, and Λ'_τ is the sum of p_τ copies of Λ_τ.

As a result, $\wedge^{p_\tau}(a)v_\tau = e^{p_\tau \mu_\tau}(a)v_\tau$ and so the new weights are of the form $\lambda = p_\tau \mu_\tau - \sum_{\beta \in \Delta} c_\beta \beta$, where $c_\beta \geq 0$ is an integer. If $\lambda = p_\tau \mu_\tau$ then $U_\lambda = \mathbb{C}v_\tau$. Hence, there exists some i, $1 \leq i \leq r$, such that $U_\lambda \subset W_i = V^\wedge$. Consequently, $W_i = V^\wedge$ is generated by $\wedge^{p_\tau}[\tau(G)]v_\tau$. Clearly, this defines a new representation $\tau^\wedge$ of G in $V^\wedge$ with the same properties as τ and with highest weight $p_\tau \mu_\tau$. From Proposition 4.19, it follows that $I_{\tau^\wedge} = I_\tau^\perp$ and $P_{\tau^\wedge} = P_\tau$. Moreover, $\dim V_{\tau^\wedge} = 1$. Since $\tau^\wedge$ is analytic, $\tau^\wedge(G) \cdot \bar{v}_\tau$ is an analytic submanifold of $P(V^\wedge)$ and $G/P_\tau = \tau^\wedge(G) \cdot \bar{v}_\tau$ is compact. □

4.25. Remark. The above proposition shows that one can always assume $\dim V_\tau = 1$ if τ is irreducible, replacing τ by $\tau^\wedge$ if necessary. This does not change the parabolic subgroup P_τ and multiplies the highest weight by an integer.

4.26. Corollary. *If the weight μ_τ is generic, i.e., satisfies $\langle \mu_\tau, \alpha\rangle > 0$ for every $\alpha \in \Delta$, then $P_\tau = P$ is a standard minimal parabolic subgroup and the corresponding tangent representation of $\mathfrak{g}$ is faithful.*

Proof. Since μ_τ is generic, $I_\tau^\perp = \emptyset$, and so it follows from Proposition 4.19 that $P_\tau = P$. Clearly, μ_τ is not orthogonal to any component of Δ and, so, from Proposition 4.17, the tangent representation of $\mathfrak{g}$ is faithful. □

4.27. Proposition. *For any standard parabolic subgroup Q of G, there exists an irreducible representation τ of G with V_τ one-dimensional and such that $Q = P_\tau$. The kernel of the tangent representation is $\mathfrak{g}^{I^\perp(\tau)}$, the largest semisimple ideal of $\mathfrak{g}$ contained in the Lie algebra of P_τ. In particular, it is faithful if and only if Q does not contain any simple factor of G. (Recall from Proposition 4.17 that $I^\perp(\tau)$ is the sum of the components of Δ which are orthogonal to μ_τ).*

Proof. Given $I \subset \Delta$, the equations $\langle \mu, \alpha\rangle = 0$ for all $\alpha \in I$ define a rational linear subspace $\mathfrak{a}_I^*$ of $\mathfrak{a}^*$, i.e., $\mathfrak{a}_I^* \cap a^*(\mathbb{Z})$ is a lattice in $\mathfrak{a}_I^*$. Moreover, the intersection of the positive Weyl chamber of $\mathfrak{a}^*$ with $\mathfrak{a}_I^*$ is $\{\mu \mid \langle \mu, \alpha\rangle > 0 \text{ for all } \alpha \notin I\}$. As a result, this intersection is an open simplicial cone of $\mathfrak{a}_I^*$ and so $\mathfrak{a}_I^*$ contains a dominant weight μ. By Proposition 4.15, there exists an irreducible representation τ of $\mathfrak{g}$ and an integer p such that $\mu_\tau = p\mu$. As pointed out in Remark 4.7, τ can be lifted to a representation $\tilde{\tau}$ of a semisimple covering group $\tilde{G}$ of G with a finite center F. This is now shown to determine a representation of G.

First, note that, because of Schur's lemma, the elements of $\tilde{\tau}(F)$ are scalar multiples of the identity matrix. Since $\tilde{\tau}(F)$ is finite, if $\gamma \in F$ then $\tilde{\tau}(\gamma)^r = Id$ with $r = \text{card}(F)$.

Now consider the tensor product $\tilde{\tau}^{\otimes r}$ and observe that $\tilde{\tau}^{\otimes r}(F) = \{Id\}$. Take an irreducible component of $\tilde{\tau}^{\otimes r}$ with highest weight $pr\mu$. This gives a representation of the adjoint group $Ad(\tilde{G}) = Ad(G)$and, hence, a representation $\tilde{\tau}$ of G itself. It follows from Remark 4.25 that one can replace $\tilde{\tau}$ by $\tilde{\tau}^\wedge$, with the result that the new highest weight is an integer multiple of μ, and the dimension of $V_{\tilde{\tau}^\wedge}$ is one. Hence, $I_{\tilde{\tau}^\wedge} = I_\tau^\perp$ and $I(\tilde{\tau}^\wedge) = I^\perp(\tau)$.

Proposition 4.17 implies that the kernel of the tangent representation of $\tilde{\tau}^\wedge$ is $\mathfrak{g}^{I^\perp(\tau)}$. It follows from Proposition 4.19 that $Q = P_\tau = P_{\tilde{\tau}^\wedge}$. Furthermore, the definition of $\mathfrak{g}^{I^\perp(\tau)}$ and Proposition 4.19 imply that $\mathfrak{g}^{I^\perp(\tau)}$ is the product of the simple factors of $\mathfrak{g}$ contained in the Lie algebra of P_τ.

It follows from this that $\tilde{\tau}^\wedge$ is faithful. Hence, the representation $\tilde{\tau}^\wedge$ is a representation that satisfies the required conditions. □

4.28. Proposition. *Let τ be an irreducible representation of G on a complex vector space V. Then the compact G-orbits in $P(V)$ are the orbits of the points of the projective subspace $P(V_\tau)$. If the dimension of* $\dim V_\tau$ *is one, and $x, y \in G \cdot \bar{v}_\tau$, then there exists a sequence $(g_n) \subset G$ such that* $\lim_n g_n \cdot x = \lim_n g_n \cdot y = \bar{v}_\tau$.

Proof. Lemma 4.22 and Remark 4.23 imply that P_τ acts on V_τ by homotheties and isometries. Hence, for every $\bar{v} \in P(V_\tau)$, the orbit $\tau(P_\tau) \cdot \bar{v}$ is compact. The orbit $\tau(G) \cdot \bar{v} \subset P(V)$ is also compact because G/P_τ is compact.

Conversely, suppose that the orbit $\tau(G) \cdot \bar{v}$ is compact for some $\bar{v} \in P(V)$ and let $a \in \exp \mathfrak{a}^+$. Then, for every weight $\mu \neq \mu_\tau$ of τ, it follows from formula (4.9) that $e^{\mu}(a) < e^{\mu_\tau}(a)$.

Let π_λ denote the projection of V on $V_\lambda, \lambda \in \Lambda$, associated with the direct sum $V = \oplus_{\lambda \in \Lambda} V_\lambda$. Then, for every $x \in V, x = \sum_{\lambda \in \Lambda} \pi_\lambda x$ and $\tau(a^n)x = \sum_{\lambda \in \Lambda} e^{n\lambda}(a)\pi_\lambda x$. This implies that $e^{-n\mu_\tau}(a)\tau(a^n)x = \pi_{\mu_\tau} x + \sum_{\lambda \neq \mu_\tau} e^{n(\lambda - \mu_\tau)}(a)\pi_\lambda x$. As a result, $\lim_n e^{-n\mu_\tau}(a)\tau(a^n)x = \pi_{\mu_\tau} x$ since $a \in \exp \mathfrak{a}^+$. Hence, the sequence $e^{-n\mu_\tau}(a)\tau(a^n)$ converges to the projection π_τ.

It follows from the irreducibility of τ that $\tau(G) \cdot \bar{v}$ is not contained in the proper projective subspace $P(\ker \pi_\tau)$. If $\bar{w} \in \tau(G) \cdot \bar{v} \setminus P(\ker \pi_\tau)$ then $\tau(a^n) \cdot \bar{w}$ converges to $\bar{w}' \in P(V_\tau)$ as $\lim_n \tau(a^n) \cdot \bar{w} = \pi_\tau \cdot \bar{w} = \bar{w}' \in P(V_\tau)$.

Since $\tau(G) \cdot \bar{w} = \tau(G) \cdot \bar{v}$ is compact, it follows that $\bar{w}' \in \tau(G) \cdot \bar{w} = \tau(G) \cdot \bar{v} = \tau(G) \cdot \bar{w}'$. This proves the first part of the proposition.

If $\dim V_\tau = 1$, then $\pi_\tau(V) = \mathbb{C}v_\tau$ and, for $x, y \in P(V)/P(\ker \pi_\tau)$, it follows that $\lim_n a^n \cdot x = \lim_n a^n \cdot y = \bar{v}_\tau$. To complete the proof of the second statement, it suffices to show that if x or y belong to $P(\ker \pi_\tau)$, then there exists $g \in G$ with $g \cdot x \notin P(\ker \pi_\tau)$, and $g \cdot y \notin P(\ker \pi_\tau)$. Then, if $g_n = a^n g$, it follows that $\lim_n g_n \cdot x = \lim_n g_n \cdot y = \bar{v}_\tau$.

In order to show the existence of g, for $z \in G \cdot \bar{v}_\tau$, let A_z denote $\{g \in G \mid g \cdot z \in P(\ker \pi_\tau)\}$. If $G = A_x \cup A_y$, then one of the closed sets A_x or A_y has non-empty interior. If, for example, $U \neq \emptyset$ is an open subset of A_x, then $U \cdot x$ is an open subset of $G \cdot \bar{v}_\tau$ contained in $P(\ker \pi_\tau)$. Since the condition that z belongs to $P(\ker \pi_\tau)$ is analytic and satisfied on an open subset of the analytic, connected manifold $G \cdot \bar{v}_\tau$, it is satisfied everywhere. Hence, $G \cdot \bar{v}_\tau \subset P(\ker \pi_\tau)$ and so $G \cdot v_\tau \subset \ker \pi_\tau$.This is a contradiction since τ is irreducible and $G \cdot v_\tau$ generates V. It follows that $A_x \cup A_y \neq G$. Hence, there exists $g \in G$ such that $g \cdot x \notin P(\ker \pi_\tau)$ and $g \cdot y \notin P(\ker \pi_\tau)$. □

4.29. Theorem. *Let Q be a closed subgroup of G. Then Q is parabolic if and only if there exists an irreducible representation τ of G on V such that the dimension of V_τ is one and Q is the stabilizer of a point in the orbit of $\bar{v}_\tau \in P(V)$.*

Moreover, the unique compact orbit of G *in* $P(V)$ *is* $G \cdot \bar{v}_\tau$. *For any two points* x, y *in the compact orbit* $G \cdot \bar{v}_\tau$, *there exists a sequence* $g_n \in G$ *such that* $\lim_n g_n \cdot x = \lim_n g_n \cdot y = \bar{v}_\tau$

Proof. If Q is a standard parabolic subgroup, the desired representation exists in view of Proposition 4.27 with $Q = P_\tau$. In general, one can find a standard parabolic subgroup P^I and $g \in G$ such that $Q = gP^Ig^{-1} = gP_\tau g^{-1}$. It follows that Q is the stabilizer of $g \cdot \bar{v}_\tau \in G \cdot \bar{v}_\tau \subset P(V)$.

Conversely, if τ is a representation with the given properties, then Q is conjugate to the stabilizer of $\bar{v}_\tau$. Hence, Q is conjugate to the parabolic group P_τ (see Proposition 4.18). The last assertions follow from Proposition 4.28. □

4.30. Remarks. (1) The action of a group H on a space is said to be proximal if the last statement of Theorem 4.29 is satisfied (see Definition 9.32). The property of proximality, which motivates Definition 4.45, is of fundamental importance in the general theory of boundaries (see Proposition 9.33). below.

(2) Let V and τ be as in Theorem 4.29. Let $G^a \subset GL(V)$ be the smallest algebraic subgroup of $GL(V)$ containing $\tau(G)$ and let P^a_τ be the smallest algebraic subgroup containing $\tau(P_\tau)$. Then it can be shown that G^a is a semisimple complex Lie group and G^a/P^a is a projective algebraic manifold, which is isomorphic in the algebraic sense to $\tau(G^a) \cdot \bar{v}_\tau$. Hence, P^a is a parabolic subgroup of G^a in the sense of algebraic group theory [B8] and $\tau(P_\tau) = P^a \cap \tau(G)$. This shows the connection with parabolic subgroups in algebraic group theory. In the present real analytic context, in view of Propositions 4.27 and 4.29, a parabolic subgroup Q of a connected semisimple group could be defined as a subgroup such that G/Q is isomorphic to a compact analytic submanifold of some projective space on which G acts by projective transformations.

(3) It is well known (see [G1, pp. 57–66]) that parabolic subgroups and their related properties can be discussed without requiring that the group be connected. In particular if H is semisimple with finite center and H has a finite number of connected components, the group $P = MAN$ may be defined as in Proposition 2.4 and a closed subgroup Q of H is said to be parabolic if one of its conjugates contain MAN. Here AN is a maximal solvable subgroup in H_0, the connected component of H, such that the group $Ad(AN)$ is represented by real triangular matrices. The group M is defined to be the centralizer of A in a maximal compact subgroup K of H. It is known that $H = KH_0$ (see [G1]).

The spaces H/Q are called flag manifolds. The parabolic subgroups Q of H correspond to those of the connected component $H_0 = G$ since $Q = MQ_0$ where $Q_0 = Q \cap H_0 = Q \cap G$ is a parabolic subgroup of G. This follows from the fact that $H = MH_0 = MG$ [G1, p. 60]. Hence the flag manifolds of H are the flag manifolds of G, and so one can, in principle, reduce questions about parabolic subgroups of H to corresponding questions for the parabolic subgroups of $H_0 = G$. This is the point of view taken in this book.

With slight modifications the results of this section remain valid. In particular, Theorem 4.29 remains valid with the following modification: if H is a semisimple Lie group with a finite number of components and finite center, the representation τ in 4.29 is not only irreducible but its restriction to the connected component H_0 of identity is irreducible.

Instead of irreducibility of a representation τ of H, one has to use the notion of strong irreducibility, that is irreducibility of $\tau(H_0)$. In this study one can always replace H by the linear group $Ad(H)$ and consider the subgroup $\hat{H}$ of real points of the smallest algebraic group which contains $Ad(H)$. Then $\tau(H)$ is of finite index in $\hat{H}$ and one can use the results of algebraic group theory [B11]. Hence, the above concept of a parabolic subgroup contains both the corresponding notion of algebraic group theory relative to the real field context and the usual concept (see [G1] for details).

4.31. Example. Let G denote the linear group $\mathrm{SL}(n, \mathbb{R})$ and consider its natural representation τ on $\mathbb{R}^n$ and the representations τ_p for $1 \leq p < n$ in the wedge product $\wedge^p \mathbb{R}^n$ determined by

$$\tau_p(g)(v_1 \wedge \cdots \wedge v_p) = \tau(g)v_1 \wedge \cdots \wedge \tau(g)v_p.$$

Let τ_p denote $\wedge^p \tau$ and $\tau_1 = \tau$. It is well known that, for p odd, τ_p is faithful and irreducible and that every irreducible representation of G can be obtained as an irreducible component of the decomposition of the tensor products of the τ_p [F2, p. 221–237].

The subgroup $K = SO(n)$ is maximal compact and $\mathfrak{sl}(n, \mathbb{R}) = \mathfrak{so}(n) \oplus \mathfrak{p}$, where $\mathfrak{p}$ is the set of symmetric matrices of trace zero. The diagonal matrices of trace zero form a maximal abelian subalgebra $\mathfrak{a}$ and if one defines the positive Weyl chamber to be the set of diagonal matrices with strictly decreasing eigenvalues, then the corresponding minimal parabolic subgroup $P = MAN$ is the upper triangular subgroup. The weight space $V_{\tau_p} \subset \wedge^p \mathbb{R}^n$ is the line generated by $e_1 \wedge e_2 \wedge \cdots \wedge e_p$. Hence, P_{τ_p} is the set of block triangular matrices of the form

$$g = \begin{pmatrix} A & B \\ O & C \end{pmatrix}$$

with $\dim A = p, \dim C = n - p$.

The standard parabolic subgroups P^I are subgroups of upper block-triangular matrices, where the block structure is related to the set $I \subset \Delta$.

The flag manifold G/P_{τ_1} is the projective space P^{n-1}. The flag manifold G/P_{τ_p} is the canonical projection in $P(\wedge^p \mathbb{R}^n)$ of the set of decomposable p-vectors, hence, the manifold of p-dimensional subspaces. A representation τ for which $P_\tau = P$ may be obtained as follows. Let W denote the irreducible component of the tensor product $\mathbb{R}^n \otimes \wedge^2 \mathbb{R}^n \otimes \cdots \otimes \wedge^{n-1} \mathbb{R}^n$ which contains

$E_1 \otimes \cdots \otimes E_{n-1}$, where E_p denotes the p-vector $e_1 \wedge \cdots \wedge e_p \in \wedge^p \mathbb{R}^n$. The natural representation τ of G in W is clearly such that $P_\tau = P$. Here the flag manifolds can be interpreted in an elementary way in terms of flags in $\mathbb{R}^n$ [B8, pp. 77–80], which explains why they are referred to in this way.

Satake Compactifications

Let $\tau : G \to \mathrm{PSL}(V)$ be an irreducible faithful projective representation of G in a complex vector space V of dimension n.

Let $\mathfrak{g} = \mathfrak{k} \oplus \mathfrak{p}$ be a Cartan decomposition of $\mathfrak{g}$ for which $\mathfrak{k}$ is the Lie algebra of K.

4.32. Proposition. *There exists a Hermitian scalar product ϕ on V such that, if u^* denotes the adjoint of $u \in \mathrm{End}(V)$ with respect to ϕ, for any $Y \in \mathfrak{g}$, $\tau(\theta(Y)) = -\tau(Y)^*$, i.e., relative to any orthonormal base of V, it follows that $\tau(\theta(Y)) = -{}^t\overline{\tau(Y)}$, where θ is the Cartan involution defined by the Cartan decomposition $\mathfrak{g} = \mathfrak{k} \oplus \mathfrak{p}$.*

Furthermore, the scalar product ϕ is unique up to a positive coefficient.

Proof. Let $\mathfrak{g}_\mathbb{C}$ be the complexification of $\mathfrak{g}$ and $\tilde{\mathfrak{g}} = \mathfrak{k} + i\mathfrak{p}$ its compact real form corresponding to the Cartan decomposition of $\mathfrak{g}$. Let $\tilde{G} \subset \mathrm{SL}(V)$ denote the compact subgroup of $\mathrm{SL}(V)$ whose Lie algebra equals $\tilde{\mathfrak{g}}$. Then, given a scalar product $\langle x, y \rangle$ on V, by integration over $\tilde{G}$, one obtains the $\tilde{G}$-invariant scalar product ϕ on V, where

$$\phi(x, y) = \int_{\tilde{G}} \langle kx, ky \rangle dk.$$

If $Y = U + Z \in \mathfrak{g}$ and $X = U + iZ \in \tilde{\mathfrak{g}}$, with $U \in \mathfrak{k}$ and $Z \in \mathfrak{p}$, the $\tilde{G}$-invariance of ϕ implies that $\tau(X) = -\tau(X)^*$. Hence, $\tau(U) = -\tau(U)^*$ if $U \in \mathfrak{k}$, and $\tau(Z) = \tau(Z)^*$ if $Z \in \mathfrak{p}$. As a result, $\theta(Y) = U - Z$ implies that

$$\tau(\theta(Y)) = \tau(U - Z) = -\tau(U + Z)^* = -\tau(Y)^*.$$

Denote by ψ another Hermitian scalar product relative to which $\tau(\theta(Y)) = -\tau(Y)^*$ if $Y \in \mathfrak{g}$. It is clear from the above calculation that this condition is equivalent to the $\tilde{G}$-invariance of ψ. If $\lambda \in \mathbb{R}$, the sesquilinear form $\lambda\phi - \psi$ is also $\tilde{G}$-invariant. Since ϕ, ψ are Hermitian and ϕ is non-degenerate, for some $\lambda \in \mathbb{R}$, $\det(\lambda\phi - \psi) = 0$ and, hence, $\lambda\phi - \psi$ is degenerate. The kernel of $\lambda\phi - \psi$ is a non-trivial subspace W of V that is $\tilde{G}$-invariant. Since $\tau(\tilde{\mathfrak{g}})$ generates $\tau(\mathfrak{g}_\mathbb{C})$, it follows that $\tau(\tilde{\mathfrak{g}})$ is irreducible. This implies that $W = V$ and so $\psi = \lambda\phi$. □

4.33. Let $\mathcal{H}_n$ denote the vector space of $n \times n$ Hermitian matrices and $P(\mathcal{H}_n)$ the associated real projective space, which can be identified with $\mathrm{SL}(n, \mathbb{C})/\mathrm{SU}(n)$. Let $\mathcal{P}_n$ be the set of elements of determinant 1 in the open cone of positive definite Hermitian matrices in $\mathcal{H}_n$.

4.34. Corollary. *Identify V with $\mathbb{C}^n$ by using an orthonormal basis of V with respect to the Hermitian scalar product in Proposition 4.32. Then for any $g \in G$, $\tau(g)\tau(g)^*$ is well-defined and belongs to $\mathcal{P}_n$.*

Proof. First note that the map $\bar{g} \in \mathrm{PSL}(n, \mathbb{C}) \to gg^* \in \mathcal{P}_n$ is well-defined since the different lifts of $\bar{g}$ in $\mathrm{SL}(n, \mathbb{C})$ differ by roots of unity. It then suffices to prove that $\tau(k)\tau(k)^* = Id$ for $k \in K$. If $k = \exp X$, with $X \in \mathfrak{k}$, then $\exp \tau(X) \in \mathrm{SL}(n, \mathbb{C})$ is a lift of $\tau(k)$. By Proposition 4.32, $\exp \tau(X) \exp \tau(X)^* = Id$, and, hence, $\tau(k)\tau(k)^* = Id$. □

4.35. Lemma. *Let τ_X denote the map $g \cdot o \in X \to \tau(g)\tau(g)^* \in \mathcal{P}_n$. It is G-equivariant and injective.*

Proof. Clearly, for $h, g \in G$, $\tau(hg)\tau(hg)^* = \tau(h)\tau(g)\tau(g)^*\tau(h)^*$. Hence, $g \cdot o \to \tau(g)\tau(g)^*$ is a G-equivariant map from G/K to the homogeneous space $\mathrm{SL}(n, \mathbb{C})/\mathrm{SU}(n)$.

If $g \in G$, $g \notin K$ then, using the fact that $G = \exp \mathfrak{p} K$ (see [H2, Theorem 1.1 (iii), p. 252]), let $g = \exp X\, k$, where $k \in K$ and $X \neq 0 \in \mathfrak{p}$. Since, by Proposition 4.32, $\exp \tau(X) \in \mathrm{SL}(n, \mathbb{C})$ is a Hermitian matrix, it follows that $\tau_X(g \cdot 0) = \tau(g)^t\overline{\tau(g)} = \tau(p)^t\overline{\tau(p)} = \exp 2\tau(X) \neq Id$. Hence, $\tau_X : X \to \mathcal{P}_n$ is injective. □

4.36. Lemma. *The map $\tau_X : X \to \mathcal{P}_n$ is a topological embedding.*

Proof. It is shown in [H2, Theorem 1.1, p. 252] that the map $Y \to \exp Y \cdot o$ of $\mathfrak{p}$ to X is a homeomorphism. Since the map $g \to \tau(g)$ is continuous, it follows that $\exp Y \cdot o \to \exp \tau(Y) \exp \tau(Y)^* = \exp 2\tau(Y)$ is continuous.

Conversely, let (Y_n) be a sequence in $\mathfrak{p}$ and $Y \in \mathfrak{p}$ such that $\tau_X(\exp Y_n \cdot o) \to \tau_X(\exp Y \cdot o)$ or, equivalently, such that $\exp 2\tau(Y_n) \to \exp 2\tau(Y)$. It suffices, in view of the fact that $\mathfrak{p}$ and X are homeomorphic under $Y \to \exp Y \cdot o$, to show that $Y_n \to Y$. Since τ is faithful, it is enough, by considering convergent subsequences, to show that (Y_n) is bounded in $\mathfrak{p}$.

If Z is Hermitian, it follows that $\exp \|Z\|$ is the maximum of $\|\exp(Z)\|$ and $\|\exp(-Z)\|$. Since $\tau(Y_n)$ and $\tau(Y)$ are Hermitian, it follows from this that if $\exp 2\tau(Y_n) \to \exp 2\tau(Y)$ then $\|Y_n\|$ is bounded in $\mathfrak{p}$. □

4.37. Remark. Because of the uniqueness of the Hermitian scalar product ϕ on V in Proposition 4.32, this embedding $\tau_X : X \to \mathcal{P}_n$ depends only on the representation τ. The image $\tau_X(X)$ is a totally geodesic submanifold in the ambient symmetric space $\mathcal{P}_n$ and any totally geodesic embedding of X into $\mathcal{P}_n$ is realized this way, as mentioned in §4.1 (see Satake [S1, §2] for details).

4.38. The restriction of the projection $\pi : \mathcal{H}_n \to P(\mathcal{H}_n)$ to $\mathcal{P}_n$ is clearly an embedding. Composed with the embedding $\tau_X : X \to \mathcal{P}_n$, this gives an embedding $X \to P(\mathcal{H}_n)$,which will be denoted by τ.

4.39. Definition. The closure of $\tau(X)$ in $P(\mathcal{H}_n)$ is called the **Satake compactification associated with the representation** τ. It will be denoted by $\overline{X}^S_\tau$.

Two Satake compactification $\overline{X}^S_{\tau_1}$ and $\overline{X}^S_{\tau_2}$ are isomorphic if their highest weights μ_{τ_1} and μ_{τ_2} lie in the same Weyl chamber face of $\mathfrak{a}^*_+$ (see [S1, Theorem 2 in §4] and Appendix B). As a result, when τ varies, there are only finitely many different Satake compactifications. When μ_τ is generic, i.e., when it belongs to the interior of $\mathfrak{a}^*_+$, the compactification $\overline{X}^S_\tau$ is maximal in the sense that it dominates all other Satake compactifications. For this reason, $\overline{X}^S_\tau$ will be denoted as $\overline{X}^S_{\max}$ when the representation is generic. The existence of such a representation follows from Proposition 4.15.

This **maximal Satake compactification** $\overline{X}^S_{\max}$ is one of the main subjects of this book. Using Satake's characterization (see [S1, p. 100]), it will now be shown to coincide with the dual cell compactification that was defined in Chapter III. In view of Theorem 3.38, this also gives another characterization of the maximal Satake compactification as that compactification in which all the fundamental sequences converge and in which their limits agree if and only if their formal limits agree. This characterization will be used throughout this book to identify various realizations of the maximal Satake compactification, including Furstenberg's maximal compactification.

In view of the polar decomposition, it is clear that, for any G-compactification of the symmetric space X, it is of crucial importance to identify which sequences (a_n) in the closure of the positive Weyl chamber $A^+ \cdot o$ converge in the compactification. In other words, in the case of a Satake compactification $\overline{X}^S_\tau$, a crucial step in determining it is the identification of the closure of a Weyl chamber in $\overline{X}^S_\tau$ (see [S1, p. 94]).

Assume for the rest of this chapter that τ is a generic representation of G, i.e., the highest weight μ_τ belongs to the interior of $\mathfrak{a}^*_+$. Let μ be any weight of τ. Then there exist non-negative integers c_α, $\alpha \in \Delta$, such that

$$\mu = \mu_\tau - \sum_{\alpha\in\Delta} c_\alpha \alpha.$$

Define the **support** of μ, denoted by $\mathrm{Supp}(\mu)$, to be $\{\alpha \in \Delta \mid c_\alpha > 0\}$.

The following result (a special case of [S1, Lemma 7 in 2.3]) is the key to determining the convergence of sequences $\tau(a_n)$. (A proof is given in Appendix B, see Lemma B.8.)

4.40 Lemma. *Given a representation τ, every subset of Δ is the support $\mathrm{Supp}(\mu)$ of some weight μ.*

With the aid of Lemma 4.40, the next lemma shows that there is a connection between fundamental sequences and sequences $(a_n \cdot o)$ in $A^+ \cdot o$ that converge in $\overline{X}^S_{\max}$.

4.41. Lemma. *Let H_m be a sequence in the closed positive Weyl chamber $\overline{\mathfrak{a}^+} \subset \mathfrak{a}$ such that $\|H_n\| \to \infty$. Let $I = \{\alpha \in \Delta \mid \limsup_{m\to+\infty} \alpha(H_m) < +\infty\}$. Then $\exp H_m \cdot o \in X$ is convergent in the Satake compactification $\overline{X}_\tau^S$ if and only if for every $\alpha \in I$, $\lim_{m\to+\infty} \alpha(H_m)$ exists, is finite and for every $\alpha \notin I, \lim_m \alpha(H_m) = +\infty$. In other words $(a_m \cdot o)$, where $a_m = \exp H_m$, is C_I-fundamental (see Definition 3.35).*

Proof. Since $H_m \in \mathfrak{a}$, one can assume that $\tau(\exp H_m)$ is a diagonal matrix:

$$\tau(\exp H_m) = (\exp \mu_1(H_m), \dots, \exp \mu_n(H_m)) \in \mathrm{PSL}(n, \mathbb{C}),$$

where $\mu_1, \dots, \mu_n$ are the weights of τ and $\exp \mu_i(H_m)$ is the ith diagonal entry. Hence,

$$\begin{aligned}\tau(\exp H_m \cdot o) &= \tau(\exp H_m)\tau(\exp H_m)^* \\ &= (\exp 2\mu_1(H_m), \dots, \exp 2\mu_n(H_m)) \in \mathcal{P}_n.\end{aligned}$$

For simplicity, assume μ_1 is the highest weight μ_τ. Write

$$\mu_i = \mu_\tau - \sum_{\alpha\in\Delta} c_{i,\alpha}\alpha = \mu_1 - \sum_{\alpha\in\Delta} c_{i,\alpha}\alpha,$$

where each $c_{i,\alpha}$ is a non-negative integer. Embedding the sequence in $P(\mathcal{H}_n)$, it follows that, in $P(\mathcal{H}_n)$,

$$\tau(\exp H_m \cdot o) = (1, \exp -2\sum_{\alpha\in\Delta} c_{2,\alpha}\alpha(H_m), \dots, \exp -2\sum_{\alpha\in\Delta} c_{n,\alpha}\alpha(H_m)).$$

Let I be the subset of Δ that is bounded on (H_m). Then

$$\limsup_{m\to+\infty} \sum_{\alpha\in\Delta} c_{m,\alpha}\alpha(H_m) < +\infty \text{ if and only if } \mathrm{supp}(\mu_i) \subset I.$$

Hence, it follows from Lemma 4.40 that $\tau(\exp H_j \cdot o)$ is convergent in $P(\mathcal{H}_n)$ if and only if, for every $\alpha \in I$, $\lim_{j\to+\infty} \alpha(H_j)$ exists and is finite and, for every $\alpha \notin I, \lim_j \alpha(H_j) = +\infty$. Since $\overline{X}_\tau^S$ is the closure in $P(\mathcal{H}_n)$ of $\tau(X)$, the lemma follows. □

This lemma says that the closure of the positive Weyl chamber in $\overline{X}_\tau^S$ does not depend on the particular generic highest weight μ_τ, but only on the fact that it lies in $\mathfrak{a}_+^*$.

Satake gave the following characterization of $\overline{X}_\tau$ which involves the orbital structure of the compactification and includes the convergence result given in Lemma 4.41.

4.42. Proposition. (See Satake *[S1, p. 100]*) *Different generic representations τ of G define the same Satake compactification $\overline{X}^S_\tau$ which dominates all other Satake compactifications and is therefore denoted by $\overline{X}^S_{\max}$. The maximal Satake compactification $\overline{X}^S_{\max}$ is characterized by the following properties:*

(1) *$\overline{X}^S_{\max}$ is a G-compactification.*
(2) *For any subset $I \subset \Delta$, let X^I be the symmetric space associated to the parabolic subgroup P^I (see § 2.13). There exists a P^I-equivariant embedding ι_I of X^I into $\overline{X}^S_{\max}$ with image X^I_∞, and*

$$\overline{X}^S_{\max} = \cup_{I \subset \Delta} G \cdot X^I_\infty.$$

(3) *For any two subsets $I_1, I_2 \in \Delta$, if $gX^{I_1}_\infty \cap X^{I_2}_\infty \neq \emptyset$, then $I_1 = I_2 = I$ and $g \in P^I$.*
(4) *For any $I \subset \Delta$, identify $\overline{A^{I,+}} \cdot o \subset X^I$ with the subset $\iota_I(\overline{A^{I,+}} \cdot o) = \overline{A^{I,+}} \cdot \iota_I(o)$ of $\overline{X}^S_{\max}$ by (2) above. Then the closure of $\overline{A^+} \cdot o$ in $\overline{X}^S_{\max}$ is equal to the union $\cup_{I \subset \Delta} \overline{A^{I,+}} \cdot \iota_I(o)$. Furthermore, a sequence $(\exp H_n \cdot o) \subset \overline{A^{I_1,+}} \cdot o$ converges to $\exp H^{I_2} \cdot \iota_{I_2}(o) \in \overline{A^{I_2,+}} \cdot \iota_{I_2}(o)$ if and only if $I_2 \subset I_1$, $\lim_{n \to \infty} \alpha(H_n) = \alpha(H^{I_2})$ for $\alpha \in I_2$, and $\lim_{n \to \infty} \alpha(H_n) = +\infty$ for $\alpha \in I_1 \setminus I_2$.*

A new proof of Satake's result is given in Chapter IX. It makes essential use of Furstenberg's maximal compactification. Other results of Satake concerning the non-maximal compactifications are proved in Appendix B.

4.43. Theorem. *The maximal Satake compactification $\overline{X}^S_{\max}$ and the dual cell compactification $X \cup \Delta^*(X)$ (see Definition 3.40) are G-isomorphic compactifications of X.*

Proof. If $(y_n) \subset A \cdot o$ is a C_I-fundamental sequence with formal limit $(C_I(\infty), a^I \cdot o)$ in $X \cup \Delta^*(X)$ (see Definition 3.24), it follows from Proposition 4.42.(4) that (y_n) converges to $a^I \cdot \iota_I(o) \in X^I$ in $\overline{X}^S_{\max}$. Since $\overline{X}^S_{\max}$ is a G-compactification, it follows that if (y_n) is a C_I-fundamental sequence with a formal limit $(k \cdot C_I(\infty), ka^I \cdot o)$, then (y_n) converges to the limit $ka^I \cdot \iota_I(o) \in k \cdot X^I_\infty \subset \overline{X}^S_{\max}$. Therefore, it follows from Theorem 3.38 and Lemma 3.28 that there is a continuous, surjective G-equivariant map from $X \cup \Delta^*(X)$ to $\overline{X}^S_{\max}$.

Since both $X \cup \Delta^*(X)$ and $\overline{X}^S_{\max}$ are compact Hausdorff spaces, to show that this map is a homeomorphism, it suffices to show that it is also injective. Let $(y_n), (y'_n)$ be two fundamental sequences with different formal limits $(k \cdot C_I(\infty), ka^I \cdot o), (k' \cdot C_{I'}(\infty), k'a'^I \cdot o)$. There are two cases: (1) $k \cdot C_I(\infty) \neq k' \cdot C_{I'}(\infty)$ and (2) $k \cdot C_I(\infty) = k' \cdot C_{I'}(\infty)$, but $ka^I \cdot \iota_I(o)$, $k'a'^I \cdot \iota_I(o)$ are different points in $k \cdot X^I_\infty = k' \cdot X^{I'}_\infty$. In Case (1), the boundary spaces $k \cdot X^I_\infty$ and $k' \cdot X^{I'}_\infty$ are disjoint in $\overline{X}^S_{\max}$ by Proposition 4.42.(3). Hence, the limits of (y_n) and (y'_n) in $\overline{X}^S_{\max}$ are distinct. In Case (2), $I = I'$ and (y_n) and (y'_n) converge in $\overline{X}^S_{\max}$ to different points $ka^I \cdot \iota_I(o)$, $k'a'^I \cdot \iota_I(o)$ in the same boundary space $k \cdot X^I_\infty$. Therefore, in both cases, the sequences (y_n) and (y'_n) have different limits in $\overline{X}^S_{\max}$. By Theorem 3.38 again, this proves that the map from $X \cup \Delta^*(X)$ to $\overline{X}^S_{\max}$ is injective. This completes the proof of Theorem 4.43. □

4.44. Example. Consider the case of $G = \mathrm{PSL}(n, \mathbb{C})$ and $X = \mathcal{P}_n = \mathrm{SL}(n, C)/\mathrm{SU}(n)$. Let $\mathfrak{a}$ be the subalgebra consisting of the diagonal matrixes $H = (e_1(H), \cdots, e_n(H))$ in $\mathfrak{sl}(n, \mathbb{C})$, where $H \in \mathfrak{a}$ and e_i are the coordinate functions on $\mathfrak{a}$. Choose as simple roots $\alpha_1 = e_1 - e_2, \cdots, \alpha_{n-1} = e_{n-1} - e_n$. If the representation τ is the standard projective representation $\tau_1 = Id : G \to \mathrm{PSL}(n, \mathbb{C})$, then $\mu_\tau = e_1$ and so does not belong to the interior of the Weyl chamber if $n > 2$ as $\langle e_1, e_2 - e_3 \rangle = 0$. Hence, the compactification $\overline{X}^S_{\tau_1}$ is not maximal if $n > 2$.

In order to get the maximal Satake compactification $\overline{X}^S_{\max}$, take τ to be the irreducible component of the tensor product $\tau_1 \otimes \wedge^2 \tau_1 \otimes \cdots \wedge^{n-1} \tau_1$ of the standard representation τ_0 whose highest weight μ_τ is the sum $ne_1 + (n-1)e_2 + \cdots + e_n$ of the representations $\wedge^k \tau_1$, $1 \le k \le n-1$. Then, for any simple root α_i, $\langle \mu_\tau, \alpha_i \rangle = 1$. Hence, $\overline{X}^S_\tau$ is the maximal Satake compactification. (See Fulton–Harris [F2, §15.3] for detailed explanations of the decomposition of the tensor product.)

Furstenberg Compactifications

Furstenberg [F3] used the natural affine action of G on the compact convex set of probability measures on the Furstenberg boundary to define a compactification of X. Moore [M8] generalized this method to define compactifications using other Furstenberg type boundaries and showed that they are the same as the Satake compactifications defined above.

In [F3, Chap. 1] Furstenberg introduced the following concept.

4.45. Definition. A compact homogeneous space Y of G is called a **boundary of** G, or a G**-boundary**, if, for every probability measure μ on Y, there is a sequence (g_n) such that $g_n \cdot \mu$ converges to a measure supported by a point.

If Y is a boundary of G and if μ is a barycenter of two Dirac measures δ_x, δ_y, i.e., $\mu = \frac{1}{2}(\delta_x + \delta_y)$, it follows from the definition that $\lim_n g_n \cdot x = \lim_n g_n \cdot y = z$, for some sequence $(G_n) \subset G$. Hence, the action of G on Y is proximal (see Remark 4.30). It also follows from Theorem 4.29 that the action of G on $G/Q = G \cdot \bar{v}_\tau$ is proximal if Q is a parabolic subgroup. A general result (Proposition 9.31) , due to Margulis, shows that this implies that G/Q is a boundary if Q is parabolic. Here a direct proof of this fact is given using the Bruhat decomposition. It will be shown in Theorem 9.37, that the construction of Theorem 4.29 gives the complete list of boundaries for G. As a result, there are only 2^r non-isomorphic boundaries of G. In other words, as Moore showed in [M8], every G-boundary is isomorphic to one of the form G/P^I, where P^I is a standard parabolic subgroup.

Furthermore, the existence of realizations of boundaries as compact analytic projective manifolds on which G acts by projective transformations is a very important property which provides a link between Satake compactifications and Furstenberg compactification. Additional details are given in chapter IX and Appendix B.

4.46. Let $\mathcal{M}_1(G/P^I)$ denote the set of probability measures on the boundary G/P^I. In addition, let $\mathcal{M}_1(\mathcal{F})$ also denote $\mathcal{M}_1(G/P)$. It follows from the Iwasawa decomposition that K acts transitively on G/P^I. Hence, there is a unique K-invariant probability measure on G/P^I, which will be denoted by $\overline{m}$.

Then the map $g \in G \to g \cdot \overline{m} \in \mathcal{M}_1(G/P^I)$ induces a continuous map $X = G/K \to \mathcal{M}_1(G/P^I) : gK \to g \cdot \overline{m}$, which is denoted by φ^I. It is injective if and only if K is the stabilizer in G of $\overline{m}$.

A key fact on which the Furstenberg approach is based can be formulated as follows. For related statements using projective realizations of boundaries see [F5], [G7], [G8], and [G15].

The Bruhat cell decomposition $G/P = \cup_{w \in W} \overline{N} w \cdot \dot{e} = \cup_{w \in W} \overline{N}^w w \cdot \dot{e}$ (see Proposition 2.20) of G/P, where $\overline{N}^w = \overline{N} \cap w\overline{N}w^{-1}$, can be used to characterize the fundamental sequences $a_n \in \overline{A^+}$.

4.47. Lemma. *Given the above Bruhat decomposition of G/P, let $(a_n) \subset \overline{A^+}$ denote a sequence converging to infinity. Then a_n is fundamental if and only if for any $x \in G/P$, the limit $a_n \cdot x$ exists.*

Furthermore, if $(a_n) \subset \overline{A^+}$ is I-canonical and $x = \eta w \cdot \dot{e} \in G/P$, with $\eta \in \overline{N}^w$ and $\eta = \eta^I(x)\eta_I(x)$, where $\eta^I(x) \in \overline{N}^I$ and $\eta_I(x) \in \overline{N}_I$, and $w = w(x)$, then $\lim_n a_n \cdot x = \eta^I(x) w(x) \cdot \dot{e}$.

Proof. Note that, if $a \in A$, then $a\eta w \cdot \dot{e} = a\eta a^{-1} a w \cdot \dot{e} = (a\eta a^{-1})\, w \cdot \dot{e}$ because $w^{-1} a w \in A$. Furthermore, if $\eta^I \in \overline{N}^I$ and $\eta_I \in \overline{N}_I$, then $a\eta a^{-1} = (a\eta^I a^{-1})(a\eta_I a^{-1})$. In particular, if $a \in A_I$, then $a\eta a^{-1} = \eta^I(a\eta_I a^{-1})$, as $\overline{N}^I \subset G^I$ which centralizes A_I.

Identify $\overline{N}^I$ and $\overline{N}_I$ with $\overline{\mathfrak{n}}^I$ and $\overline{\mathfrak{n}}_I$ under the exponential map. Then conjugation by $a \in A$ corresponds to the action of a on the corresponding Lie algebra by the diagonal linear map $Ad(a)$. The eigenvalues of $Ad(a)$ restricted to $\overline{\mathfrak{n}}_I$ are $e^{-\beta}(a) = e^{-\beta(\log a)}$ for $\beta \in \Sigma_I^+$

Assume that (a_n) is a C_I-canonical sequence (also referred to as I-canonical in Definition 7.19). It follows that $\lim_n e^{-\beta}(a_n) = 0$ if $\beta \in \Sigma_I^+$. Hence, $\lim_n a_n \eta_I a_n^{-1} = e$. It follows that if $x = \eta w \cdot \dot{e} \in G/P$, where

$\eta = \eta^I \eta_I$, then $\lim_n a_n.x = \eta^I w \cdot \dot{e}$ if $\eta = \eta^I \eta_I$. This proves the last assertion.

If a_n is I-fundamental, then $a_n = a_n^I a_{n,I}$, where $a_n^I \to a^I \in \overline{A^{I,+}} = \exp \overline{\mathfrak{a}^{I,+}}$ and $(a_{n,I})$ is I-canonical. Hence, $\lim_n a_n \cdot x = a^I \lim_n a_{n,I} \cdot x = a^I \eta^I w \cdot \dot{e}$ if $x = \eta w \cdot \dot{e}$.

Conversely, if $a_n \cdot x$ converges for every $x \in G/P$, let $x = \eta \cdot \dot{e}$ with $\eta \in \overline{N}$. Then, since $a_n \cdot x = a_n \eta a_n^{-1} \cdot \dot{e}$, and because of the identification of $\overline{N}$ with the "large cell" $\overline{N} \cdot \dot{e}$, the sequence $a_n \eta a_n^{-1}$ converges for every $\eta \in \overline{N}$. The diagonal form of the action of $Ad(a_n)$ on $\overline{\mathfrak{n}}$ implies that $e^{-\alpha}(a_n)$ converges for every $\alpha \in \Sigma^+$ and, hence, for every $\alpha \in \Delta$. Let I be the set of roots $\alpha \in \Delta$ such that $\lim_n e^{-\alpha}(a_n) > 0$. Then a_n is I-fundamental. □

It follows immediately from this lemma that under the action of an I-canonical sequence, every probability ν on G/P has a weak limit.

4.48. Lemma. *If $\nu \in \mathcal{M}_1(G/P)$ and (a_n) is I-canonical the sequence of probability measures $a_n \cdot \nu$ converges weakly to the image of ν under the Borel map $x \to \eta^I(x) w(x) \cdot \dot{e}$.*

Furthermore, the compact space G/P is a G-boundary.

Proof. The first assertion follows from Lemma 4.47 and dominated convergence. If ϕ is continuous on G/P, then $a_n \cdot \nu(\phi) = \int \phi(a_n \cdot x) d\nu(x)$ and so $\lim_n a_n \cdot \nu(\phi) = \int \phi[\eta^I(x) w(x) \cdot \dot{e}] d\nu(x)$. In other words, the sequence $a_n \cdot \nu$ converges to the image of ν under the Borel map $x \to \eta^I(x) w(x) \cdot \dot{e}$. Note that this map is continuous on the open dense set $\overline{N} \cdot \dot{e} \subset G/P$ as well as on each of the sets $\overline{N}^w w \cdot \dot{e}$.

Fix a neighborhood V of e in G. The set $F_w = \{g \in V \mid gw \cdot \dot{e} \notin \overline{N} \cdot \dot{e}\}$ is closed with no interior points, since the complement of $\overline{N} \cdot \dot{e}$ in G/P is a finite union of submanifolds of codimension at least one, and since, for every $w \in W$, the map $g \to gw \cdot \dot{e}$ is open. Let $F = \cup_{w \in W} F_w$. Then F has no interior points and $F \neq V$. In other words, there exists $h \in V$ such that, for every $w \in W, hw \cdot \dot{e} \in \overline{N} \cdot \dot{e}$. If V is sufficiently small then, in addition, $B = \cup_{w \in W} hVw \cdot \dot{e} \subset \overline{N} \cdot \dot{e}$. If $a \in A^+$ then (a^n) is $\emptyset$-canonical. Hence, $\lim_n a^n \cdot (B) = \dot{e}$. It follows from the first assertion that $\lim_n a^n \cdot \nu = \sum_{w \in W} \nu(\overline{N} w \cdot \dot{e}) \delta_{w \cdot \dot{e}} = \sum_{w \in W} u_w \delta_{w \cdot \dot{e}}$, with $u_w \geq 0$ and $\sum_{w \in W} u_w = 1$. As a result, it follows that $\lim_n ha^n \cdot \nu = \sum_{w \in W} u_w \delta_{hw \cdot \dot{e}}$ and $\lim_n ha^n \nu(B) = 1$

Since $a^n ha^n (a^n \cdot B) = ha^n \cdot \nu(B)$, this implies that $\lim_n a^n ha^n \cdot \nu(a^n B) = 1$. Hence, $\lim_n a^n ha^n \cdot \nu(\mathcal{F} \setminus a^n B) = 0$. If ϕ is a continuous function whose support does not contain $\dot{e}$, then $\lim_n a^n ha^n \cdot \nu(\phi) = 0$. Hence, the sequence $a^n ha^n \cdot \nu$ converges to $\delta_{\dot{e}}$. Consequently, G/P is a G-boundary. □

It was observed in § 4.46 that, provided K is the stabilizer in G of the K-invariant measure $\overline{m}$, the symmetric space X can be identified, as a set, with the G-orbit of $\overline{m}$ by the map φ^I. A boundary G/P^I with this property is said to be a **faithful boundary**. Faithfulness of the boundary is equivalent to a condition involving the structure of the set Δ of simple roots as stated in the next result.

4.49. Proposition. (See Moore [M8]) *The map* $\varphi^I : X \to \mathcal{M}_1(G/P^I)$ *is injective if and only if the complement* $\Delta \setminus I$ *has non-empty intersection with every connected component of* Δ. *If this condition holds, then* φ^I *is a topological embedding.*

A modification follows of Moore's original argument (see [M8, Theorem 4 and Lemma 9]), that explains this non-empty intersection condition. It is based on the next lemma.

Note that, by Proposition 4.18 and Theorem 4.27, it follows that if $I \subset \Delta$ equals $I_\tau^\perp$ for some representation τ, i.e., if $P^I = P^{I_\tau^\perp} = P_\tau$, then the boundary G/P^I is faithful if and only if the tangent representation of τ is faithful.

4.50. Lemma. *If* G *is simple, then* G/P^I *is a faithful boundary if and only if* P^I *is a proper parabolic subgroup of* G, *i.e.,* I *is a proper subset of* Δ.

Proof. If G/P^I is faithful, it clearly cannot be a single point. Hence, P^I is a proper parabolic subgroup.

On the other hand, suppose P^I is a proper parabolic subgroup. If φ^I is not injective, then there exists $g \in G$, $g \notin K$ such that $g \cdot \overline{m} = \overline{m}$. Then, for any open U in $\overline{N_I}$, $g \cdot \overline{m}(U) = \overline{m}(U)$. If $g = k_1 a\, k_2$, where $k_1, k_2 \in K$, and $\log a = H \in \mathfrak{a}^+$, then $H \neq 0$, and $a \cdot \overline{m}(U) = \overline{m}(a^{-1} \cdot U) = \overline{m}(U)$.

Now let $\overline{P^I}$ be the parabolic subgroup opposite to P^I, i.e., the Lie algebra of the unipotent radical $\overline{N_I}$ of $\overline{P^I}$ is equal to $\overline{\mathfrak{n}_I} = \theta(\mathfrak{n}_I)$. Consider the Bruhat cell decomposition of G/P^I (see Proposition 2.20) in which $\overline{N_I}$ is mapped diffeomorphically onto the large cell — an open subset of G/P^I of full measure with respect to $\overline{m}$.

If $\beta(H) = 0$, then for every $t \in \mathbb{R}$ and $\beta \in \Sigma_I^+$, the action of $\exp tH$ on $\overline{N_I} \cdot \dot{e}$ is trivial because it is given by a diagonal matrix with eigenvalues $e^{-t\beta(H)}$. Hence, because $\overline{N_I} \cdot \dot{e}$ is dense in G/P^I, the action of $\exp tH$ on G/P^I is trivial. Therefore, the one parameter subgroup $\exp tH$ is contained in the kernel of the action of G on G/P^I, which is a normal subgroup of G. Since $\mathfrak{g}$ is simple this implies that $\exp tH = e$ and so $H = 0$. This is a contradiction.

As a result, for some $\beta \in \Sigma_I^+$, it follows that $e^{\beta(H)} > 1$. Hence, $a = \exp H$ contracts $\overline{N_I}$. Suppose $O \subset \overline{\mathfrak{n}_I}$ is a box of center e associated with a basis of eigenvectors for $Ad(a)$. Let $U = \exp O$. Then $a \cdot U \subset U$ and the volume (in the group $\overline{N}$) of $a \cdot U$ is at most $e^{-\beta(H)}$ times the volume of U. This implies that $\overline{m}(a \cdot U) < \overline{m}(U)$. This contradiction shows that φ^I is injective. □

Proof of Proposition 4.49.

Using the above lemma, one sees easily that the non-empty intersection property of Proposition 4.49 holds. Recall that one may assume that G has a trivial center. Then G can be written as a product of simple Lie subgroups: $G = G_1 \times \cdots \times G_k$, and $X = X_1 \times \cdots \times X_k$. Any boundary G/P^I of G is the product of boundaries of the G_i. Then it is clear that the boundary G/P^I is faithful if and only if every boundary of G_i, $i = 1, \ldots, k$, is faithful. By the above lemma, this condition is exactly equivalent to requiring that I contain no connected component of Δ. (Recall that these components correspond to the factor Lie algebras $\mathfrak{g}_i$ of $\mathfrak{g}$.)

In order to show that φ^I is an embedding consider a sequence $g_n \in G$ such that $\lim_n g_n \cdot \overline{m} = \overline{m}$. Using a polar decomposition, one can suppose $g_n = a_n = \exp H_n$ with $H_n \in \overline{\mathfrak{a}^+}$. Using a subsequence, one can also assume that the limits of the sequences $e^{-\beta}(a_n)$ exist and are between 0 and 1 for every $\beta \in \Sigma^+$. The linear maps $Ad\, a_n$ of $\overline{\mathfrak{n}}_I$ into itself are diagonal with coefficients $e^{-\beta}(a_n)$, where $\beta \in \Sigma^+$. It follows that $\lim_n Ad\, a_n = u$ exists, as an endomorphism of $\overline{\mathfrak{n}}_I$.

In view of the identification of $\overline{\mathfrak{n}}_I$ with the open dense set $\overline{N}_I \cdot \dot{e}$ of G/P^I, the measure $\overline{m}$ can be lifted to $\overline{\mathfrak{n}}_I$ as a measure equivalent to Lebesgue measure and, so, $u(\overline{m}) = \overline{m}$. If, for some $\beta \in \Sigma_I^+$, one has $\lim_n e^{-\beta}(a_n) = 0$, the linear map u degenerates and $u(\overline{m})$ is concentrated on a proper linear subspace of $\overline{\mathfrak{n}}_I$. This contradicts $u(\overline{m}) = \overline{m}$; hence, $\lim_n e^{-\beta}(a_n) > 0$ for every $\beta \in \Sigma_I^+$. It follows that $\lim_n \beta(H_n) < +\infty$. If (H_n) is converging to infinity, consider the sequence $\frac{H_n}{\|H_n\|}$. One can assume that $\lim_n \frac{H_n}{\|H_n\|} = H$ exists with $\|H\| = 1$. Hence, $\beta(H) = \lim_n \beta(\frac{H_n}{\|H_n\|}) = 0$ for every $\beta \in \Sigma_I^+$. This implies that $b = \exp H$ acts trivially on $\overline{N}_I \cdot \dot{e}$, hence, on G/P^I, as the large cell is dense. Then $b \cdot \overline{m} = \overline{m}$ and the injectivity of φ^I implies $b = e$, and $H = 0$, which contradicts $\|H\| = 1$. As a result H_n is bounded and $a_n = \exp H_n$ converges to $a \in \overline{A^+}$ with $a \cdot \overline{m} = \overline{m}$. Since φ^I is injective, $a = e$ and, so, $\lim_n a_n = e$. □

4.51. Definition. Let G/P^I be a faithful boundary. The closure of the image $\varphi^I(X)$ in $\mathcal{M}_1(G/P^I)$ is called a **Furstenberg compactification** of X.

Remarks. (1) It is clear from the definition of φ^I and Proposition 4.49 that $\varphi^I(X) \subset \mathcal{M}_1(G/P^I)$ is isomorphic to X as a G-space. Furthermore, since $\overline{\varphi^I(X)}$ contains Dirac measures, it contains the boundary G/P^I as an isomorphic copy. The map $x \to \delta_x \in \overline{\varphi^I(X)}$ embedding G/P^I into $\mathcal{M}_1(G/P^I)$ is clearly continuous. Hence, it is an isomorphism because G/P^I is compact. It follows easily from the definition of a boundary that the image of G/P^I is the unique compact G-orbit in $\mathcal{M}_1 = \mathcal{M}_1(G/P^I)$.

(2) Clearly, the boundary G/P projects onto the boundary G/P^I. Since $\pi(G{\cdot}m) = G{\cdot}\overline{m}$, the Furstenberg compactification $\overline{\varphi^{\emptyset}(X)}$ dominates $\overline{\varphi^I(X)}$. For this reason, it is called the **maximal Furstenberg compactification**.

In this book, the boundary G/P is referred to as simply the **Furstenberg boundary** rather than as the maximal Furstenberg boundary.

Let $\overline{\varphi^{\emptyset}(X)}$ be denoted by $\overline{X}^F_{\max}$. As in the case of the maximal Satake compactification, since $\overline{X}^F_{\max}$ is a G-compactification, it is of key importance to understand the closure of $\overline{A^+} \cdot o$ in this compactification. The following lemma is the analog, for $\overline{X}^F_{\max}$, of Lemma 4.41.

4.52. Lemma. *Let $a_n = \exp H_n$ and assume that $(a_n) \subset \overline{A^+}$ converges to infinity. Denote by I the set of roots bounded on (H_n). Then, $(a_n \cdot m) \subset \mathcal{M}_1(G/P)$ converges if and only if (a_n) is I-fundamental.*

Proof. It was already proved in Lemma 4.48 that, if (a_n) is I-canonical, the sequence $a_n \cdot m$ converges.

Because $m(\overline{N}_I \cdot \dot{e}) = 1$, the image of the measure m by the map $x \to \eta^I(x) \cdot \dot{e}$ is well defined and will be denoted $\eta^I(m)$. It follows that $\lim_n a_n \cdot m = \eta^I(m)$. If (a_n) is C_I-fundamental and $a_n = a^I a_{n,I}$ with $a^I = \lim_n a^I_n$ and $(a_{n,I})$ I-canonical, then $a_n \cdot m$ converges to $a^I \cdot \eta^I(m)$.

Conversely, assume that $\lim_n a_n \cdot m = \nu$ where $(a_n) \subset \overline{A^+}$ is converging to infinity. Then (a_n) has fundamental subsequences. To show that (a_n) is itself fundamental, it therefore suffices to show that if $(a_n^{I_1} a_n^1)$ and $(a_n^{I_2} a^2)$ are fundamental sequences, where $a_n^{I_i} \to a^{I_i}$ and (a_n^i) is I_i-canonical for $i = 1$ or 2, then $a^{I_1} \cdot \eta^{I_1}(m) = a^{I_2} \cdot \eta^{I_2}(m)$ implies that $I_1 = I_2$ and $a^{I_1} = a^{I_2}$.

If $a \in A^I$, then a acts on $\overline{N}^I$ as an automorphism and the measure m, and the map η^I can be considered on $\overline{N}$ rather than on $\overline{N} \cdot \dot{e} \subset G/P$. With this identification $\eta^I(m)$ is equivalent to a Haar measure on $\overline{N}^I$ as is $a \cdot \eta^I(m)$.

Consequently, if $a^{I_1} \cdot \eta^{I_1}(m) = a^{I_2} \cdot \eta^{I_2}(m)$, the equality of the supports of these measures in $\overline{N}$ implies that $\overline{N}^{I_1} = \overline{N}^{I_2}$ and, hence, $I_1 = I_2 = I$. Let $a_i = a^{I_i}, i = 1, 2$. It remains to show that if $a = a_2^{-1} a_1 \in A^I$ and $a \cdot \eta^I(m) = \eta^I(m)$, then $a = e$.

If $Ad(a) \neq Id$ then one of the diagonal maps $Ad(a)$ or $Ad(a^{-1})$ has an eigenvalue greater than unity. One may assume that this is so for $Ad(a)$. Then, for almost every $x \in \overline{\mathfrak{n}}^I$ (in the sense of Lebesgue measure on $\overline{\mathfrak{n}}^I$), one has $\lim_n \|Ad(a^n)(x)\| = \infty$.

If ϕ is a continuous function with compact support on $\overline{N}^I$ it follows that $\lim_n \phi(a^n \cdot y) = 0$ for almost every $y \in \overline{N}^I$. Hence, by dominated convergence, $\lim_n (a^n \cdot \eta^I(m))(\phi) = \lim_n \eta^I(m)(\phi \circ a^n) = 0$. This contradicts the assumption that $a \cdot \eta^I(m) = \eta^I(m)$. Hence, $a_1 = a_2$. Consequently, a sequence (a_n), converging to infinity, is fundamental if $\lim_n a_n \cdot m$ exists. □

In view of the isomorphism of $X \cup \Delta^*(X)$ and $\overline{X}^S$, proved in Theorem 4.43, and the uniqueness of the extension of G-action to the boundaries of these compactifications, Lemma 4.41 and Lemma 4.51 suggest the first part of following result. A proof will be given later in Chapter IX (see Corollary 9.53).

4.53. Proposition. *The maximal Furstenberg compactification* $\overline{X}^F_{\max}$ *is isomorphic to the maximal Satake compactification* $\overline{X}^S_{\max}$. *Furthermore, if* τ *is a faithful irreducible representation such that the parabolic subgroup* P_τ *is equal to the parabolic subgroup* P^I, *then the compactifications* $\overline{X}^S_\tau$ *and* $\overline{\varphi^I(X)} \subset \mathcal{M}_1(G/P^I)$ *are isomorphic.*

Because of this proposition, the maximal Satake compactification and the maximal Furstenberg compactification are referred to as the **maximal Satake–Furstenberg compactification** and denoted by $\overline{X}^{SF}_{\max}$. Since other non-maximal Satake–Furstenberg compactifications are not used in the identification of the Martin compactification, $\overline{X}^{SF}_{\max}$ will be denoted by $\overline{X}^{SF}$ for convenience. When specific reference to either one of these two compactifications is made, $\overline{X}^{SF}$ will often be denoted as either $\overline{X}^S$ or $\overline{X}^F$. Proposition 4.53 was originally proved by Moore [M8, Theorem 8]. The projective representation of G/P given in Theorem 4.29 leads to another self-contained proof of this result together with the characterization of convergent sequences (see Theorems 9.52 and 9.55). More detailed discussions of the boundary theory and the maximal Furstenberg compactification are to be found in Chapter IX. Non-maximal Furstenberg compactifications, together with non-maximal Satake compactifications, are studied in Appendix B where the second part of Proposition 4.53 is proved.

CHAPTER V

THE KARPELEVIČ COMPACTIFICATION

This compactification was introduced by Karpelevič in [K3]. The original inductive definition of $\overline{X}^K$ is recalled in § 5.3. By examining the closure of a flat $A\cdot o$ in $\overline{X}^K$, a non-inductive characterization of the closure $\overline{A^+ \cdot o}^K$ of $A^+ \cdot o$ is obtained (see Theorem 5.6). The nature of the Karpelevič topology restricted to the flat is clarified by the introduction of the class of K-fundamental sequences. Using this concept, one shows that $\overline{A\cdot o}^K$ is isomorphic to a compactification of $\mathfrak{a}$ determined by its polyhedral structure. This compactification of $\mathfrak{a}$, referred to as the Karpelevič compactification of $\mathfrak{a}$, is used to give a new proof that the Karpelevič topology is compact. Lemma 5.26, Proposition 5.27, and Corollary 5.28 explain the relations between the Karpelevič compactification and the conic and dual cell compactifications. Finally, in Remark 5.32 a new way to define the Karpelevič compactification is presented. It consists of fitting together the Karpelevič compactifications of the flats $kA\cdot o$, $k \in K$, in exactly the same way that the dual cell compactification is obtained from the polyhedral compactifications of the flats $kA\cdot o$.

THE KARPELEVIČ COMPACTIFICATION

§ **5.1.** Karpelevič [K3] used the geometry of geodesics and induction on the rank of the space X to define a compactification $\overline{X}^K$ of X. He was motivated by the need to parametrize the minimal Martin boundary and to study the boundary behavior of harmonic functions on X. It is known, and also follows from the results of this book, that this compactification is too big for the purpose of parametrization when the rank is greater than two.

In what follows, the definition of $\overline{X}^K$ is recalled, and a non-inductive characterization of the topology is given in terms of convergent sequences in the positive Weyl chamber $\overline{\mathfrak{a}}^+$ (see Theorem 5.6).

§ **5.2.** The Karpelevič compactification $\overline{X}^K$ is constructed by induction on the rank of X (see Karpelevič [K3, §13]). If the rank is equal to one, $\overline{X}^K$ is defined to be the conic compactification $X \cup X(\infty) = \overline{X}^c$. If X has rank two the Karpelevič compactification $\overline{X}^K$ dominates the conic compactification $\overline{X}^c$, i.e., there is a continuous map $\varphi_c : \overline{X}^K \to \overline{X}^c$ that is the identity on X (see Lemma 5.26). This means that if a sequence (y_n) converges in $\overline{X}^K$ it converges in $\overline{X}^c$ to a limiting direction z. Because the rank of X is two, either z is in a 1-simplex of the Tits building $\Delta(X)$ or it is in a 0-simplex. In the first case the fiber $\varphi_c^{-1}(\{z\})$ is a point and in the second it is $\overline{X_z}^K$, that is, by definition, the conic compactification of X_z, the (rank one) boundary symmetric space associated with z.

Note that, for any symmetric space X of non-compact type, one may define the **boundary symmetric space for** $z \in X(\infty)$ to be $X_z = k \cdot X^I$ when $z \in k \cdot C_I(\infty)$. Then X_z is the cell in $\Delta^*(X)$ dual to the simplex $k \cdot C_I(\infty) \in \Delta(X)$ corresponding to the parabolic subgroup $P_z = kP^I k^{-1}$ that is the stabilizer of z (see § 3.34). Note that the simplex $k \cdot C_\emptyset(\infty) = k \cdot \mathfrak{a}^+(\infty)$ corresponds to the Weyl chamber $kA^+ \cdot o$ and the boundary symmetric space X_z, for $z \in k \cdot C_\emptyset(\infty)$, is then a point.

To summarize, in the case of a rank two symmetric space X, a sequence (y_n) converges in $\overline{X}^K$ if it has a limiting direction z and its projection onto X_z converges in $\overline{X_z}^K$, equivalently, has a limiting direction in $\overline{X_z}^K$. The ideal boundary of $\overline{X}^K$ is then $\cup_{z \in X(\infty)} \overline{X_z}^K$.

This serves to motivate the first step in defining the Karpelevič compactification $\overline{X}^K$ of X, namely, the definition of the ideal boundary. Assume that for any symmetric space Y of non-compact type with rank less than the rank of X, the compactification $\overline{Y}^K$ has been defined. Then, for each $z \in X(\infty)$, the Karpelevič compactification $\overline{X_z}^K$ is defined, as it is clear that X_z has rank less than the rank of X. The union of these Karpelevič compactifications at infinity is the ideal boundary.

5.3. Definition. (See Karpelevič [K3, p. 119]) The **Karpelevič compactification** $\overline{X}^K$ is defined set-theoretically to be

$$\overline{X}^K = X \cup (\cup_{z \in X(\infty)} \overline{X_z}^K).$$

By this inductive definition, every point in the boundary of $\overline{X}^K$ is given by an ordered pair (z_0, y_0), where $z_0 \in X(\infty)$ and $y_0 \in \overline{X_{z_0}}^K$. If y_0 is in the Karpelevič boundary of X_{z_0}, it is represented by an ordered pair (z_1, y_1) with $z_1 \in X_{z_0}(\infty)$ and $y_1 \in \overline{X_{z_1}}^K$. If y_0 is in X_{z_0}, the point (z_0, y_0) will be said to be of level 1. As a result, the generic point of the ideal boundary of $\overline{X}^K$ is given by a sequence $(X_0, z_0; X_1, z_1; X_2, z_2; \cdots; X_\ell, z_\ell; x)$, where $X_0 = X$, $z_i \in X_i(\infty)$, $X_{i+1} = (X_i)_{z_i}$ is the boundary symmetric space of X_i at z_i, and the last point x belongs to $X_{\ell+1} \stackrel{\text{def}}{=} X_{z_\ell}$ if the rank of X_ℓ is greater than 1 (see eq. (13.4.11) in [K3]). When the rank of X_ℓ equals one, all the boundary symmetric spaces X_{z_ℓ} degenerate to one point as $\overline{X_\ell}^K = \overline{X_\ell}^c = X_\ell \cup X_\ell(\infty)$. In this case, the generic point will be denoted by a sequence $(X_0, z_0; X_1, z_1;\ X_2, z_2; \cdots; X_\ell, z_\ell)$. The parameter $\ell + 1$ will be called the **level** of the point.

Equivalently, suppressing explicit reference to the boundary symmetric spaces that are involved, a point y_0 in the Karpelevič boundary is of the form $(z_0, z_1, \ldots, z_\ell, x)$, where

$$\begin{aligned} z_0 &= k_0 \cdot D_0(\infty), \;\; D_0 \in C_{I_1}, k_0 \in K, \\ z_1 &= k_0 k_1 \cdot D_1(\infty), \;\; D_1 \in C_{I_2}^{I_1}, k_1 \in K^{I_1} M, \\ &\vdots \qquad \vdots \\ z_\ell &= k_0 k_1 \cdots k_\ell \cdot D_\ell, \;\; D_\ell \in C_I^{I_\ell}, k_\ell \in K^{I_\ell} M, \\ x &\in k_0 k_1 \cdots k_\ell \cdot X^I. \end{aligned}$$

Here, the cone $C_{I_{i+1}}^{I_i}$ is the projection of $C_{I_{i+1}}$ onto $\mathfrak{a}^{I_i}$ (see Lemma 3.37) and D_i is a **direction** (i.e., a vector of unit length) in the cone $C_{I_{i+1}}^{I_i}$.

At the ith level, the direction is given by a direction in the image under $k_0 k_1 \cdots k_i$ of the ith symmetric space X^{I_i}, where $I_i \supset I_{i+1}$ for $0 \leq i \leq \ell - 1$, with $I_0 = \Delta$ and $I \stackrel{\text{def}}{=} I_{\ell+1}$. If the rank of X_ℓ equals 1, then the generic point will also be denoted by $(z_0, z_1, \ldots, z_\ell)$.

Let $\partial \overline{X}^K \stackrel{\text{def}}{=} \cup_{z \in X(\infty)} \overline{X}_z^K$ — a disjoint union — denote the ideal boundary of the Karpelevič compactification. The compact group K acts on the set $\partial \overline{X}^K$ as follows. If $(z_0, z_1, \ldots, z_\ell, x) \in \partial \overline{X}^K$, then $k \cdot (z_0, z_1, \ldots, z_\ell, x) \stackrel{\text{def}}{=} (k \cdot z_0, k \cdot z_1, \ldots, k \cdot z_\ell, k \cdot x)$. This formula makes sense since, for any $z \in X(\infty)$, one has $k \cdot X_z = X_{k \cdot z}$ and, hence, $k \cdot \overline{X}^K = \overline{k \cdot X}^K$. As a result, if $(z, y) \in \partial \overline{X}^K$, then $k \cdot y \in \overline{X}_{k \cdot z}^K$, i.e., $(k \cdot z, k \cdot y) \in \partial \overline{X}^K$.

§ **5.4.** Karpelevič defined the topology on the set $\overline{X}^K$ in two steps. The first step involves putting a topology on the ideal boundary $\partial \overline{X}^K$. Then, using this topology on the boundary, one can describe the basic open neighborhoods of the ideal boundary points.

While one can simplify the following discussion by defining the topology so as to be obviously K-invariant, Karpelevič will be followed here and the topology defined so as to be G-invariant. This requires that G act on $\overline{X}^K$. For this it is necessary to show that, for all $g \in G$ and $z \in X(\infty)$, one has $g \cdot X_z = X_{g \cdot z}$, which is evident if $g \in K$.

If $[\gamma] = z$, this result follows from the intrinsic (and, hence, G-invariant) description of the boundary space X_z [K3, §8.3, p. 92] in terms of the finite space defined by the corresponding f-bundle F of geodesics γ' where $\gamma' \in F$ if and only if $\lim_{t \to \infty} d(\gamma(t), \gamma'(t)) < \infty$. Since the group G acts on $\overline{X}^c$, as pointed out in § 3.1, and, from the intrinsic formulation of X_z, one sees that $g \cdot X_z = X_{g \cdot z}$ for all $g \in G$ and $z \in X(\infty)$, it follows that G acts on $\overline{X}^K$.

The points z of the conical boundary may be classified by their **degree of singularity**: if $z \in k \cdot C_I(\infty)$, i.e., if $z = k \cdot D(\infty)$ where $D \in C_I$ is a unit vector, it will be said to have degree of singularity I. To discuss some of the details of topology introduced by Karpelevič, it is important to understand a certain parametrization of $X(\infty)$ near a point $D(\infty)$ when

$D \in C_I$. Recall that, by the Cartan decomposition, $X(\infty)$ is the K-orbit of the boundary of $\mathfrak{a}(\infty)$, even of $\overline{\mathfrak{a}^+}(\infty)$.

In terms of the flat $A \cdot o$ one sees that the image under the exponential map of $C_I(\infty) \times O^I(\infty)$, where $O^I \subset \mathfrak{a}^I$ is a neighborhood of 0, gives a neighborhood of any $z \in C_I(\infty)$ in the boundary $(A \cdot o)(\infty) \subset X(\infty)$ of the flat $A \cdot o$. The subgroup $K^I M$ rotates this neighborhood, leaving $C_I(\infty)$ pointwise fixed and generating the set $C_I(\infty) \times (U^I \cdot o)(\infty)$, where U^I is the neighborhood of $o \in X^I = G^I \cdot o$ equal to $K^I \exp O^I \cdot o$. To finally obtain a neighborhood in $X(\infty)$ of $D(\infty)$, one acts on this new object by K, or more usefully by $\exp W$, where $W \subset \mathfrak{k} \cap (\mathfrak{k}^I + \mathfrak{m})^\perp$ is a neighborhood of 0. As a result, a boundary point $z \in X(\infty)$ near to $D(\infty) \in C_I(\infty)$ is of the form $z = \exp w \cdot k_1 \cdot \exp(D' + H_2)(\infty)$, where $w \in W, k_1 \in K^I M, D' \in C_I$, and $H_2 \in O^I$ is small.

One consequence of this is that, for any $z_0 \in X(\infty)$, there is an open neighborhood V_0 of $z_0 = k_0 \cdot D(\infty)$, where $D \in C_I$, such that the degree J of singularity of any $z \in V_0$ is a subset of I. In this case, V_0 will be said to have **degree of singularity of at most** I.

Let V_0 be an open subset of $X(\infty)$ with a degree of singularity of at most I and let $z_0 = k_0 \cdot D(\infty) \in V_0$ with $D \in C_I$. Then, for each $z \in V_0$, there is a canonical map π_{z,z_0} of X_z into X_{z_0}: if $z = k \cdot D(\infty), D \in C_J$, then $X_z = k \cdot X^J \subset k \cdot X^I$ is mapped by $k_0 k^{-1}$ to $X_{z_0} = k_0 \cdot X^I$.

Karpelevič [K3, p. 121] describes the topology of $\partial \overline{X}^K$ as follows. Let $(z_0, y_0) \in \overline{X}^K_{z_0}$, where $z_0 = k_0 \cdot D(\infty), D \in C_I$. Choose a neighborhood V_0 of z_0 in $X(\infty)$ small enough so that it has degree of singularity of at most I. Then, by the inductive assumption and the properties of the Karpelevič compactifications, for any $z_0 \in V_0$, the canonical map π_{z,z_0} determines a map $\overline{\pi}^K_{z,z_0} : \overline{X}^K_z \to \overline{X}^K_{z_0}$, whose image is $\overline{X}^K_{z_0}$ if $\pi_{z,z_0}(X_z) = X_{z_0}$ or is $\overline{X}^K_{z'} \subset \partial \overline{X}^K_{z_0}$ if $\pi_{z,z_0}(X_z) = X_{z'}$ is a boundary symmetric space of X_{z_0}. For any neighborhood V_1 of y_0 in $\overline{X}^K_{z_0}$, define a subset

$$V = \{(z, y) \in \partial \overline{X}^K \mid z \in V_0, \ \overline{\pi}^K_{z,z_0}(y) \in V_1\}.$$

Let V^G be a neighborhood of the identity in G. Then the set

$$U = \{g \cdot (z, y) \mid g \in V^G, (z, y) \in V\}$$

is a neighborhood of (z_0, y_0) in $\partial \overline{X}^K$. Note that it is at this point that the formulation makes use of the action of G on $\overline{X}^K$.

Remarks. While the above definition of π_{z,z_0} is clear, it is not so evident how to define $\overline{\pi}^K_{z,z_0}$, the issue being the determination of the point in $X_{z_0}(\infty)$ that corresponds to $z \in X(\infty)$. For simplicity, assume that $k_0 = e$. Then, if $z = k_1 \cdot D_J(\infty)$ with $k_1 \in K^I M$ and $D_J \in C_J$, $J \subset I$, the corresponding point $z^I \in X_{z_0}(\infty) = X^I(\infty)$ is $k_1 \cdot D^I_J(\infty)$, where D^I_J is the direction (in C^I_J) of the projection of D_J onto $\mathfrak{a}^I$ (recall that $\mathfrak{a}_J = \mathfrak{a}^I_J \oplus \mathfrak{a}_I$). This defines the boundary point for all $z \in St(C_I(\infty))$, where $St(C_I(\infty))$ denotes the union of the simplices $C(\infty)$ in $\Delta(X)$ for which $\overline{C(\infty)} \supset C_I(\infty)$. To see this, observe that for a simplex $C(\infty) = k \cdot C_J(\infty)$ of $\Delta(X)$, one has $\overline{C(\infty)} \cap C_I(\infty) \neq \emptyset$ if and only if $\overline{C(\infty)} \supset C_I(\infty)$. In terms of parabolic subgroups, this implies that $kP^J k^{-1} \subset P^I$. Hence, $J \subset I$ and $k \in K^I M$, i.e., $St(C_I(\infty)) = \{z \in X(\infty) \mid \text{ for some } J \subset I, k_1 \in K^I M,\ z \in k_1 \cdot C_J(\infty)\}$.

The K-orbit of $St(C_I(\infty))$ is open as its complement is the K-orbit of the union of the simplices $C_J(\infty)$ with $J \not\subset I$. Furthermore, $K \cdot St(C_I(\infty))$ is the disjoint union of the sets $k \cdot St(C_I(\infty))$, where k is unique modulo $K^I M$ (by an easy application of Proposition 2.18). If $k \notin K^I M$ and $z = kk_1 \cdot D_J(\infty), J \subset I$ and $k_1 \in K^I M$ define z^I to be $k_1 \cdot D_J^I(\infty)$.

§ **5.5.** The second step involves attaching the boundary $\partial \overline{X}^K$ to X. Let $z_0 \in X(\infty)$ have a neighborhood that has a degree of singularity of at most I. Its parabolic subgroup P_{z_0} has a Langlands decomposition $P_{z_0} = N_{z_0} A_{z_0} M_{z_0}$ (see Corollary 2.16). Let G_{z_0} be the boundary semisimple Lie group associated to z_0, as in § 2.13. If $z_0 \in C_I(\infty)$, $G_{z_0} = G^I$; otherwise G_{z_0} is conjugate to G^I for some I. Then the orbit $A_{z_0} G_{z_0} \cdot o$ of $A_{z_0} G_{z_0}$ in $X = G/K$ is denoted by $\mathcal{R}_0$ in [K3, §9.2], also denoted by $\mathcal{R}$ in [K3, §13.8].

This manifold $\mathcal{R}_0$ is isometric to $A_{z_0} \times X_{z_0}$ as $x = a_{z_0} \cdot x^{z_0}$ with $a_{z_0} \in A_{z_0}$ and $x^{z_0} \in X_{z_0}$ (see § 7.6). (In the terminology of [K3, p. 52], $\mathcal{R}_0$ is a realization of the set of n-bundles of geodesics in the f-bundle of geodesics corresponding to the point z_0— Karpelevič refers to this as the finite space associated with the f-bundle.)

Let γ be the unique geodesic in X that passes through the basepoint o and belongs to the class z_0. Let $\pi_{z_0} : X \to X_{z_0}$ be the projection associated to the generalized horospherical decomposition $X = N_{z_0} A_{z_0} X_{z_0}$ (see § 7.6) induced from the Langlands decomposition of P_{z_0}. Denote the restriction of π_{z_0} to $\mathcal{R}_0$ also by π_{z_0} (this is the projection $\mathcal{R}_0 = A_{z_0} \times X_{z_0} \to X_{z_0}$).

Let ε be a small positive number and W_0 be a neighborhood of (z_0, y_0) in $\partial \overline{X}^K$. Following Karpelevič [K3, p. 127], let $V^{\mathcal{R}_0}$ denote the set of points in $\mathcal{R}_0$ such that

(1) $\pi_{z_0}(x) \in W_0$,
(2) $d(o, x) > \frac{1}{\varepsilon}$,
(3) $\angle(\gamma, \overline{ox}) < \varepsilon$.

where $d(\cdot, \cdot)$ is the distance function on X, $\overline{ox}$ is the geodesic ray from o to x, and $\angle(\gamma, \overline{ox})$ is the angle at o between γ and $\overline{ox}$.

Let V^G be a neighborhood of the identity in G. Define

$$U = \{g \cdot x \mid g \in V^G, x \in V^{\mathcal{R}_0}\}.$$

Then the union $U \cup W_0 \subset \overline{X}^K$ is a neighborhood of (z_0, y_0) in $\overline{X}^K$.

Remarks. (1) The set $V^{\mathcal{R}_0}$ is in effect a basic neighborhood of (z_0, y_0) in the closure of $\mathcal{R}_0$ in $\overline{X}^K$. Recall that $\mathcal{R}_0 = k_0 \cdot A_I \cdot X^I$ if $z_0 \in k_0 \cdot C_I$. The set U is obtained by rotating this neighborhood by k close to k_0. In order to determine the sequences that converge in $\overline{X}^K$, it will suffice to work in the flat $A \cdot o$. The intersection with $A \cdot o$ of a basic neighborhood in $\overline{A \cdot o}^K$ of a boundary point (z, y) is given by the points x in a cone (in the flat) about the direction $z \in Ad(k_j)C_I$, with $d(x, o)$ large, whose projection onto $X^I \subset \partial\overline{X}^K$ lies in a neighborhood of y in the compactified boundary symmetric space $\overline{X^I}^K$ (see Definition 5.17).

(2) It is clear from the definition of this topology on $X \cup \partial\overline{X}^K = \overline{X}^K$ that the G-action (and, hence, the K-action defined earlier) is continuous.

Convergence in the Karpelevič topology restricted to a flat

To get a better picture of, and characterize, the topology of $\overline{X}^K$, to give a new proof that it is compact, and to relate it to other compactifications, it is important to give a non-inductive characterization of the topology in terms of convergent sequences in the positive Weyl chamber. The first thing to observe is that the sequences in the flat $A \cdot o$ that converge in $\overline{X}^K$ have a very special form.

5.6. Theorem. *Let $A^+ \cdot o = \exp \mathfrak{a}^+$ be the image under the exponential map of the positive Weyl chamber, and let $\overline{A^+ \cdot o}$ be its closure, equivalently, the image under the exponential map of $\overline{\mathfrak{a}^+}$. A sequence $\exp(H_n) \cdot o \subset \overline{A^+ \cdot o}$, converging to infinity, is convergent in $\overline{X}^K$ if and only if there is an ordered partition $J_0 \cup J_1 \cup \cdots \cup J_\ell \cup I$ of the set Δ of simple roots such that*

(1) *For $\alpha \notin I, \alpha(H_n) \to +\infty$.*
(2) *For $\alpha \in I$, $\lim_{n\to+\infty} \alpha(H_n)$ exists and is finite (I could be empty).*
(3) *For α, $\beta \in J_i, 0 \le i \le \ell$, $\lim_{n\to+\infty} \frac{\alpha(H_n)}{\beta(H_n)}$ exists, and is positive and finite.*
(4) *For $\alpha \in J_i$, $\beta \in J_j$, $0 \le i < j \le \ell$, $\lim_{n\to+\infty} \frac{\beta(H_n)}{\alpha(H_n)} = 0$.*

One may explain this theorem in the following way. Let $(X_0, z_0; X_1, z_1; \cdots; X_\ell, z_\ell; x)$ be a boundary point in $\overline{X}^K$. This point can be reached from the interior X in k steps. First, one moves out to infinity in the direction of $z_0 \in X(\infty)$. On the boundary space X_1, one chooses another direction and moves to infinity in the direction of $z_1 \in X_1(\infty)$. Keeping on doing this, one eventually reaches x. If one approximates this process by points in the interior X, this movement in different stages corresponds to a division of Δ according to the different rates at which $\alpha(H_n)$ goes to infinity. The reason is that at the ith stage, the direction is determined by the terms $\alpha(H_n)$ with highest rate for $\alpha \in I_i \stackrel{\text{def}}{=} J_i \cup J_{i+1} \cup \cdots \cup J_\ell \cup I$.

The following four lemmas give an inductive proof of Theorem 5.6.

5.7. Lemma. *Let $(H_n) \subset \overline{\mathfrak{a}^+}$. Assume that $\exp(H_n) \cdot o$ converges to a point boundary (z_0, y_0), where z_0 is an interior point of the simplex $C_I(\infty) \subset X(\infty)$ for some $I \subset \Delta$, and $y_0 \in X_{z_0}$. Then*

(1) *for any $\alpha, \beta \in \Delta \setminus I$, $\lim_{n\to+\infty} \frac{\alpha(H_n)}{\beta(H_n)}$ exists and is positive and finite; and*

(2) *for any $\delta \in I$, $\lim_{n\to+\infty} \delta(H_n)$ exists and is finite.*

Equivalently, the sequence (H_n) is C_I-fundamental and has a limiting direction in C_I.

Proof. By the definition of the topology of $\overline{X}^K$, in particular, the condition (3) in the definition of $V^{\mathcal{R}_0}$ in § 5.5, it is clear that the directions $\frac{1}{\|H_n\|} H_n$ converge to an interior point D of the simplex $\{H \in C_I \mid |H| = 1\}$. This simplex can be naturally identified with $C_I(\infty)$. Under this identification, this limit corresponds to z_0, i.e., $z_0 = D(\infty)$. For any root α, it follows that $\frac{\alpha(H_n)}{\|H_n\|} \to \alpha(D)$. Since $C_I = \{H \in \mathfrak{a} \mid \alpha(H) > 0 \text{ if } \alpha \in \Delta \setminus I \text{ and } \delta(H) = 0 \text{ if } \delta \in I\}$, this implies that, for any $\alpha, \beta \in \Delta \setminus I$, $\lim_{n\to+\infty} \frac{\alpha(H_n)}{\beta(H_n)}$ exists and is positive, finite and, for any $\delta \in I$, $\lim_{n\to+\infty} \frac{\delta(H_n)}{\beta(H_n)} = 0$.

Since y_0 is an interior point of X_{z_0}, the condition (1) in the definition of $V^{\mathcal{R}_0}$ following Lemma 5.5 implies that $\pi_{z_0}(\exp(H_n) \cdot o)$ eventually belongs to X_{z_0} and converges to y_0. This projection amounts to projecting H_n onto $\mathfrak{a}^I$, since if one decomposes H_n as $H_{n,I} + H_n^I$, with $H_{n,I} \in \mathfrak{a}_I$ and $H_n^I \in \mathfrak{a}^I$ (see § 2.7), then $\pi_{z_0}(\exp(H_n) \cdot o) = \exp(H_n^I) \cdot o$. Since for any $\delta \in I$, $\delta(H_n) = \delta(H_n^I)$, it follows that $\lim_{n\to+\infty} \delta(H_n)$ exists and is finite. □

5.8. Lemma. *If the sequence $(\exp(H_n) \cdot o)$ is convergent in $\overline{X}^K$, then it satisfies the conditions of Theorem 5.6.*

Proof. Let $(X_0, z_0; X_1, z_1; \cdots; X_\ell, z_\ell; x)$ be the limit of $(\exp(H_n) \cdot o)$. This lemma is proved by induction on the level $\ell + 1$. If the level is 1, it has been proved in the previous lemma. Assume that the lemma is true for any convergent sequence in the Karpelevič compactification of any symmetric space whose limit has level strictly less than $\ell + 1$ and that the level of the limit point is $\ell + 1$. Since the sequence (H_n) has a limiting direction, by the first part of the proof of the previous lemma there exists a subset $J_0 \subset \Delta$ such that, for $\alpha, \beta \in J_0$, $\lim_{n\to+\infty} \frac{\alpha(H_n)}{\beta(H_n)}$ exists and is positive, finite and, for any $\delta \in \Delta \setminus J_0$, $\lim_{n\to+\infty} \frac{\delta(H_n)}{\beta(H_n)} = 0$.

By the condition (1) on $V^{\mathcal{R}}$ following Lemma 5.5 , $\pi_{z_o}(\exp(H_n) \cdot o)$ converges to the boundary point $y_0 = (X_1, z_1; \ldots; X_\ell, z_\ell; x)$ in $\overline{X}^K_{z_0}$. If $I_1 = \Delta \setminus J_0$, then $X_1 = X_{z_0} = k_0 \cdot X^{I_1}$. Furthermore, $\pi_{z_1}(\exp(H_n) \cdot o) = \exp(H_n^1) \cdot o$, where H_n^1 is the component of H_n in $\mathfrak{a}^{I_1}$. Since the set $\Delta \setminus J_0$ forms the set of simple roots of X_1 relative to the positive Weyl chamber $\mathfrak{a}^{I_1,+}$, it follows from the induction hypothesis that there exists an ordered partition $J_1 \cup J_2 \cup \cdots \cup J_\ell \cup I$ of I_1 such that the conditions in Theorem 5.6, applied to $\overline{X}^K_{z_1}$, are satisfied for the sequence (H_n^1). To finish the proof, note that for any $\alpha \in I_1$, $\alpha(H_n) = \alpha(H_n^1)$. □

These two lemmas prove that the conditions stated in Theorem 5.6 are necessary in order that a sequence $(\exp H_n \cdot o) \subset \overline{A^+ \cdot o}$ converge in $\overline{X}^K$.

5.9. Lemma. *If the sequence* $(\exp(H_n) \cdot o)$ *satisfies the condition of Theorem 5.6 with* $\ell = 0$, *then it converges in* $\overline{X}^K$ *to the point* $(D_0(\infty), \exp H^I \cdot o)$, *where* $D_0 \in C_I$ *is the limiting direction of the sequence* (H_n).

In other words, if (H_n) *is a* C_I*-fundamental sequence whose limiting direction is in* C_I, *then* $\exp H_n \cdot o$ *converges in* $\overline{X}^K$.

Proof. It suffices that, for any $V^{\mathcal{R}_0}$ associated to $(X_0, D_0(\infty); \exp H^I \cdot o)$ (see Remark (1) in § 5.5), $\exp(H_n) \cdot o$ eventually belongs to $V^{\mathcal{R}_0}$.

For any two simple roots α and β in J_0, $\lim_{n\to\infty} \alpha(H_n)/\beta(H_n)$ exists and is positive. This implies that the sequence (H_n) has a limiting direction $D_0 \in C_I$ (see the lemma following Proposition 5.12).

Let $z_0 = D_0(\infty)$. Since $\pi_{z_0}(\exp(H_n) \cdot o) = \exp(H_n^1) \cdot o$, where H_n^1 is the projection of H_n onto $\mathfrak{a}^I$, it follows that H_n^1 converges to H^I. Therefore, $\exp(H_n \cdot o)$ also satisfies the conditions (1) and (2) defining $V^{\mathcal{R}_0}$ and so the sequence eventually belongs to $V^{\mathcal{R}_0}$. □

By a similar induction to that used in the proof of Lemma 5.8, one can prove the following lemma and, hence, finish the proof of Theorem 5.6.

5.10. Lemma. *If the sequence* $(\exp(H_n) \cdot o)$ *satisfies the conditions in Theorem 5.6, then it converges in* $\overline{X}^K$ *to the point* (z_0, y_0), *where*

(1) $z_0 = D_0(\infty)$ *and* $D_0 \in C_{I_1}$ *is the limiting direction of the sequence* (H_n),

(2) y_0 *is the point in* $\overline{X}^K_{z_0}$ *to which the sequence* $\pi_{z_o}(\exp H \cdot o)$ *converges.*

Theorem 5.6 shows that the Karpelevič compactification of X selects a special class of sequences on the closure of a positive Weyl chamber. It will now be shown how to associate a compactification of the flat $A \cdot o$, or, equivalently, of $\mathfrak{a}$ with this class of sequences. As will be evident in what follows, it appears that this compactification is completely determined by the polyhedral cone decomposition of $\mathfrak{a}$, and can be defined for any vector space equipped with such a decomposition, provided each cone is simplicial.

The Karpelevič compactification of $\mathfrak{a}$

5.11. Definition. A sequence (H_n) in $\overline{\mathfrak{a}^+}$, converging to infinity, will be called a **Karpelevič fundamental sequence** or K**-fundamental sequence** if there is an ordered partition $J_0 \cup J_1 \cup \cdots \cup J_\ell \cup I$ of the set Δ of simple roots such that

(1) $\alpha \in I$ if and only if $\lim_n \alpha(H_n)$ exists and is finite;

and for each i, $0 \le i \le \ell$,

(2) $\lim_n \frac{\alpha_1(H_n)}{\alpha_2(H_n)}$ exists and is bounded away from zero for any two roots $\alpha_1, \alpha_2 \in J_i$; and

(3) $\lim_n \frac{\alpha'(H_n)}{\alpha(H_n)} = 0$ if $\alpha \in J_i$ and $\alpha' \in J_j, 0 \le i < j \le \ell$.

Note that one may have $I = \emptyset$, but that the sets J_i are all non-empty.

Given an ordered partition $J_0 \cup J_1 \cup \cdots \cup J_\ell \cup I$ of Δ, let $I_i = J_i \cup \cdots \cup J_\ell \cup I$, for $0 \le i \le \ell$. Then $I_0 = \Delta$ and $\mathfrak{a} = \mathfrak{a}^{I_0} \supset \mathfrak{a}^{I_1} \supset \cdots \supset \mathfrak{a}^{I_\ell} \supset \mathfrak{a}^I$. Conversely, a flag of this type, associated with the basis of simple roots, determines an ordered partition of Δ.

To such a flag one may associate an orthogonal decomposition of $\mathfrak{a}$, namely,

$$\mathfrak{a} = \mathfrak{a}_{I_1} \oplus \mathfrak{a}_{I_2}^{I_1} \oplus \mathfrak{a}_{I_3}^{I_2} \oplus \cdots \oplus \mathfrak{a}_I^{I_\ell} \oplus \mathfrak{a}^I,$$

where $\mathfrak{a}_{I_{i+1}}^{I_i} \stackrel{\text{def}}{=} \mathfrak{a}^{I_i} \cap \mathfrak{a}_{I_{i+1}}$. This is a consequence of the fact that for each i, $0 \le i \le \ell$, $\mathfrak{a}^{I_i} = \mathfrak{a}_{I_{i+1}}^{I_i} \oplus a^{I_{i+1}}$, where $I_{\ell+1}$ denotes I.

In terms of this splitting of $\mathfrak{a}$, to each factor one may associate a cone that is in the closure $\overline{\mathfrak{a}^{I_i,+}}$ of the positive Weyl chamber in $\mathfrak{a}^{I_i}$. (Recall that the positive Weyl chamber in $\mathfrak{a}^{I_i}$ is the projection onto $\mathfrak{a}^{I_i}$ of the closure of the positive Weyl chamber $\overline{\mathfrak{a}^+}$ in $\mathfrak{a}$.) Namely, for each i, $0 \le i \le \ell$, define $C_{I_{i+1}}^{I_i}$ to be the projection of $C_{I_{i+1}}$ onto $\mathfrak{a}^{I_i}$, equivalently onto $\mathfrak{a}_{I_{i+1}}^{I_i}$. Then $C_{I_{i+1}}^{I_i} = \{H \in \overline{\mathfrak{a}^{I_i,+}} \mid \alpha(H) = 0, \text{ for all } \alpha \in J_i\}$ and $C_{I_1}^{I_0} = C_{I_1}$. (See Lemma 3.37.)

Recall that if D is a unit vector in $\mathfrak{a}$, it is called a direction.

5.12. Proposition. *Let (H_n) be a K-fundamental sequence in $\overline{\mathfrak{a}^+}$ with associated ordered partition $J_0 \cup J_1 \cup \cdots \cup J_k \cup I$ of Δ. Then, for each i, $0 \le i \le \ell$, there is a unique direction $D_i \in C_{I_{i+1}}^{I_i}$ and a unique point $H^I \in \mathfrak{a}^I$ such that*

(1) $\lim_n \frac{\alpha(H_n)}{\|H_n\|} = \alpha(D_i)$ *for all $\alpha \in J_i$; and*

(2) $\lim_n \alpha(H_n) = \alpha(H^I)$ *if $\alpha \in I$.*

In case $I = \emptyset$, $\mathfrak{a}^\emptyset = \{0\}$ and $H^\emptyset = 0$.

Proof. There is at most one point $H \in \mathfrak{a}^I$ such that $\alpha(H) = \lim_n \alpha(H_n)$ for all $\alpha \in I$ and the existence of such a point follows from the fact that the $H_\alpha, \alpha \in I$, form a basis of $\mathfrak{a}^I$.

The directions $\frac{H_n}{\|H_n\|}$ have a convergent subsequence $\frac{H_{n_k}}{\|H_{n_k}\|}$. It follows that if D is the limit of a convergent subsequence then, for all $\alpha \in \Delta, \alpha(D)$ $= \lim_k \frac{\alpha(H_{n_k})}{\|H_{n_k}\|}$. As a result, if $\alpha_1, \alpha_2 \in J_0$, $\frac{\alpha_1(D)}{\alpha_2(D)} = \lim_n \frac{\alpha_1(H_n)}{\alpha_2(H_n)}$. It follows from Definition 5.11 that $\alpha(D) = 0$ if and only if $\alpha \in I_1$. As a result, $D \in C_{I_1}$ and D is unique by the lemma following this proposition. Hence, the directions $\frac{H_n}{\|H_n\|}$ converge to a (primary) direction $D_0 \in C_{I_1}$.

Project the sequence onto $\mathfrak{a}^{I_1}$. The same argument shows that the projected sequence has a limiting direction $D_1 \in C_{I_2}^{I_1}$. Continuing in this way, by projecting the sequence at the ith stage onto $\mathfrak{a}^{I_i}$, one obtains the directions D_i. □

Lemma. *If $\alpha(D)/\beta(D) = \alpha(D')/\beta(D')$ for all $\alpha, \beta \in \Delta\backslash I$ and D, D' are directions in C_I, then $D = D'$.*

Proof. Let $\Delta\backslash I = \{\alpha_1, \alpha_2, \ldots, \alpha_k\}$. Map H to $(\alpha_1(H), \alpha_2(H), \ldots, \alpha_k(H)) \in \mathbb{R}^k$. Let $c_i \stackrel{\text{def}}{=} \frac{\alpha_i(D)}{\alpha_1(D)}$, $2 \leq i \leq k$. Then D maps to $\alpha_1(D)(1, c_2, \ldots, c_k)$ and D' maps to $\alpha_1(D')(1, c_2, \ldots, c_k)$. Under this map, the unit sphere in $\mathfrak{a}_I$ maps to the boundary of an ellipsoid in $\mathbb{R}^k$ and so $\alpha_1(D) = \alpha_1(D')$, which implies that $\alpha_i(D) = \alpha_i(D')$ for all i, $1 \leq i \leq k$. □

5.13. Lemma. *Every sequence on $\overline{\mathfrak{a}^+}$ that converges to infinity has a K-fundamental subsequence.*

Proof. By § 7.17 one may assume that the sequence is C_I-fundamental (see Definition 3.24). If (H_n) is C_I-fundamental it has a subsequence (H_{n_k}) that has a limiting direction $D_0 \in C_{I_1} \subset \overline{C_I}$: note that $I_1 \supset I$. In other words, $\lim_k \frac{\alpha(H_{n_k})}{\|H_{n_k}\|} = \alpha(D_0)$ for any simple root α. As a result, one has that

(1) $\lim_k \alpha(H_{n_k})$ exists if $\alpha \in I$; and
(2) $\lim_k \frac{\alpha_1(H_{n_k})}{\alpha_2(H_{n_k})}$ exists and is bounded away from zero for any two roots $\alpha_1, \alpha_2 \in J_0 \stackrel{\text{def}}{=} \Delta\backslash I_1$; and
(3) $\lim_k \frac{\alpha'(H_{n_k})}{\alpha(H_{n_k})} = 0$ if $\alpha \in J_0$ and $\alpha' \in I_1$.

If $I_1 = I$, then (H_{n_k}) is a C_I-fundamental subsequence. If not, project the sequence (H_{n_k}) onto $\mathfrak{a}^{I_1}$. It is a $C_I^{I_1}$-fundamental sequence on $\mathfrak{a}^{I_1}$, as $C_I^{I_1}$ is, by definition, the projection of C_I onto $\mathfrak{a}^{I_1}$. The set of simple roots for $\mathfrak{a}^{I_1}$ is I_1 and, as before, one extracts a further subsequence $(H_{n_{k'}})$ that has a limiting direction D_1 in $C_{I_2}^{I_1} \subset \overline{C_I^{I_1}}$, where $I \subset I_2 \subset I_1$.

Hence, for all $\alpha \in I_1$, it follows that $\lim_k \frac{\alpha(H_{n_{k'}})}{\|H_{n_{k'}}\|} = \alpha(D_1)$ for any simple root $\alpha \in I_1$. As a result, one has

(4) $\lim_{k'} \frac{\alpha_1(H_{n_{k'}})}{\alpha_2(H_{n_{k'}})}$ exists and is bounded away from zero for any two roots $\alpha_1, \alpha_2 \in J_1 \stackrel{\text{def}}{=} I_1\backslash I_2$; and
(5) $\lim_{k'} \frac{\alpha'(H_{n_{k'}})}{\alpha(H_{n_{k'}})} = 0$ if $\alpha \in J_1$ and $\alpha' \in I_2$.

Continuing in this way, in a finite number of steps, one determines an ordered partition π of Δ whose last set is I and a subsequence that is K-fundamental with π as its ordered partition. □

5.14. Definition. Given an ordered partition $J_0 \cup J_1 \cup \cdots \cup J_\ell \cup I$ of Δ, let D_i be a direction in each cone $C_{I_{i+1}}^{I_i}$ with corresponding point $D_i(\infty) \in \mathfrak{a}^{I_i}(\infty)$, the boundary of the conical compactification of $\mathfrak{a}^{I_i}$, and let $H^I \in \overline{\mathfrak{a}^{I,+}}$. A K-fundamental sequence (H_n) will be said to **converge formally**

to $(D_0(\infty), D_1(\infty), \ldots, D_\ell(\infty), H^I)$ if its associated ordered partition is the given one and

(1) $\lim_n \alpha(H_n)/\parallel H_n \parallel = \alpha(D_i)$ for all $\alpha \in J_i$;
(2) $\lim_n \alpha(H_n) = \alpha(H^I)$ for $\alpha \in I$.

The formal limit will be denoted by $(D_0(\infty), D_1(\infty), \ldots, D_\ell(\infty))$ if $I = \emptyset$ and the term $H^\emptyset = 0$ will be omitted.

In other words, Lemma 5.12 states that a K-fundamental sequence converges formally to a unique point $(D_0(\infty), D_1(\infty), \ldots, D_\ell(\infty), H^I)$ or to $(D_0(\infty), D_1(\infty), \ldots, D_\ell(\infty))$ in case $I = \emptyset$.

5.15. Proposition. *Given an ordered partition $J_0 \cup J_1 \cup \cdots \cup J_\ell \cup I$ of Δ, let D_i be a direction in each cone $C^{I_i}_{I_{i+1}}$ and let $H^I \in \overline{\mathfrak{a}^{I,+}}$. Then there is a sequence of integers $m_0 > m_1 > \cdots > m_l$ such that $\sum_{i=0}^{\ell} m_i D_i + H^I \in C_J$ if $H^I \in C^I_J$. Hence, there is a K-fundamental sequence on $\overline{\mathfrak{a}^+}$ with $(D_0(\infty), D_1(\infty), \ldots, D_\ell(\infty), H^I)$ as its formal limit.*

Proof. First note that, by the proof of Lemma 3.37, there are positive integers $m'_0 > m'_1 > \ldots m'_\ell$ such that $L = m'_o D_0 + m'_1 D_1 + \cdots + m'_\ell D_\ell \in C_I$. If $H^I \in C^I_J$, $J \subset I$, another application of this proof implies that $m'L + H^I \in C_J$ for some $m' > 0$. Let $m_i = m' m'_i$.

For each i, $0 \le i \le \ell$, let $f_i(n)$ be a strictly positive increasing function such that (i) $\lim_n f_i(n) = \infty$ and (ii) if $i < j$, then $\lim_n \dfrac{f_j(n)}{f_i(n)} = 0$, e.g., $f_i(n) = n^{\ell+1-i}$. Set $H^i_n = f_i(n) m_i D_i$ and let $H_n = \sum_{i=0}^{\ell} H^i_n + H^I$. This sequence is K-fundamental and $(D_0(\infty), D_1(\infty), \ldots, D_\ell(\infty), H^I)$ is its formal limit. □

Two K-fundamental sequences are said to be **equivalent** if they determine the same ordered partitions and if they converge formally to the same point $(D_0(\infty), D_1(\infty), \ldots, D_\ell(\infty), H^I)$. Let $\mathcal{K}(\overline{\mathfrak{a}^+})$ denote the set of points of this form that are obtained from all the possible ordered partitions of Δ. It follows from Propositions 5.12 and 5.15 that this set parametrizes the set of equivalence classes of K-fundamental sequences on $\overline{\mathfrak{a}^+}$.

5.16. Definition. A **K-fundamental sequence on $\mathfrak{a}$** is defined to be the image of a K-fundamental sequence on $\overline{\mathfrak{a}^+}$ under the action of an element of the Weyl group. More specifically, the K-fundamental sequences on $\mathfrak{a}$ are the sequences $(Ad(k_j)H_n)$ where $k_j \in M'$ is one of the representatives of W (see § 3.2) and (H_n) is K-fundamental on $\overline{\mathfrak{a}^+}$.

Consider an ordered partition $J_0 \cup J_1 \cup \cdots \cup J_k \cup I$ of Δ with associated flag $\mathfrak{a} = \mathfrak{a}^{I_0} \supset \mathfrak{a}^{I_1} \supset \cdots \supset \mathfrak{a}^{I_k} \supset \mathfrak{a}^I$, where $I_i = J_i \cup \cdots \cup J_k \cup I$ if $0 \le i \le k$. This flag is defined by the simple roots and so is associated to the positive Weyl chamber $\mathfrak{a}^+$.

If $k \in M'$, the Weyl chamber $\mathfrak{c} = Ad(k)\mathfrak{a}^+$ can also be taken to be the positive chamber, in which case the simple roots are the roots $\alpha \circ Ad(k^{-1})$, $\alpha \in \Delta$. The ordered partition $J_0 \cup J_1 \cup \cdots \cup J_k \cup I$ of Δ

carries over to an ordered partition of $\Delta \circ Ad(k^{-1})$. Since the root vector that represents $\alpha \circ Ad(k^{-1})$ is $Ad(k)H_\alpha$, it follows that the corresponding flag is $\mathfrak{a} = Ad(k)\mathfrak{a}^{I_0} \supset Ad(k)\mathfrak{a}^{I_1} \supset \cdots \supset Ad(k)\mathfrak{a}^{I_k} \supset Ad(k)\mathfrak{a}^{I}$. As a result, it follows that if (H'_n) is K-fundamental on $\overline{Ad(k)\mathfrak{a}^+}$, its formal limit is of the form $(Ad(k)D_0(\infty), Ad(k)D_1(\infty), \ldots, Ad(k)D_\ell(\infty), Ad(k)H^I)$, since, if $H_n = Ad(k^{-1})H'_n$, the sequence (H_n) is K-fundamental on $\overline{\mathfrak{a}^+}$, with $(D_0(\infty), D_1(\infty), \ldots, D_\ell(\infty), H^I)$ as its formal limit. Restated, this means that $(Ad(k)H_n)$ is a K-fundamental sequence on the chamber $\overline{Ad(k)\mathfrak{a}^+}$ with formal limit $(Ad(k)D_0(\infty), Ad(k)D_1(\infty), \ldots, Ad(k)D_\ell(\infty), Ad(k)H^I)$.

This leads one to define the action of K on formal limits by the formula

$$Ad(k)(D_0(\infty), D_1(\infty), \ldots, D_\ell(\infty), H^I) \stackrel{\text{def}}{=} (Ad(k)D_0(\infty), Ad(k)D_1(\infty), \ldots, Ad(k)D_\ell(\infty), Ad(k)H^I).$$

Alternatively,

$$k \cdot (D_0(\infty), D_1(\infty), \ldots, D_\ell(\infty), a^I) \stackrel{\text{def}}{=} (k \cdot D_0(\infty), k \cdot D_1(\infty), \ldots, k \cdot D_\ell(\infty), k \cdot a^I),$$

where $D_i(\infty) \in \mathfrak{a}^{I_i}(\infty)$, $0 \le i \le \ell$, and $a^I = \exp H^I \in \overline{A^{I,+}}$.

Consider the collection $\mathcal{K}(\mathfrak{a})$ of formal limit points of the K-fundamental sequences on all the Weyl chambers. If $\mathcal{K}(\overline{\mathfrak{a}^+})$ denotes the set of formal limit points of the K-fundamental sequences on $\overline{\mathfrak{a}^+}$, this set is

$$\{Ad(k)y \mid k \in M', \text{ where } y = (D_0(\infty), D_1(\infty), \ldots, D_\ell(\infty), H^I) \in \mathcal{K}(\overline{\mathfrak{a}^+})\}.$$

Formally, $\mathcal{K}(\mathfrak{a})$ is the set of limit points of all the K-fundamental sequences. One now introduces a topology on $\mathfrak{a} \cup \mathcal{K}(\mathfrak{a})$ for which convergence to a point of $\mathcal{K}(\mathfrak{a})$ is equivalent to formal convergence to that point. Eventually, one proves that, with this topology, $\mathfrak{a} \cup \mathcal{K}(\mathfrak{a})$ is a compact space, even a compactification of $\mathfrak{a}$, that will be called the Karpelevič compactification of $\mathfrak{a}$. This compactification will also be denoted by $\overline{\mathfrak{a}}^K$ and the ideal boundary $\mathcal{K}(\mathfrak{a})$ will, consequently, also be denoted by $\partial\overline{\mathfrak{a}}^K$.

If D is a direction in $\mathfrak{a}$ and $0 < \eta < 1$, let $\text{Cone}(D;\eta) \stackrel{\text{def}}{=} \{H \in \mathfrak{a} \mid \langle H, D\rangle > \eta \parallel H \parallel\}$, where the inner product $\langle H, D\rangle = B(H, D)$ is given by the Killing form B. In addition, if $d > 0$, let $\text{Cone}(D;\eta, d) = \text{Cone}(D;\eta) \cap \{H \mid d < \parallel H \parallel\}$.

Let $\eta, d, \epsilon > 0$ and define $O(D_0, D_1, \ldots, D_\ell, H^I;\eta, d, \epsilon)$ to be the intersection of the following sets:

$$O(D_0) = O(D_0;\eta, d) \stackrel{\text{def}}{=} \text{Cone}(D_0;\eta, d)$$

$$O(D_1) = O(D_1;\eta, d) \stackrel{\text{def}}{=} \mathfrak{a}_{I_1} \oplus \text{Cone}(D_1;\eta, d) \cap \mathfrak{a}^{I_1}$$

$$\vdots \qquad \vdots \qquad \vdots$$

$$O(D_i) = O(D_i;\eta, d) \stackrel{\text{def}}{=} \mathfrak{a}_{I_i} \oplus \text{Cone}(D_i, \eta, d) \cap \mathfrak{a}^{I_i}$$

$$\vdots \qquad \vdots \qquad \vdots$$

$$O(D_\ell) = O(D_\ell; \eta, d) \stackrel{\text{def}}{=} \mathfrak{a}_{I_\ell} \oplus \mathrm{Cone}(D_\ell; \eta, d) \cap \mathfrak{a}^{I_\ell} \text{ and}$$

$$N(H^I) = N(H^I; \epsilon) \stackrel{\text{def}}{=} \mathfrak{a}_I \oplus B^\perp(H^I; \epsilon),$$

i.e., $O(D_0, D_1, \ldots, D_\ell, H^I; \eta, d, \epsilon) \stackrel{\text{def}}{=} (\cap_{i=0}^{\ell} O(D_i)) \cap N(H^I)$. Recall that $B^\perp(H^I; \epsilon)$ denotes the ball in $\mathfrak{a}_I^\perp = \mathfrak{a}^I$ with center H^I and radius ϵ.

5.17. Definition. The **trace of a basic neighborhood** of $(D_0(\infty), D_1(\infty), \ldots, D_\ell(\infty), H^I)$ is defined to be $O(D_0, D_1, \ldots, D_\ell, H^I; \eta, d, \epsilon)$. The image $Ad(k)O(D_0, D_1, \ldots, D_\ell, H^I; \eta, d, \epsilon)$ under $Ad(k)$ of the trace of a basic neighborhood of $(D_0(\infty), D_1(\infty), \ldots, D_\ell(\infty), H^I)$, is defined to be the **trace of a basic neighborhood** of $k \cdot (D_0(\infty), D_1(\infty), \ldots, D_\ell(\infty), H^I)$.

Hence, $H \in O(D_0, D_1, \ldots, D_\ell, H^I; \eta, d, \epsilon)$ if and only if, for $0 \le i \le \ell$, $H_i \in \mathrm{Cone}(D_i; \eta, d)$, where H_i is the projection of H onto $\mathfrak{a}^{I_i}$, and its projection onto $\mathfrak{a}^I$ is within ϵ of H^I.

A topology is defined on the set $\mathfrak{a} \cup \mathcal{K}(\mathfrak{a})$ by taking as basis the sets O^* where, for each open subset O of $\mathfrak{a}$, O^* is the union of O and all the points in $\mathcal{K}(\mathfrak{a})$ for which the trace of some basic neighborhood of the point is a subset of O.

In other words, $Ad(k)(D_0(\infty), D_1(\infty), \ldots, D_\ell(\infty), H^I) \in O^*$ if and only if, for some $\eta, d, \epsilon > 0$, one has $Ad(k)O(D_0, D_1, \ldots, D_\ell, H^I; \eta, d, \epsilon) \subset O$. In particular, $Ad(k)O(D_0, D_1, \ldots, D_k, H^I; \eta, d, \epsilon)^*$ contains a point of $\mathcal{K}(\mathfrak{a})$ if and only if $Ad(k)O(D_0, D_1, \ldots, D_k, H^I; \eta, d, \epsilon)$ contains the trace of some basic neighborhood of that point. It is clear that M' acts on the topological space $\mathfrak{a} \cup \mathcal{K}(\mathfrak{a})$.

5.18. Proposition. *If $Ad(k)(D_0(\infty), D_1(\infty), \ldots, D_\ell(\infty), H^I) \in \mathcal{K}(\mathfrak{a})$, and $k \in M'$, the open sets $Ad(k)O(D_0, D_1, \ldots, D_\ell, H^I; (1-\frac{1}{n}), n, \frac{1}{n})^*$ form a neighborhood base for $Ad(k)(D_0(\infty), D_1(\infty), \ldots, D_\ell(\infty), H^I)$. Further, a K-fundamental sequence converges formally to a point of $\mathcal{K}(\mathfrak{a})$ if and only if it converges to the same point relative to the topology on $\mathfrak{a} \cup \mathcal{K}(\mathfrak{a})$. Consequently, the topology is Hausdorff.*

Proof. Let $O(\eta, d, \epsilon)$ denote $O(D_0, D_1, \ldots, D_k, H^I; \eta, d, \epsilon)$. From the definition of $Ad(k)O(D_0, D_1, \ldots, D_k, H^I; \eta, d, \epsilon)^* = Ad(k)O(\eta, d, \epsilon)^*$, it is obvious that it contains $Ad(k)(D_0(\infty), D_1(\infty), \ldots, D_\ell(\infty), H^I)$ and, hence, is a neighborhood of that point. It is also clear from the definition that $Ad(k)O(\eta, d, \epsilon)$ and $Ad(k)O(\eta, d, \epsilon)^*$ decrease as η increases to 1, d increases to infinity, and ϵ decreases to zero. This proves the first statement.

Let $Ad(k)(D_0(\infty), D_1(\infty), \ldots, D_\ell(\infty), H^I)$ denote a point of $\mathcal{K}(\mathfrak{a})$. To discuss convergence at this point, since it is clear that M' acts on the topological space $\mathfrak{a} \cup \mathcal{K}(\mathfrak{a})$, it suffices to consider the case when $k = e$.

Assume that $(D_0(\infty), D_1(\infty), \ldots, D_\ell(\infty), H^I)$ is the formal limit of the sequence (H_n). The limiting direction of the sequence is therefore D_0. Thus, for large n, $H_n \in \mathrm{Cone}(D_0; \eta, d) = O(D_0; \eta, d)$.

Project the sequence onto $\mathfrak{a}^{I_1}$. The limiting direction of the projected sequence is then D_1 and so, for large n, one has $H_n^{I_1} \in O(D_1, \eta, d)$. Continuing to project, in turn, on the subspaces $\mathfrak{a}^{I_i}$, it follows that for each $i, \leq i \leq \ell$, one has $H_n^{I_i} \in O(D_i; \eta, d)$ for large n.

Finally, projecting the sequence onto $\mathfrak{a}^I$, it follows that the projected sequence converges to H^I. Hence, for large n, one has $H_n^I \in N(H^I; \epsilon)$.

Consequently, if $(D_0(\infty), D_1(\infty), \ldots, D_\ell(\infty), H^I)$, is the formal limit of (H_n), for any $\eta \in (0,1)$ and $d, \epsilon > 0$, the sequence is eventually in $O(D_0, D_1, \ldots, D_\ell, H^I; \eta, d, \epsilon)$.

Conversely, assume that a K-fundamental sequence $(Ad(k_j)(H_n))$ converges topologically to $y = (D_0(\infty), D_1(\infty), \ldots, D_\ell(\infty), H^I)$. Let $y_o = (D_0^o(\infty), D_1^o(\infty), \ldots, D_{\ell^o}^o(\infty), {H_o}^{I^o})$ denote the formal limit of the sequence (H_n).

Let $m_0 > m_1 > \cdots m_\ell$, as in Proposition 5.15, be such that $\sum_{i=0}^{\ell} m_i D_i + H^I \in C_J, J \subset I$. The following lemma implies that (i) $y = y_o$ and (ii) that $k_j \in K^J M$. Hence, $Ad(k_j)D_i = D_i$, $0 \leq i \leq \ell$ and $Ad(k_j)H^I = H^I$. Consequently, the formal limit $Ad(k_j)(D_0(\infty), D_1(\infty), \ldots, D_\ell(\infty), H^I)$ of $(Ad(k_j)H_n)$ equals its topological limit y as $Ad(k_j)y = y$.

Lemma 5.19. *Let $k_j \in W$ and (H_n) be a K-fundamental sequence on $\overline{\mathfrak{a}^+}$. Let $y_o = (D_0^o(\infty), D_1^o(\infty), \ldots, D_{\ell^o}^o(\infty), {H_o}^{I^o})$ denote the formal limit of the sequence (H_n). Let $y = (D_0(\infty), D_1(\infty), \ldots, D_\ell(\infty), H^I)$ be the limit of $(Ad(k_j)H_n)$ in $\mathfrak{a} \cup \mathcal{K}(\mathfrak{a})$. Then,*

(1) *$y_o = y$,*
(2) *$k_j \in K^J M \cap M'$, and is in M if $J = \emptyset$.*

In case $I = \emptyset$, one has $k_j \in M$ in (2).

Continuation of the proof of Proposition 5.18.

Let $(D_0^o(\infty), D_1^o(\infty), \ldots, D_{\ell^o}^o(\infty), {H_o}^{I^o})$ and $Ad(k)(D_0(\infty), D_1(\infty), \ldots, D_\ell(\infty), H^I)$ be two points y_o and $Ad(k)y$ of $\mathcal{K}(\mathfrak{a})$ that do not have disjoint neighborhoods. Then, by the first part of the proof, there is a sequence on $\mathfrak{a}$ that converges topologically to both points. Hence, for some k_j, by Lemma 5.13, there is a K-fundamental subsequence $(Ad(k_j)(H_n))$ that converges topologically to both points. As a result, Lemma 5.19 implies that (i) $k^{-1}k_j \in K^J M$ where $\sum_{i=0}^{\ell} m_i D_i + H^I \in C_J$ (see Proposition 5.15), (ii) that (H_n) is K-fundamental on $\overline{\mathfrak{a}^+}$ with formal limit $y = (D_0(\infty), D_1(\infty), \ldots, D_\ell(\infty), H^I)$, and (iii) that (H_n) also converges formally to $y_o = (D_0^o(\infty), D_1^o(\infty), \ldots, D_{\ell^o}^o(\infty), {H_o}^{I^o})$. In other words, $y = y_0$. From (ii), it also follows by Lemma 5.19 that $k_j \in K^J M$, with the result that $k \in K^J M$. Consequently, $Ad(k)y = y = y_o$. $\square$

Proof of Lemma 5.19.

The sequence $(Ad(k_j)H_n)$ is eventually in $O(D_0, D_1, \ldots, D_\ell, H^I; (1 - \frac{1}{n}), n, \frac{1}{n})$ for any integer n. As a result, the (primary) direction of this sequence is D_0. Projecting it successively on the subspaces $\mathfrak{a}^{I_i}$ of the flag associated with $(D_0(\infty), D_1(\infty), \ldots, D_\ell(\infty), H^I)$, one sees that the corresponding directions are the directions D_i and that only the roots in I are bounded on this sequence. It follows that the projection of the sequence onto $\mathfrak{a}^I$ converges to H^I.

Since $(D_0^o(\infty), D_1^o(\infty), \ldots, D_{\ell^o}^o(\infty), H_o{}^{I^o})$ is the formal limit of the K-fundamental sequence (H_n), D_0^o is the limiting direction of this sequence, i.e., $Ad(k_j)D_0^o = D_0$ and so $Ad(k_j)C_{I^o} = C_I$. Proposition 3.4 implies that $I_1 = I_1^o$, and so the first proper subspace $\mathfrak{a}^{I_1^o}$ in the flag corresponding to $(D_0^o(\infty), D_1^o(\infty), \ldots, D_\ell^o(\infty), H_o{}^{I^o})$ coincides with $\mathfrak{a}^{I_1}$. Furthermore, since $k_j \in K^{I_1}$, $Ad(k_j)$ leaves $\mathfrak{a}_{I_1}$ pointwise fixed and so $D_0 = D_0^o$.

Project the sequence (H_n) onto the subspace $\mathfrak{a}^{I_1}$. It follows for similar reasons that $Ad(k_j)D_1^o = D_1$ and $\mathfrak{a}^{I_2^o} = \mathfrak{a}^{I_2}$. Continuing in this way, it follows that $\ell = \ell^o$ and $Ad(k_j)D_i^o = D_i$ for all i, $0 \le i \le \ell$, and $I = I^o$. To see this, first assume that $\ell \le \ell^o$. Then one has $I \supset I^o$ and $Ad(k_j)C_I = C_I$ as $Ad(k_j)\sum_{i=0}^{\ell} m_i D_i = \sum_{i=0}^{\ell} m_i D_i$, where the m_i are as in Proposition 5.15. As a result, $k_j \in K^I M$. Since the projection of the sequence $(Ad(k_j)H_n)$ onto $\mathfrak{a}^I$ is bounded, this implies that the projection of the sequence (H_n) onto $\mathfrak{a}^I$ is bounded. Hence, $\mathfrak{a}^I = \mathfrak{a}^{I^o}$, i.e., $I = I^o$. If on the other hand, $\ell^o \le \ell$, for similar reasons, $k_j \in K^{I^o}M$ and, hence, as before, $\mathfrak{a}^{I^o} \supset \mathfrak{a}^I$ and so $I^o = I$.

The projection onto $\mathfrak{a}^I$ commutes with $Ad(k_j)$. As a result, $Ad(k_j)H^I = H_o^I$ since the projections onto $\mathfrak{a}^I$ of the sequences $(Ad(k_j)H_n)$ and (H_n) converge to H^I and H_o^I respectively. It follows from the Cartan decomposition, applied to the symmetric space X^I, that $H^I = H_o^I$.

It was shown earlier that there exist $m_0 > m_1 > \cdots > m_\ell$ such that $H' = \sum_{i=0}^{\ell} m_i D_i + H^I \in C_J^I$ (Proposition 5.15). As a result, $k_j \in K^J M$ since $Ad(k_j)H' = H'$. Note that $J = \emptyset$ if $H^I \in \mathfrak{a}^{I,+}$ and so $k_j \in M$. Finally, observe that, when $I = \emptyset$, then $\sum_{i=0}^{\ell} m_i D_i \in \mathfrak{a}^+$ and so $k_j \in M$. □

5.20. Theorem. *The topological space $\mathfrak{a} \cup \mathcal{K}(\mathfrak{a})$ is compact. When $\mathfrak{a}$ is identified with the flat $A \cdot o$ under the map $H \to \exp H \cdot o$, this compactification is isomorphic to the closure of the flat $A \cdot o$ in $\overline{X}^K$.*

The proof of this result is given in Appendix A. The basic point to establish is that the topology is regular and has a countable base.

The identification of this compactification with the closure of the flat in $\overline{X}^K$ is made as follows. If $(D_0(\infty), D_1(\infty), \ldots, D_\ell(\infty), H^I)$ is the formal limit of (H_n), then $(\exp H_n \cdot o)$ converges to $(X_0, z_0; X_1, z_1; \ldots, X_\ell, z_\ell; x)$ where $X_0 = X, z_0 = D_0(\infty), X_1 = X^{I_1}, z_1 = D_1(\infty), \ldots, X_\ell = X^{I_\ell}, z_\ell = D_\ell(\infty)$, and $x = \exp H^I \cdot o$. The K-action defined earlier then shows that a boundary point $k_j \cdot (D_0(\infty), D_1(\infty), \ldots, D_\ell(\infty), H^I)$ for $k_j \in W$ corresponds to the point $k_j \cdot (z_0, z_1, \ldots, z_\ell, x)$.

As in the case of the boundary of $\overline{X}^{SF}$, points of the Karpelevič boundary have a **polar representation**, established by the next result. This representation is related to a new construction (see Remark 5.32) of the

Karpelevič compactification $\overline{X}^K$ from the group K and the compact space $\mathfrak{a} \cup \mathcal{K}(\mathfrak{a})$.

5.21. Proposition. *If* $(z, y) = (z_0, z_1, \ldots, z_\ell, x) \in \partial\overline{X}^K$, *then*

$$(z, y) = k \cdot (D_0(\infty), D_1(\infty), \ldots, D_\ell(\infty), \exp H^I \cdot o),$$

for a unique point $(D_0(\infty), D_1(\infty), \ldots, D_\ell(\infty), H^I)$ *in* $\mathcal{K}(\overline{\mathfrak{a}^+})$. *Furthermore,* k *is unique modulo* $K^J M$ *if* $H^I \in C_J^I$ *for* $J \subsetneq I$ *and modulo* M *if* J *or* I *is the empty set.*

Proof. The result is clearly valid if the rank of X is one. Assume that it is true for any symmetric space when the rank is less than that of X.

It is clear that $z = k_0 \cdot D_0(\infty)$ for a unique direction $D_0 \in C_{I_1}$, where $z \in k_0 \cdot C_{I_1}(\infty)$. Further, k_0 is unique modulo $K^{I_1}M$ and $y \in \overline{k_0 \cdot X^{I_1}}^K$. Since $k_0^{-1} \cdot y \in \overline{X^{I_1}}^K$ it follows from the inductive assumption that $k_0^{-1} \cdot y = k_1 \cdot (D_1(\infty), \ldots, D_\ell(\infty), \exp H^I \cdot o)$ where $(D_1(\infty), \ldots, D_\ell(\infty), H^I)$ is a unique point of $\mathcal{K}(\overline{\mathfrak{a}^{I_1,+}})$, with $k_1 \in K^{I_1}M$ and k_1 unique modulo $K^J M$ if $H^I \in C_J^I$ for $J \subsetneq I$, and modulo M if J or $I = \emptyset$. Since $Ad(k_1)D_0 = D_0$, it follows that $(z, y) = k_0 k_1 \cdot (D_0(\infty), D_1(\infty), \ldots, D_\ell(\infty), \exp H^I \cdot o)$. Furthermore, since $(D_1(\infty), \ldots, D_\ell(\infty), H^I)$ is unique, it follows that $(D_0(\infty), D_1(\infty), \ldots, D_\ell(\infty), \exp H^I \cdot o)$ is unique.

Assume that, in addition, $k \cdot (D_0(\infty), D_1(\infty), \ldots, D_\ell(\infty), \exp H^I \cdot o) = (z, y)$. Let $k' = k^{-1}(k_0 k_1)$. Since $Ad(k')D_i = D_i$, $0 \le i \le \ell$ and $Ad(k')H^I = H^I$, it follows from Proposition 5.15 that $k' \in K^J M$ if $J \neq I$ and is in M if J or $I = \emptyset$. □

5.22. Definition. The **Karpelevič compactification** of $\mathfrak{a}$ is defined to be the compact space $\mathfrak{a} \cup \mathcal{K}(\mathfrak{a})$. It will denoted by $\overline{\mathfrak{a}}^K$ and $\partial\overline{\mathfrak{a}}^K$ will denote the ideal boundary $\mathcal{K}(\mathfrak{a})$.

THE KARPELEVIČ TOPOLOGY IS COMPACT

By extending the notion of a K-fundamental sequence from the flat $A \cdot o$ to X and making use of the continuity of the K-action, it is easy to prove that $\overline{X}^K$ is compact, given that $\overline{\mathfrak{a}}^K$ is compact.

5.23. Definition. A sequence (y_n) on X is said to be a **Karpelevič-fundamental sequence**, or **K-fundamental sequence, in** X if, when $y_n = k_n \exp H_n \cdot o, H_n \in \overline{\mathfrak{a}^+}$, then

(1) (k_n) converges, and
(2) (H_n) is K-fundamental on $\overline{\mathfrak{a}^+}$.

5.24. Remarks. (1) It follows from Lemma 5.13 that every sequence in X, converging to infinity, has a K-fundamental subsequence.

(2) Every K-fundamental sequence is fundamental in the sense of Definition 3.35. Furthermore, if $(D_0(\infty), D_1(\infty), \ldots, D_\ell(\infty), \exp H^I \cdot o)$ is the

limit in $\overline{X}^K$ of $(\exp H_n \cdot o)$, when (H_n) is K-fundamental on $\overline{\mathfrak{a}^+}$, then $(\exp H_n \cdot o)$ converges to $(C_I(\infty), \exp H^I \cdot o)$ in $X \cup \Delta^*(X)$. In case $I = \emptyset$, the limit point is $(\mathfrak{a}^+(\infty), 0)$ since all the simple roots converge to ∞ along the sequence (H_n).

(3) Every K-fundamental sequence converges in the topology previously defined on the set $\overline{X}^K$.

(4) It follows from (1), as in the case of the dual cell compactification $X \cup \Delta^*(X)$, that a sequence (y_n) in X, converging to infinity, converges in $\overline{X}^K$ if and only if all of its K-fundamental subsequence have the same limit.

5.25. Theorem. (See [K3, §13.7]) *The topological space $\overline{X}^K$ is compact.*

Proof. By identifying $\mathfrak{a}$ with $A \cdot o$ via the exponential map, it follows from Theorem 5.6 that the Karpelevič compactification $\overline{\mathfrak{a}}^K$ of $\mathfrak{a}$ is identified with the closure $\overline{A \cdot o}^K$ of $A \cdot o$ in $\overline{X}^K$.

The map $K \times \overline{A \cdot o}^K \to \overline{X}^K$ defined by the mapping (k, y) to $k \cdot y$ is continuous and its image is therefore compact. In fact, in view of Proposition 5.21, this image equals $\overline{X}^K$ as each point of $\partial\overline{X}^K$ has a polar representation. This proves the theorem once one establishes the continuity of the map $(k, y) \to k \cdot y$.

To verify this, it is enough to show that if $k_n \to k$ and $a_n \cdot o \to y \in \partial\overline{X}^K$, with $a_n \in A$, for $n \geq 1$, then $k_n a_n \cdot o \to k \cdot y$.

If the sequence $(a_n \cdot o)$ is in $\overline{A^+ \cdot o}$, then it follows from the definition of the basic neighborhoods of y that $k_n a_n \cdot o$ converges to $k \cdot y$.

In general, $a_n = k_{j(n)} \exp H_n \cdot o$ with $H_n \in \overline{\mathfrak{a}^+}, n \geq 1$ and $k_{j(n)} \in W, j(n) \in \{1, 2, \ldots, |W|\}$. If j is a value that occurs an infinite number of times, define $n_k(j) = n_k$ to be the integer at which j occurs for the kth time. The subsequence $(k_{n_k}(\exp H_{n_k} \cdot o))$ converges in $\overline{X}^K$ as does the subsequence $(\exp H_{n_k} \cdot o)$. Let $(D_0(\infty), D_1(\infty), \ldots, D_\ell(\infty), \exp H^I \cdot o)$ be its limit. Since $y = k_j \cdot (D_0(\infty), D_1(\infty), \ldots, D_\ell(\infty), \exp H^I \cdot o)$, it follows from Proposition 5.21 that $(D_0(\infty), D_1(\infty), \ldots, D_\ell(\infty), \exp H^I \cdot o)$ does not depend upon the value of j.

Furthermore, by what was observed before, $k_{n_k} \exp H_{n_k} \cdot o = k_{n_k} a_{n_k} \cdot o$ converges to $k \cdot y$. This shows that $k_n a_n \cdot o \to k \cdot y$. □

THE RELATION BETWEEN THE KARPELEVIČ COMPACTIFICATION AND THE CONICAL AND DUAL CELL COMPACTIFICATIONS

5.26. Lemma. *The Karpelevič compactification $\overline{X}^K$ dominates the conic compactification $X \cup X(\infty)$. More specifically, there is a unique continuous function $\varphi_c : \overline{X}^K \to \overline{X}^c$ that is the identity on X. In addition, $\varphi_c(z, y) = z$ if $(z, y) \in \partial\overline{X}^K$. As a result, the fiber $\varphi_c^{-1}(\{z\}) = \overline{X}^K_z$, for all $z \in X(\infty)$.*

Proof. If a sequence y_n in X converges to $(X_0, z_0; X_1, z_1; \cdots; X_\ell, z_\ell; x)$ in $\overline{X}^K$, then the condition (3) in the definition of $V^{\mathcal{R}_0}$ shows that (y_n)

converges to z_0 in $X \cup X(\infty)$. The continuous map φ_c is defined by $\varphi_c(z_0, z_1, \ldots, z_\ell, x) = z_0$. □

5.27. Proposition. *The Karpelevič compactification $\overline{X}^K$ dominates the compactification $X \cup \Delta^*(X)$ (see Definition 3.40). These two compactifications are isomorphic to each other if and only if X has rank one.*

More specifically, there is a unique continuous map $\varphi_{\Delta^*} : \overline{X}^K \to X \cup \Delta^*(X)$ such that $\varphi_{\Delta^*}(x) = x$ for all $x \in X$. Furthermore,

(1) *if $(z, y) = k \cdot (D_0(\infty), D_1(\infty), \ldots, D_\ell(\infty), \exp H^I \cdot o) \in \partial\overline{X}^K$, then $\varphi_{\Delta^*}((z, y)) = k \cdot (C_I(\infty), \exp H^I \cdot o)$ and*
(2) *if $(z, y) = k \cdot (D_0(\infty), D_1(\infty), \ldots, D_\ell(\infty))$, then $\varphi_{\Delta^*}((z, y)) = k \cdot (\mathfrak{a}^+(\infty), o)$.*

Proof. In view of Remark 5.24(2), it follows from Lemma 3.28 that $\overline{X}^K$ dominates $X \cup \Delta^*(X)$, i.e, there is a unique continuous map $\varphi : \overline{X}^K \to X \cup \Delta^*(X)$ that is the identity on X. Let φ_{Δ^*} denote this map. Furthermore, the proof of this lemma shows that if (y_n) is K-fundamental in X with limit (z, y) in $\overline{X}^K$ then $\varphi_{\Delta^*}((z, y))$ is the limit in $X \cup \Delta^*(X)$ of (y_n). From this fact, and the observation in Remark 5.24(2) on limit points, it follows that $\varphi_{\Delta^*}(k \cdot (D_0(\infty), D_1(\infty), \ldots, D_\ell(\infty), \exp H^I \cdot o)) = k \cdot (C_I(\infty), \exp H^I \cdot o)$ and, in the case $I = \emptyset$, $\varphi_{\Delta^*}(k \cdot (D_0(\infty), D_1(\infty), \ldots, D_\ell(\infty)) = k \cdot (\mathfrak{a}^+(\infty), o)$.

In case the rank of X is one, both compactifications are equal to the conical compactification $\overline{X}^c$. Conversely, if the natural map φ_{Δ^*} from $\overline{X}^K$ onto $X \cup \Delta^*(X)$ that is the identity on X is one-to-one, it follows that each C_I is 1-dimensional, for all proper subsets I of Δ. Hence, the rank of X is one. □

5.28. Corollary. *The Karpelevič compactification $\overline{X}^K$ dominates the compactification $\overline{X}^c \vee (X \cup \Delta^*(X))$. It is isomorphic to $\overline{X}^c \vee (X \cup \Delta^*(X))$ if and only if the rank of X is less than or equal to two.*

More explicitly, the unique continuous map $\varphi : \overline{X}^K \to \overline{X}^c \vee (X \cup \Delta^*(X))$ that is the identity on X has the property that, for each $z \in X(\infty)$, its restriction to $\overline{X}_z^K$ is the natural map of $\overline{X}_z^K$ onto $X_z \cup \Delta^*(X_z)$ given by Proposition 5.27. As a result, it is one-to-one if and only if the rank of X is less than or equal to two.

Proof. By Lemma 5.26 and Proposition 5.27, $\overline{X}^K$ dominates both $\overline{X}^c$ and $X \cup \Delta^*(X)$ and, hence, dominates $\overline{X}^c \vee (X \cup \Delta^*(X))$. The unique continuous map $\varphi : \overline{X}^K \to \overline{X}^c \vee (X \cup \Delta^*(X))$ is such that $\varphi_c = \varphi_1 \circ \varphi$, where φ_1 is the projection on the first coordinate (see Proposition 3.45). It follows from the identification of the fibers of φ_c and of φ_1 that $\varphi(\overline{X}_z^K) = X_z \cup \Delta^*(X_z)$.

Furthermore, $\varphi_{\Delta^*} = \varphi_2 \circ \varphi$, where $\varphi_2 : \overline{X}^c \vee (X \cup \Delta^*(X)) \to (X \cup \Delta^*(X))$ is given by projection on the second coordinate. This implies that $\varphi(z, y) = (z, \varphi_{\Delta^*}(z, y))$. If $z = D_0(\infty)$, and $y = k \cdot (D_1(\infty), \ldots, D_\ell(\infty), \exp H^I)$, then $\varphi_{\Delta^*}(z, y) = k \cdot (C_I(\infty), \exp H^I \cdot o) \in \Delta^*(X)$. (Note that $k \in K^{I_1} M$.) In

$\Delta^*(X)$ the symmetric spaces $k \cdot X^I$ are the "open" cells corresponding to the simplices $k \cdot C_I(\infty)$ in $X(\infty)$.

It follows from Proposition 3.44, and the continuity of the K-action that the closure in $\Delta^*(X)$ of $k{\cdot}X^I$ is naturally isomorphic to $k{\cdot}X^I \cup \Delta^*(k{\cdot}X^I) = k \cdot (X^I \cup \Delta^*(X^I))$. As a result, $k \cdot (C_I(\infty), \exp H^I \cdot o) \in \Delta^*(X)$ is naturally identified with the point $k \cdot (C_I^{I_1}(\infty), \exp H^I \cdot o) \in \Delta^*(k \cdot X^{I_1})$, where $D_0 \in C_{I_1}$. Hence, the restriction of φ to $\overline{X}^K_{D_0(\infty)}$ can then be viewed as the natural map of this Karpelevič compactification onto $X_{D_0(\infty)} \cup \Delta^*(X_{D_0(\infty)})$ given by Proposition 5.27. Using the K-action, the result follows for any $z \in X(\infty)$.

Consequently, φ is one-to-one if and only if, for all $z \in X(\infty)$, the Karpelevič compactification of X_z is isomorphic to the dual cell compactification of X_z. By Proposition 5.27, this is the case if and only if the rank of X_z is less than or equal to one, i.e., if and only if the rank of X is less than or equal to two. ☐

5.29. Example. The following is a simple example of a sequence that converges in $\overline{X}^c \vee (X \cup \Delta^*(X))$ but does not converge in $\overline{X}^K$. Assume that the rank of X is equal to three and the simple roots are $\alpha_1, \alpha_2, \alpha_3$. Let (H_n) be a sequence in the positive chamber $\mathfrak{a}^+$ such that, for $n \geq 1, \alpha_1(H_{2n}) = n^3, \alpha_2(H_{2n}) = n^2, \alpha_3(H_{2n}) = n$ and $\alpha_1(H_{2n-1}) = n^3, \alpha_2(H_{2n-1}) = n$, $\alpha_3(H_{2n-1}) = n$. Since $\lim_{n\to\infty} \frac{1}{\|H_n\|} H_n$ exists, it follows that $(\exp(H_n) \cdot o)$ is convergent in $\overline{X}^c$. It is also convergent in $X \cup \Delta^*(X)$ since it is an $\mathfrak{a}^+$-fundamental sequence in the sense of Definition 3.24. Therefore, it is convergent in $\overline{X}^c \vee (X \cup \Delta^*(X))$. However, the condition in Theorem 5.6 is not satisfied and, hence, $(\exp(H_n) \cdot o)$ is not convergent in $\overline{X}^K$.

A CHARACTERIZATION OF THE KARPELEVIČ COMPACTIFICATION

It is evident from the definition of a K-fundamental sequence on X and Definition 5.14 that one may define the concept of a formal limit independently of the definition of the space $\overline{X}^K$. Namely, $(k_n \exp H_n \cdot o)$ has a **formal limit** $k \cdot (D_0(\infty), D_1(\infty), \ldots, D_\ell(\infty), H^I)$ if $k_n \to k$ and $(D_0(\infty), D_1(\infty), \ldots, D_\ell(\infty), H^I)$ is the formal limit of the sequence (H_n).

5.30. Theorem. *There is a unique compactification $\overline{X}$ of X such that*

(1) *every K-fundamental sequence on X converges in $\overline{X}$,*
(2) *two K-fundamental sequences converge to the same point in $\overline{X}$ if and only if they have the same formal limits.*

This compactification is isomorphic to the Karpelevič compactification $\overline{X}^K$ of X.

Proof. The Karpelevič compactification has these two properties. It follows from Lemma 3.28, as in the proofs of Theorems 3.29 and 3.38, that any two compactifications that satisfy these two properties are isomorphic. ☐

Theorem 3.39 has an analogue in the case of the Karpelevič compactification that characterizes it as a K-compactification of X for which the closure of a flat $kA \cdot o$ is its Karpelevič compactification.

5.31. Theorem. *The compactification $\overline{X}$ of X determined by Theorem 5.30 has the following properties:*

(1) *it is a K-compactification (see Definition 3.27);*
(2) *the closure of a flat $kA \cdot o$ in $\overline{X}$ is the Karpelevič compactification of that flat; and*
(3) *the closure of the intersection of two flats $k_1 A \cdot o$ and $k_2 A \cdot o$ is the intersection of their closures, i.e., of their Karpelevič compactifications.*

These three properties characterize this compactification.

Proof. If $y \in \overline{X} \backslash X$, define $k \cdot y$ as the limit of $k \cdot y_n$ if (y_n) is K-fundamental and converges to y. This defines a continuous action of K. For a fixed $k \in K$, the sequences $k\ell_n \exp H_n \cdot o$, with $\ell_n \in M'$ that are fundamental in the sense of Definition 5.23 are exactly the fundamental sequences of the flat $kA \cdot o$. It follows from the identification of $\mathfrak{a}$ with $kA \cdot o$ that the Karpelevič compactification of $\mathfrak{a}$ may be identified with the closure of $kA \cdot o$ in $\overline{X}^K$.

To verify (3), let $(H_n(1))$ and $(H_n(2))$ be two K-fundamental sequences in $\overline{\mathfrak{a}^+}$ with formal limits $y_1 = (D_0^1(\infty), D_1^1(\infty), \dots, D_{\ell_1}^1(\infty), H^{I^1})$ and $y_2 = (D_0^2(\infty), D_1^2(\infty), \dots, D_{\ell_2}^2(\infty), H^{I^2})$. Assume that the formal limits of the sequences $(k_1 \exp H_n(1) \cdot o)$ and $(k_2 \exp H_n(2) \cdot o)$ agree, in other words, that $k_1 \cdot y_1 = k_2 \cdot y_2$.

Note that this implies $\ell_1 = \ell_2 = \ell, D_i^1(\infty) = D_i^2(\infty)$, $0 \leq i \leq \ell$, $I^1 = I^2 = I$ with $H^{I^1} = H^{I^2} = H^I$, and $k_2^{-1} k_1 \in K^J M$ if $H^I \in C_J^I$. It is enough to show that there is a K-fundamental sequence in $(k_1 A \cdot o) \cap (k_2 A \cdot o)$ (equivalently, in $Ad(k_1)\mathfrak{a} \cap Ad(k_2)\mathfrak{a}$) that has the same formal limit.

Let $H_n = H^I + H_{n,J}(1)$, where H_J is the projection of H onto $\mathfrak{a}_J$. Then $(k_1 \exp H_n \cdot o)$ is a K-fundamental sequence in the intersection of the flats $(k_1 A \cdot o)$ and $(k_2 A \cdot o)$ (equivalently, $(Ad(k_1) H_n)$ is a K-fundamental sequence in $Ad(k_1)\mathfrak{a} \cap Ad(k_2)\mathfrak{a}$) whose formal limit is the formal limit of $(k_1 \exp H_n(1) \cdot o)$ and of $(k_2 \exp H_n(2) \cdot o)$.

Consider a compactification of X that satisfies (1), (2) and (3). From (2), it follows that K-fundamental sequences of the form $(k\ell_n \exp H_n \cdot o)$, with $\ell_n \in M'$, all converge. Property (1) implies that if a sequence (y_n) converges and (k_n) converges, then $(k_n \cdot y_n)$ converges. Hence, all the K-fundamental sequences converge. It remains to verify condition (2) of Theorem 5.30, i.e., that their limits are determined by their formal limits.

Suppose that $(k_n \exp H_n \cdot o)$ and $(k_n' \exp H_n' \cdot o)$ are two K-fundamental sequences in the sense of Definition 5.23. The sequences (H_n) and (H_n') are also K-fundamental. Let $y = (D_0(\infty), D_1(\infty), \dots, D_\ell(\infty), \exp H^I \cdot o)$ and $y' = (D_0'(\infty), D_1'(\infty), \dots, D_{\ell'}'(\infty), \exp H'^{I'} \cdot o)$ denote their formal limits.

By (2) and Proposition 5.18, the sequences (H_n) and $(H_{n'})$ converge to the same limit in the compactification $\overline{\mathfrak{a}}^K$ if and only if their formal limits coincide.

Let $k = \lim_n k_n$ and $k' = \lim_n k'_n$. Assume that $(k \exp H_n \cdot o)$ and $(k' \exp H'_n \cdot o)$ converge to the same point x_0 in the given compactification of X. It follows from (3) that there is a K-fundamental sequence in $(kA \cdot o) \cap (k'A \cdot o)$ that also converges to x_0. In order to conclude from (2) that $(k \exp H_n \cdot o)$ and $(k' \exp H'_n \cdot o)$ have the same formal limit, one needs to observe that Lemma 3.36(2) implies that a K-fundamental sequence $(k_n \exp H_n \cdot o)$ in a flat $kA \cdot o$ has the form $k \cdot y_n$, where (y_n) is K-fundamental in $A \cdot o$.

Conversely, assume that the formal limits of two K-fundamental sequences $(k \exp H_n \cdot o)$ and $(k' \exp H'_n \cdot o)$ are the same, i.e., assume that $k \cdot y = k' \cdot y'$. It follows that $\ell = \ell', I = I'$, $D_i(\infty) = D'_i(\infty)$, $0 \le i \le \ell$, and $H^I = H^{I'}$. Furthermore, $k^{-1}k' \in K^J M$ if $H^I \in C^I_J$ (and is in M if $I = \emptyset$). As a result $C_J \subset Ad(k^{-1}k')\mathfrak{a} \cap \mathfrak{a}$, and there is a K-fundamental sequence (y_n) in $(kA \cdot o) \cap (k'A' \cdot o)$ that has the common formal limit. It suffices to let $y_n = k \exp H_n \cdot o$, where $H_n = \sum_{i=0}^{\ell} H^i_n + H^I$ is defined as in the proof of Proposition 5.15.

It follows from (2) and the earlier observation about the application of Lemma 3.36(2) to K-fundamental sequences that the sequences $(k \exp H_n \cdot o)$ and $(k' \exp H'_n \cdot o)$ converge to the same point in the compactification. □

5.32. Remark. In Appendix A another construction of the Karpelevič compactification $\overline{X}^K$ of X is given. It parallels the construction of the dual cell compactification from the polyhedral compactification and the action of K on the flat $A \cdot o$. Namely, one defines the obvious equivalence relation $\sim_K$ on $K \times \overline{A \cdot o}^K$ by setting $(k_1, y_1) \sim_K (k_2, y_2)$ if $k_1 \cdot y_1 = k_2 \cdot y_2$. One proves that this is a closed equivalence relation and, hence, that the quotient space is compact (see Theorem A.9).

As in the case of the dual cell compactification, one embeds X in this quotient space by mapping $x = ka \cdot o$, $a \in A$, to the equivalence class $[k, a]$ of (k, a). The resulting compactification of X has the characteristic properties stated in Theorem 5.31. As a result, this gives a new construction of $\overline{X}^K$.

CHAPTER VI

MARTIN COMPACTIFICATIONS

This chapter presents some general facts about the Martin compactification of X and its connection to the asymptotic behavior of the Brownian motion on X. For an expanded version of this chapter see Taylor [T5].

The Martin compactification

§ **6.1.** Let $G(z,w)$ denote the Green function, with Dirichlet boundary conditions, of the Laplacian for the unit disk D. It follows from Green's identities that, for any continuous function h on $\overline{D}$, harmonic on D, $h(z)$ equals the integral, with respect to arc length, of h times the inward normal derivative of the Green function $w \to G(z,w)$ over the boundary ∂D. Up to a constant, this normal derivative equals $\frac{1-|z|^2}{|z-b|^2}$, the Poisson kernel $P(z,b)$ of D.

It is well-known that, as a result, every positive harmonic function h on the unit disk D is determined by a unique measure μ on the unit circle ∂D and the Poisson kernel, i.e., for all $z \in D$,

$$h(z) = \int_{\partial D} P(z,b)d\mu(b).$$

Since the Poisson kernel is a normal derivative at the boundary of the Green function, it is clear that, for b on the unit circle, the Poisson kernel $P(z,b) = \lim_{w\to b} \dfrac{G(z,w)}{G(0,w)}$. This observation, less well-known, was used by Martin to determine a Poisson kernel for any bounded domain Ω in $\mathbb{R}^n$, regardless of the smoothness of the boundary.

In 1943, Martin showed in [M5] that, as y tends to infinity in Ω, the set of possible limits of the normalized Green function $x \to \frac{G(x,y)}{G(x_0,y)}$, where G is the Green function of the Laplacian for Ω, with Dirichlet boundary conditions, and $x_0 \in \Omega$ is a reference point, gives a family of functions on Ω that can be used as a Poisson kernel.

If $(y_n) \subset \Omega$ is a sequence that tends to infinity, i.e., if it is eventually outside any compact subset of Ω, it follows that, for any relatively compact subdomain Ω_1 containing x_0 with $\overline{\Omega_1} \subset \Omega$, the functions $x \to K(x,y) = \frac{G(x,y_n)}{G(x_0,y_n)}$ are all eventually positive, harmonic, with value 1 at x_0. From Harnack's inequality and the local uniform equicontinuity of such functions, it follows that the convex set $\mathcal{H}_1(\Omega)$ of positive harmonic functions on Ω that take the value 1 at x_0 is compact in the topology of uniform convergence on compact sets. Using an exhaustion of Ω by such relatively compact subdomains and a diagonal procedure, it is not hard to see that, for some subsequence (y_{n_k}), the functions $K(x, y_{n_k})$ converge uniformly on compact subsets to a positive harmonic function.

The set of limit functions is a subset of the convex compact set $\mathcal{H}_1(\Omega)$. Martin showed that this set could be viewed as a boundary $\partial\Omega$ of Ω (denoted by Δ in [M5]) and that, with respect to a topology induced by the limit functions, $\Omega \cup \partial\Omega \stackrel{\text{def}}{=} \tilde{\Omega}$ is compact. If $b \in \partial\Omega$, let $K(x,b) = \lim_n K(x, y_n)$ when $y_n \to b$. The set of functions $x \to K(x,b), b \in \partial\Omega$, is the set of limit functions. Martin showed that, for every positive harmonic function h on Ω, there is a measure μ on $\partial\Omega$ such that, for all $x \in \Omega$,

$$h(x) = \int_{\partial\Omega} K(x,b)d\mu(b).$$

One way to view $\partial\Omega$ is as the set of equivalence classes of sequences in Ω that converge to infinity where two such sequences are said to be equivalent if they give rise to the same limit function. Hence, the compactification is determined by knowing the equivalence classes of the sequences that converge to infinity and converge in $\tilde{\Omega}$.

Martin also showed that $\tilde{\Omega}$ is metrizable and that there is a G_δ subset $\partial_e\Omega$ of $\partial\Omega$, that could be identified with the set of extremal elements of the compact convex set $\mathcal{H}_1(\Omega)$. Limit functions h in this set (denoted by Δ_1 by Martin) are usually called minimal functions since they have the characteristic property that if h' is harmonic and $0 \leq h' \leq h$, then $h' = \lambda h$ for some $\lambda \in [0,1]$. The measure μ, which represents h in the sense that $h(x) = \int_{\partial\Omega} K(x,b)d\mu(b)$, can be assumed to be carried by $\partial_e\Omega$, in which case it is unique.

While Martin computed some examples of $\tilde{\Omega}$, general examples other than simply connected plane domains, which are conformally equivalent to the unit disc, were not computable simply because, in general, it is not possible to say much about the Green function $G(x,y)$. In 1970 Hunt–Wheeden [H6] showed, using techniques of harmonic analysis, that for any bounded Lipschitz domain Ω in $\mathbb{R}^n$, the Martin compactification $\tilde{\Omega}$ equals its closure $\overline{\Omega}$ in $\mathbb{R}^n$ and all boundary points are minimal. There is now a literature on this subject for domains in $\mathbb{R}^n$ (see the bibliographies of Bass–Burdzy [B7] and Cranston–Salisbury [C6]).

It is clear that Martin's ideas carry over to the case of the Laplace–Beltrami operator L on a Riemannian manifold X, and also to $L + \lambda Id$, whenever the Green function with Dirichlet conditions at infinity exists. Hence, to such an operator $L+\lambda Id$, one can associate a Martin compactification of the manifold. It is, therefore, a natural problem to try to describe this compactification in terms of the geometry of the manifold. In the case of symmetric spaces of non-compact type, Dynkin [D4] was the first to compute these compactifications when the manifold X equals $\mathrm{SL}(n,\mathbb{C})/\mathrm{SU}(n)$, the space of positive definite Hermitian matrices of determinant one. He obtained the complete list of limit functions, but did not describe the compactifications in geometric terms. He was followed by Nolde [N] and then

Olshanetsky [O1]. For general symmetric spaces, the main paper in this area is the seminal paper by Karpelevič [K3] in which, among other things, he obtained a description of the complete list of minimal functions for each Martin compactification of a general symmetric space of non-compact type.

In the following two chapters of this book, the Martin compactifications are computed and described geometrically for all symmetric spaces X of non-compact type. While X can be expressed as the left coset space G/K, it is also important to make use of the fact that a symmetric space X of non-compact type can be identified with the solvable group $S = NA$ in view of the Iwasawa decomposition $G = KAN$ of G. The Riemannian metric on X then determines a metric on S and the Laplace–Beltrami operator L can then be viewed as a left invariant differential operator on this group.

As pointed out in the introduction, the bottom λ_0 of the L^2-spectrum of $-L$ determines the set of λ for which positive solutions exist for $L + \lambda u = 0$ (see Guivarc'h [G13], Sullivan [S4], and Taylor [T3]). This number is well-known for symmetric spaces to be equal to $\|\rho\|^2$. It may be computed from the harmonic analysis of $L^2(X)$, in particular from the spectral resolution of the Laplace–Beltrami operator given by the Fourier–Helgason transform (see Helgason [H4, p. 223, eq. (6) and p. 227]). Since the eigenvalues of the exponential functions that appear in the Fourier–Helgason transform are greater than or equal to $\|\rho\|^2$ with infimum equal to $\|\rho\|^2$, it follows that the bottom of the spectrum λ_0 equals this number.

This number can also be computed, without using harmonic analysis, by using the fact that if $Lu + \lambda u = 0$, viewed as a differential operator on the group $S = NA$, has a positive global solution, then it has one that is N-left invariant. Given this it follows from eq. (7.9), that $\lambda \leq \|\rho\|^2$. This is because the formula for the Laplacian acting on N-left invariant functions reduces the question to asking for which $\lambda \in \mathbb{R}$ does

$$\sum_{i=1}^{r} \frac{\partial^2 u}{\partial x_i^2} + 2 \sum_{i=1}^{r} h_i \frac{\partial u}{\partial x_i} + \lambda u = 0$$

have a positive global solution. Letting $u(x) = e^{-h \cdot x} f(x)$, this is equivalent to asking for what $\lambda \in \mathbb{R}$ does the equation

$$\sum_{i=1}^{r} \frac{\partial^2 f}{\partial x_i^2} + (\lambda - \|h\|^2) f = 0$$

have a positive solution. Clearly, this is the case if and only if $\lambda \leq \|h\|^2$. (The vector h corresponds to H_ρ whose length is $\|\rho\|^2$.)

The reason that $Lu + \lambda u = 0$ has a N-left invariant solution if it has a positive solution is because the group S has the fixed line property (see Conze–Guivarc'h [C5] and Example 11.33). Karpelevič established the existence of N-left invariant solutions without explicit use of this property (see [K3, p. 179]). His simple argument is given in Appendix A.

For each $\lambda \leq \lambda_0$, $L + \lambda Id$ admits a Green function $G^\lambda(x, y)$ as shown by Karpelevič [K3, p. 174, Theorem 16.6.1] and also by the estimates of Anker–Ji [A6][1] . Hence, in the symmetric space case, for each $\lambda \leq \lambda_0$, the operator $L + \lambda Id$ determines a Martin compactification.

Returning to the general theory of the Martin compactification, if $L + \lambda_0 Id$ admits a positive global solution then two possibilities occur. Either there is, up to a constant multiple, a unique positive solution, or the operator admits a Green function $G^\lambda(x, y)$ with Dirichlet conditions at infinity. In the first case, the Martin compactification $\tilde{X}(\lambda)$ is defined to be the one-point compactification of X. In the case where the Green function G^λ exists (which is so if X is a symmetric space of non-compact type for any $\lambda \leq \lambda_0$), if x_0 is a base point, the **Martin compactification** $\tilde{X}(\lambda)$ is the unique compactification $\tilde{X}$ of X with the following formal properties:

(1) the functions $y \to K^\lambda(x, y) \stackrel{\text{def}}{=} \dfrac{G^\lambda(x, y)}{G^\lambda(x_0, y)}$ extend continuously to $\tilde{X}$ for each $x \in X$; and

(2) the extended functions separate the points of the ideal boundary $\tilde{X} \backslash X$ (see, for example, Taylor [T1], Constantinescu–Cornea [C3]).

The function $K^\lambda(x, y)$ is called the **Martin kernel**. (Note that since X is locally compact it is (necessarily) open and dense in $\tilde{X}(\lambda)$.)

The Martin compactification is well-known to be metrizable and may be obtained by completing X in the metric given by

$$d(y_1, y_2) = \int_{B(x_0;r)} \frac{|K^\lambda(x, y_1) - K^\lambda(x, y_2)|}{1 + |K^\lambda(x, y_1) - K^\lambda(x, y_2)|} \mu(dx),$$

where $B(x_0; r)$ is the ball of radius r with center x_0 and μ is the volume measure (see Martin [M5]). Note that, by Harnack's inequality, the completion does not depend upon the value of r.

Hence, to say that a sequence $(y_n) \subset X$ converges to a point of the ideal boundary of $\tilde{X}(\lambda)$ is equivalent to saying that either it is Cauchy relative to the above metric or that the sequence of functions $x \longrightarrow K^\lambda(x, y_n)$ converge uniformly on the compact subsets of X to a positive function h for which $Lh + \lambda h = 0$. As a result, to describe the Martin compactification $\tilde{X}(\lambda)$ amounts to finding, in as explicit a fashion as possible, the sequences, converging to infinity, that converge and their equivalence classes. In the case of a symmetric space of non-compact type, this will be done in terms of the corresponding explicit results for the dual cell compactification and the conic compactification. In other words, the equivalence classes of convergent sequences for the Martin compactifications are determined by the equivalence classes for these two geometric compactifications.

The Martin compactification $\tilde{X}(\lambda)$ will also be denoted by $X \cup \partial X(\lambda)$, where $\partial X(\lambda)$ denotes the ideal boundary, which will be referred to as the

[1] It is not hard to show that if $\lambda \leq \lambda_0$ then $Lu + \lambda u = 0$ has at least two distinct solutions. This also implies that the Green function exists (see [T5] for details).

Martin boundary. This ideal boundary is the set of functions h in the compact convex set $\mathcal{H}_1^\lambda = \mathcal{H}_1^\lambda(X)$, the positive functions u for which $Lu + \lambda u = 0$ and $u(x_0) = 1$, that are limit functions, i.e., $h(x) = \lim_n K^\lambda(x, y_n)$ for some sequence $(y_n) \subset X$ that converges to infinity. The convex set $\mathcal{H}_1^\lambda$ is compact in the topology of uniform convergence on compact sets and the topology of the ideal boundary is the subspace topology. A basic open neighborhood $O(h; C_0, C, \epsilon)$ of a limit function $h = K^\lambda(\cdot, b) \in \partial X(\lambda)$, corresponding to the compact subsets C_0, C of X and $\epsilon > 0$ is

$$O(h; C_0, C, \epsilon) = \{y \in \tilde{X}(\lambda) \backslash C_0 \mid |K^\lambda(x, y) - K^\lambda(x, b)| < \epsilon \text{ for all } x \in C\},$$

where $K^\lambda(x, y) = \lim_n K^\lambda(x, y_n)$ if $y_n \to y \in \partial X(\lambda)$.

Martin's arguments in [M5] show that every positive solution u of $Lu + \lambda u = 0$ is represented by a unique positive measure μ of total mass $u(x_0)$ carried by the G_δ-set of extreme points of $\mathcal{H}_1^\lambda$. This measure will be called the **representing measure** for u. It is well-known that the extreme points of $\mathcal{H}_1^\lambda$ all belong to the set of limit functions, and the corresponding subset of the Martin boundary, called the set of **minimal points**, will be denoted by $\partial_e X(\lambda)$. Hence, if $x \to K^\lambda(x, b)$ is the limit function corresponding to a minimal point $b \in \partial_e X(\lambda)$ (sometimes called a **kernel function at** b), the connection between representing measure and function is given by

$$(*) \qquad u(x) = \int_{\partial_e X(\lambda)} K^\lambda(x, b) d\mu(b).$$

In general (and certainly in the case of a symmetric space of non-compact type with rank greater than one), not every boundary point is minimal. The limit function associated with a minimal point is a so-called minimal function, where, as in the classical case, a positive solution u of $Lu + \lambda u = 0$ is said to be **minimal** if, for any other solution $v, 0 \le v \le u$ implies that $v = \gamma u, \gamma \in [0, 1]$.

When the manifold X is a symmetric space of non-compact type, the reference point x_0 will be taken to be $o = K \in G/K$. From now on the case of a general Riemannian manifold will not be considered and the manifold X will be assumed to be an arbitrary symmetric space of non-compact type. In the rank one case, the Martin compactifications all coincide with the conic compactification. This follows, when $\lambda < \lambda_0$, from the general results of Ancona [A3] on the Martin compactifications of Cartan–Hadamard manifolds with pinched negative curvature. It also follows by explicit computation, for $\lambda \le \lambda_0$, since the asymptotics of the Green functions are known (see, for example, Lyons–MacGibbon–Taylor [L3]).

When $\lambda = 0$, it follows from eq. (*) that the bounded solutions of $Lu = 0$ are in bijection with $L^\infty(\mu_1)$, where μ_1 is the measure that represents the constant function 1. Let E denote the support of this measure. Then $(E, \mathfrak{B}(E), \mu_1)$ is a measure space $(E, \mathfrak{E}, \mu)$ for which $L^\infty(E, \mathfrak{E}, \mu)$ is isomorphic with the set of bounded solutions. Such a measure space is called a **Poisson boundary** (see Furstenberg [F3], Kaimanovič–Vershik [K1]). While a Poisson boundary is unique up to a measure preserving isomorphism, one may always realize a Poisson boundary as a subset of the Martin boundary of X. It will be shown later, by identifying the measure μ_1 representing 1, that $G/P = K/M = \mathcal{F}$ — the so-called Furstenberg boundary — is in fact a Poisson boundary, the measure $\mu = \mu_1$ being the unique K-invariant probability on $\mathcal{F}$.

Convergence of Brownian motion

It is fairly well-known that second order elliptic differential operators can be viewed as the infinitesimal generators of diffusions. In the context of a Riemannian manifold, the Brownian motion is the diffusion associated with the Laplace–Beltrami operator L

In 1966, Šur [S5] proved a general result on the convergence of a diffusion to the points on the Poisson boundary (viewed as the support of μ_1). His result is proved for a diffusion on an open subset of $\mathbb{R}^n$ whose infinitesimal generator L is a second order elliptic operator with smooth enough coefficients. This proof applies without change to a Riemannian manifold and its Laplace–Beltrami operator L (see [T5]).

6.2. Proposition. (Šur [S5]) *Let (X_t) be a Brownian motion on $X = G/K$ started at x. Then a.s. $t \to X_t(\omega)$ converges in the Martin compactification $X \cup \partial X(0)$ to a minimal point in the support of the measure μ_1 representing the constant function 1.*

This result of Šur is used later in §8.30, after identifying the support of μ_1, to prove a convergence theorem for Brownian motion in polar coordinates due to Malliavin–Malliavin [M2].

Extension of the group action to the Martin compactification

§ 6.3. Let $\phi : X \to X$ be a Riemannian isometry. Then, for any smooth function f on X, one has $L(f \circ \phi) = (Lf) \circ \phi$. Hence, if $G_y^\lambda(x) = G^\lambda(x, y)$, it follows that $G_{\phi(y)}^\lambda \circ \phi = G_y^\lambda$. Consequently,

$$(\dagger) \qquad \frac{G^\lambda(x,y)}{G^\lambda(o,y)} = \frac{G^\lambda(\phi(x),\phi(y))}{G^\lambda(\phi(o),\phi(y))} = \frac{G^\lambda(\phi(x),\phi(y))}{G^\lambda(o,\phi(y))} \times \frac{G^\lambda(o,\phi(y))}{G^\lambda(\phi(o),\phi(y))}.$$

Therefore, a sequence (y_n) converges in the Martin compactification $X \cup \partial X(\lambda)$ if and only if the sequence $(\phi(y_n))$ converges in the same compactification. As a result, the isometry ϕ extends to the Martin compactification as an isomorphism (see Definition 3.27) of $X \cup \partial X(\lambda)$ with itself, mapping the limit of the sequence (y_n) to the limit of the sequence $\phi(y_n)$.

Since $X = G/K$ is a left G-space, and L is left invariant, it follows from this observation that the group G acts isomorphically on the Martin compactifications $X \cup \partial X(\lambda)$, $\lambda \leq \lambda_0$. On the Martin boundary, viewed as a set of limit functions, this action can be made more explicit. It agrees with a natural left action (due to Dynkin [D4]) of G on the set of functions f on X normalized to take the value 1 at o: namely, define $S_g f$ by setting $S_g f(x) = f(g^{-1} \cdot x)/f(g^{-1} \cdot o)$. This observation about the G-action is a consequence of the next result.

6.4. Proposition. *If $(y_n) \subset X$ converges in $X \cup \partial X(\lambda)$ to h, then $(g \cdot y_n)$ converges to $S_g h$. Further, if $(g_n) \subset G$ converges to g, then $(g_n \cdot y_n)$ also converges to $S_g h$. In addition, if $(h_n) \subset \partial X(\lambda)$ converges to h, then $S_{g_n} h_n$ converges to $S_g h$.*

Hence, G acts continuously on the Martin compactification $X \cup \partial X(\lambda)$.

Proof. First, note that Eq. (†) implies that

$$K^\lambda(x, g \cdot y_n) = \frac{G^\lambda(g^{-1} \cdot x, y_n)}{G^\lambda(o, y_n)} \times \frac{G^\lambda(o, y_n)}{G^\lambda(g^{-1} \cdot o, y_n)} \to \frac{h(g^{-1} \cdot x)}{h(g^{-1} \cdot o)} = S_g h(x)$$

as n tends to ∞.

Harnack's inequality implies that $K^\lambda(\cdot, y_n) \to h$ uniformly on compact sets. Hence,

$$\frac{G^\lambda(g_n^{-1} \cdot x, y_n)}{G^\lambda(o, y_n)} \to h(g^{-1} \cdot x) \text{ and } \frac{G^\lambda(g_n^{-1} \cdot o, y_n)}{G^\lambda(o, y_n)} \to h(g^{-1} \cdot o).$$

This proves the second statement.

Let (h_n) be a sequence in the Martin boundary that converges to h. It follows from the definition that $h_n(x) \to h(x)$ for all $x \in X$. Since $h_n(o) = h(o) = 1$, local uniform equicontinuity implies that $h_n \to h$ uniformly on compact sets. As a result, if $g_n \to g$, it follows that $S_{g_n} h_n \to S_g h$. Hence, the G-action is continuous. □

The Martin compactification for a random walk

§ **6.5.** There is a close connection between diffusions and Markov chains. A Markov chain is given by one transition kernel $\check{P}$ rather than a family of transition kernels — the heat kernel for the manifold in the case of Brownian motion. When this Markov kernel is defined on an H-space, where H is a group, and is H-left invariant, the associated random process is called a random walk and is determined by a probability measure on the space. In the case of a symmetric space X, random walks on X correspond to random walks on G that are given by K-bi-invariant probabilities p. Random walks determine harmonic functions, the functions that satisfy $\check{P}h = h * p = h$, and there is a corresponding theory of Martin compactification. When the state space is discrete, this theory is fairly simple.

In Chapter XI, a new formulation of this compactification procedure is given, using what is referred to as Martin's method. When the K-bi-invariant probability p on H has a continuous density with compact support

S for which $H = \cup_{n>0} S^n$, it is shown that the classical arguments of Martin can be used to define a Martin compactification $\tilde{H}(p,r)$ for any constant $r > 0$ such that there are positive solutions f to the convolution equation $u * p = ru$.

It turns out that the set of all minimal eigenfunctions u of the Laplacian L, for all $\lambda \leq \lambda_0$, coincides with the set of all minimal functions u for which a convolution equation $u * \bar{p} = ru$ holds, where $\bar{p}$ is a suitable K-invariant probability on $X = G/K$. (This is a consequence of Theorems 13.23 and 13.33.) The requirement on the probability $\bar{p}$ is that its lift to G as a K-bi-invariant probability p on G has a density with compact support S for which $G = \cup_{n>0} S^n$.

The random walk on G associated with p determines a family of Martin compactifications $\tilde{G}(p,r)$ for all $r \geq r(p)$, where $r(p)$, the bottom of the positive spectrum is determined by the behavior of the convolution powers p^n at $e \in G$ (see Theorem 11.17). These compactifications of G also determine Martin compactifications $\tilde{X}(p,r)$ of X relative to the corresponding random walk on X (see Theorem 14.24).

The fact that Martin compactifications can be defined, and in some cases computed, for random walks opens the way to the consideration, in the later part of this book, of the Martin boundaries associated with convolution equations. This topic carries over, for example, to the study of random walks on Euclidean buildings.

For additional details on some of the technical questions involved in this chapter, see the expository article by Taylor [T5].

CHAPTER VII

THE MARTIN COMPACTIFICATION $X \cup \partial X(\lambda_0)$

The Martin compactifications associated with the Laplace–Beltrami operator L are determined by the conical compactification and the Martin compactification $X \cup \partial X(\lambda_0)$ corresponding to the bottom of the positive spectrum (see Theorem 8.21).

To compute a Martin compactification, in principle, some kind of asymptotic behavior of the Green function is needed in order to identify the limit functions. This is not always the case as, in fact, a number of computations have been done that make no use of such asymptotics, e.g., [H6]. In particular, it will be shown here that, for the case of $X \cup \partial X(\lambda_0)$, the asymptotics can be dispensed with. The structure of the Lie algebra $\mathfrak{g}$ of G determines the compactification and it turns out that this Martin compactification is isomorphic to the dual cell compactification (see Theorem 7.33). This is due to the fact that if a sequence in X goes to infinity through a Weyl chamber or a Weyl chamber face, in the sense of Definition 3.24, then it converges in the Martin compactification. (This convergence to infinity through a Weyl chamber or a Weyl chamber face means that the distance to all the walls also goes to infinity.) Hence, the limit functions of this type are in one-to-one correspondence with the simplices of the Tits building $\Delta(X)$ (see Definition 3.13). The remaining limit functions are obtained by the action of G and they are shown to correspond to the points in $\Delta^*(X)$.

The key to computing $\partial X(\lambda_0)$, without making use of information about the asymptotics of the Green function G^{λ_0} of $L + \lambda_0 Id$, is to view L as a left invariant differential operator on $S = NA$. This enables one to obtain explicit formulas as to how it acts on N-invariant and N_I-invariant functions. To do this one is obliged to introduce the (generalized) horospherical coordinates of X associated with parabolic subgroups of G. These formulas also play an important role in computing the Martin kernels for $\lambda < \lambda_0$ in Chapter VIII.

The Laplacian in horocyclic coordinates

§ **7.1.** As pointed out in § 3.1, a symmetric space X of non-compact type is homeomorphic to the Euclidean space $\mathfrak{p}$ under the map $Y \to \exp Y \cdot o$, where the base point o is the coset K in G/K. While this map introduces global coordinates into X and is useful for the study of the geometrical boundaries of X, it is necessary to introduce another type of coordinate system to relate these boundaries to the potential theory on X. These coordinates are the so-called horocyclic coordinates. They have the crucial advantage that, in terms of them, X becomes a group and the Laplace-Beltrami operator on X is a left invariant operator.

As pointed out earlier in § 2.1, the choice of a maximal abelian subalgebra $\mathfrak{a}$ of $\mathfrak{p}$ and a positive Weyl chamber $\mathfrak{a}^+$ determines a corresponding Iwasawa decomposition $G = KAN = NAK$ of G. As a result, the symmetric space $X = G/K$ can be identified with the solvable group $S = NA$. This identification gives the **horocyclic coordinates** of a point $x = g \cdot o \in X$: if $g = nak$ the horocyclic coordinates of $x = g \cdot o = na \cdot o$ are n and a. The G-invariant metric on X corresponds to a left invariant metric on the group S. In order to identify the Laplacian with a left invariant operator on S, one first determines this metric. It is given by a positive definite quadratic form Q on the Lie algebra $\mathfrak{s}$. The map $\tau : \mathfrak{g} \to T_o(X)$ given by $\tau(Y) = \frac{d}{dt} \exp tY \cdot o_{|_{t=0}}$ is linear with kernel $\mathfrak{k}$ and maps $\mathfrak{p}$ bijectively to $T_o(X)$ as pointed out earlier in § 3.1. If $Y \in \mathfrak{p}$ one identifies $\tau(Y)$ and Y.

Since $X = NA \cdot o$, it follows that $T_o(X)$ and $\mathfrak{s} = \mathfrak{n} \oplus \mathfrak{a}$ are isomorphic under τ and if $H \in \mathfrak{a} \subset \mathfrak{s}$, then $\tau(H) = H$.

§ **7.2.** To determine Q it is useful to express this isomorphism $\mathfrak{s} \to T_o(X)$ in terms of the root spaces (see Helgason [H3], also Taylor [T2]). If $X \in \mathfrak{p}$ is orthogonal to $\mathfrak{a}$, then $X = \frac{1}{2}\sum_{\alpha>0}\{X_\alpha - \theta(X_\alpha)\}$ where $X_\alpha \in \mathfrak{g}_\alpha$ and $\alpha > 0$.

This expression for X corresponds to a decomposition of $\mathfrak{q} \overset{\text{def}}{=} \mathfrak{p} \cap \mathfrak{a}^\perp$ into the sum $\mathfrak{q} = \sum_{\alpha>0} \mathfrak{p}_\alpha$, where $\mathfrak{p}_\alpha = \{X_\alpha - \theta(X_\alpha) \mid X_\alpha \in \mathfrak{g}_\alpha\}$. There is a parallel decomposition of $\mathfrak{l} \overset{\text{def}}{=} \mathfrak{k} \cap \mathfrak{m}^\perp$ as $\mathfrak{l} = \sum_{\alpha>0} \mathfrak{k}_\alpha$ where $\mathfrak{k}_\alpha = \{X_\alpha + \theta(X_\alpha) \mid X_\alpha \in \mathfrak{g}_\alpha\}$ (see § 2.1).

Let $U = \frac{1}{2}\sum_{\alpha>0}\{X_\alpha + \theta(X_\alpha)\}$ be the vector in $\mathfrak{l}$ corresponding to $X \in \mathfrak{q}$. If $Z = \sum_{\alpha>0} X_\alpha \in \mathfrak{n}$, then $Z = X + U$ and $\tau(Z) = X$. Define Q on $\mathfrak{a}$ to coincide with $B_{|_\mathfrak{a}} = B_{\theta|_\mathfrak{a}}$. If $Z \in \mathfrak{n}$, and τ is to be an isometry, $Q(Z, Z) = B(X, X)$. As indicated in § 2.7, B_θ is positive definite on $\mathfrak{g}$. One has $B_\theta(Z, Z) = B_\theta(X, X) + B_\theta(U, U) = B(X, X) - B(U, U) = 2B(X, X)$ since, by computation, one sees that $B(X, X) = -B(U, U)$. Define $Q_{|_\mathfrak{n}}$ to be $\frac{1}{2}B_{\theta|_\mathfrak{n}}$. Then $\tau : (\mathfrak{s}, Q) \to (\mathfrak{p}, B)$ is an isometry. Hence, the group S, equipped with the left invariant metric defined by Q, is isometric to the symmetric space X under the map $na \to na \cdot o$.

§ **7.3.** An orthogonal basis for the quadratic form Q on $\mathfrak{s}$ may be constructed as follows. Let $Y_1, Y_2, \ldots, Y_q$ be a basis of $\mathfrak{n}$ that is adapted to the $\mathfrak{g}_\alpha, \alpha > 0$, i.e., each vector Y_i is in one of the $\mathfrak{g}_\alpha$. Assume this basis of $\mathfrak{n}$ is chosen as, for example, in Taylor [T2] so that (i) they are orthogonal relative to the form B_θ and (ii) the length of each vector Y_i with respect to the form B_θ is $\sqrt{2}$: this implies that $\frac{1}{2}B_\theta(Y_i, Y_i) = Q(Y_i, Y_i) = 1$. Hence the basis $Y_1, Y_2, \ldots, Y_q$ of $\mathfrak{n}$ is orthonormal relative to Q and the tangent vector $\tau(Y_i) = X_i \in \mathfrak{p}$ corresponding to Y_i has length 1. Denote by $H_1, H_2, \ldots, H_r$ a basis of $\mathfrak{a}$ that is orthonormal relative to the Killing form B. Let H_ρ be the vector in $\mathfrak{a}$ that represents one half the sum of the positive roots, i.e., if $\rho = \frac{1}{2}\sum_{\alpha>0} m_\alpha \alpha$, $m_\alpha = \dim \mathfrak{g}_\alpha$, then $B(H, H_\rho) = \rho(H)$, where $H \in \mathfrak{a}$.

Using this Q-orthonormal basis, one has the following formula for the left invariant Laplacian on S corresponding to the metric given by Q.

Proposition 7.4. (Laplacian in horocyclic coordinates, see Helgason [H3], Karpelevič [K3, Theorem 15.3.2]) *The Laplace-Beltrami operator L on X, when viewed as a left invariant operator on the group S, is given by the explicit formula*

$$L = \sum_{i=1}^{q} Y_i^2 + \sum_{j=1}^{r} H_j^2 - 2H_\rho. \tag{7.5}$$

Proof. This is a consequence of Corollary 8.2 in [T2]. □

Generalized horocyclic coordinates and the Laplacian

§ 7.6. As shown in § 2.7, corresponding to a subset I of the set $\Delta = \Delta(\mathfrak{g}, \mathfrak{a}^+)$ of simple roots, one has the semidirect decompositions $N = N^I \ltimes N_I$ and $A = A^I \times A_I$. They determine the so-called generalized horocyclic coordinates for a point $na \cdot o = x \in X$. Let $n = n_2 n_1$ with $n_2 \in N_I$, $n_1 \in N^I$, and $a = a_2 a_1$ with $a_2 \in A_I$, $a_1 \in A^I$. Since A_I commutes with N^I, $na \cdot o = n_2 n_1 a \cdot o = n_2 a_2 n_1 a_1 \cdot o = n_2 a_2 \cdot x^I$, where x^I is a point in the symmetric space X^I. The **generalized horocyclic coordinates** of x are then $n_2 = n_I, a_2 = a_I$ and x^I. As pointed out in § 2.7, the solvable group $S = NA$ is also the semidirect product $S_1 \ltimes S_2$, where $S_1 = N^I A^I$ and $S_2 = N_I A_I$. In terms of $S = S_1 \ltimes S_2$, the generalized horocyclic coordinates amount to writing $s \in S$ as $s = s_2 s_1$, associating with $s_2 \cdot o$ its horocyclic coordinates n_2 and a_2 and observing that $s_1 \cdot o = x^I$.

As observed in Corollary 2.16, the generalized horocyclic coordinates are also determined by the Langlands decomposition of the standard parabolic subgroup P^I. Since $P^I = M_I A_I N_I = N_I A_I M_I = N_I A_I G^I M$, it follows that $X = P^I \cdot o = N_I A_I \cdot G^I \cdot o = N_I A_I \cdot X^I$.

The subset I splits each of the two summands of the Lie algebra $\mathfrak{s} = \mathfrak{n} \oplus \mathfrak{a}$ of S. Hence, $\mathfrak{s} = \mathfrak{n}_I \oplus \mathfrak{n}^I \oplus \mathfrak{a}_I \oplus \mathfrak{a}^I$. The basis for $\mathfrak{n}$ used in § 7.3 can be chosen so that $Y_1, Y_2, \ldots, Y_p$ is a basis of $\mathfrak{n}^I$ that is adapted to the $\mathfrak{g}_\alpha, \alpha \in \Sigma^{I,+}$, and $Y_{p+1}, Y_{p+2}, \ldots, Y_q$ is a basis of $\mathfrak{n}_I$ adapted to the $\mathfrak{g}_\alpha, \alpha \in \Sigma_I^+$.

Similarly the basis for $\mathfrak{a}$ can be taken so that $H_1, H_2, \ldots H_s$ is a basis of $\mathfrak{a}^I$ and $H_{s+1}, H_{s+2}, \ldots, H_r$ is a basis of $\mathfrak{a}_I$.

In terms of this basis and the formula eq. (7.5) for L, it is easy to determine how L acts on functions $f(x)$ on X that do not depend on n_2 when x is expressed in generalized horocyclic coordinates, i.e., those functions f such that $f(n_2 a_2 \cdot x^I) = f(a_2 \cdot x^I)$, equivalently, $f(n_2 n_1 a \cdot o) = f(n_1 a \cdot o), a \in A$.

This is an immediate consequence of the following lemma.

Lemma 7.7. *Let $Y \in \mathfrak{n}_I$ and f be a function such that $f(n_2 n_1 a \cdot o) = f(n_1 a \cdot o)$ for all n_2, n_1, a. Then $Yf = 0$.*

Proof. By definition, $Yf(na \cdot o) = \frac{d}{dt} f(na \exp tY \cdot o)|_{t=0}$ and $na \exp tY = n_2(\exp tAd(n_1)Ad(a)Y)n_1 a$. Since $Ad(a)\mathfrak{n}_I \cup Ad(n_1)\mathfrak{n}_I \subset \mathfrak{n}_I$, the result follows. □

Let f be a function on NA such that $f(n_2n_1a) = f(n_1a)$ for all n_2, n_1, a. It follows from eq. (7.5) and Lemma 7.7 that

$$Lf = \sum_{i=1}^{p} \mathrm{Y}_i^2 f + \sum_{j=1}^{r} \mathrm{H}_j^2 f - 2\mathrm{H}_\rho f. \tag{7.8}$$

In particular, if f is a function on NA such that $f(n_2n_1a) = f(n_1a)$ for all n_2, n_1, a, then Lf has the same property, since the vector fields $Y_i, 1 \le i \le p$, and $H_j, 1 \le j \le r$, are left invariant differential operators that preserve this class of functions.

In addition, if f is a function on NA that is left N-invariant, i.e., $f(na) = f(a)$ for all $n \in N, a \in A$, then

$$Lf = \sum_{j=1}^{r} \mathrm{H}_j^2 f - 2\mathrm{H}_\rho f. \tag{7.9}$$

The significance of eq. (7.8) is clarified by the next result.

Proposition 7.10. *The natural map from $NA \to N^IA = A_IN^IA^I$ given by $n_2n_1a \to n_1a = a_2n_1a_1$ where $a = a_2a_1$, intertwines the Laplace-Beltrami operator, eq. (7.5), and the left invariant operator*

$$L' = \sum_{i=1}^{p} Y_i^2 + \sum_{j=1}^{r} H_j^2 - 2H_\rho, \tag{7.11}$$

on $A_IN^IA^I$.

It follows from § 2.13 that the group $A_IN^IA^I$, which is the direct product of A_I and the solvable group $S_1 = N^IA^I$, can be identified with the product $A_I \times X^I$ of A_I with the symmetric space X^I under the map $a_2n_1a_1 \to a_2n_1a_1 \cdot o$. The next lemma — that explains how the linear functional ρ decomposes under the splitting of $\mathfrak{a}$ as $\mathfrak{a}^I \oplus \mathfrak{a}_I$ — and its corollary show that the operator L' is the direct sum of the Laplace-Beltrami operator L^I on X^I (relative to the induced metric) and an invariant operator on A_I.

Lemma 7.12. *Let I be a set of simple roots. If $\alpha_i \in I$ and $\alpha \in \Sigma_I^+$, then $s_{\alpha_i}\alpha \in \Sigma_I^+$. Furthermore, if $2\rho_I = \sum_{\alpha \in \Sigma_I^+} m_\alpha \alpha$, where $m_\alpha = \dim \mathfrak{g}_\alpha$, then $s_{\alpha_i}\rho_I = \rho_I$.*

Proof. (See Helgason [H2, Lemma 3.11, p. 461], and the proposition in the appendix of Taylor [T2].) Since $\alpha \in \Sigma_I^+, \alpha = \sum_{j=1}^r n_j\alpha_j, n_j \ge 0$ and $n_{j_0} > 0$ for some j_0 with $\alpha_{j_0} \notin I$.

Now $s_{\alpha_i}\alpha = \alpha - a_{\alpha,\alpha_i}\alpha_i = \{n_i - a_{\alpha,\alpha_i}\}\alpha_i + \sum_{j \ne i} n_j\alpha_j$. Since the coefficients of the α_i are integers all of the same sign (cf. Helgason [H2, Theorem 3.6, p. 458]), $n_{j_0} > 0$ implies that $\beta = s_{\alpha_i}\alpha \in \Sigma_I^+$.

The fact that the reflection s_{α_i} is realized by an element of M′ — where M' is the subgroup of K that normalizes A — implies that $m_\alpha = m_\beta$. This implies the last statement. □

Corollary 7.13. (See Karpelevič [K3, Lemma 1.14.1]) *If* $\alpha_i \in I$, *then* $\rho_I(H_{\alpha_i}) = 0$. *Hence,* $\rho_{|\mathfrak{a}^I} = \rho^I$ *— where* $2\rho^I$ *denotes the sum of the positive roots of* $\mathfrak{a}^I$ *— and* $\ker(\rho_I) \supset \mathfrak{a}^I$. *In addition,*

$$\rho = \rho^I + \rho_I \text{ and } ||\rho||^2 = ||\rho^I||^2 + ||\rho_I||^2.$$

Proof. If $k \in M'$ is such that $\beta \circ Ad(k) = s_{\alpha_i}\beta$ for all $\beta \in \Sigma$, then $Ad(k)^{-1}(H_{\alpha_i}) = -H_{\alpha_i}$ since $Ad(k^{-1})H_\beta = H_{s_{\alpha_i}\beta}$. Hence, $s_{\alpha_i}\beta(H_{\alpha_i}) = -\beta(H_{\alpha_i})$. From this it follows that $\rho_I(H_{\alpha_i}) = 0$ if $\alpha_i \in I$.

The second statement follows from the observation that the vectors H_{ρ^I} and H_{ρ_I}, representing the corresponding linear functionals, are perpendicular: it is clear that $H_{\rho^I} \in \mathfrak{a}^I$ and that $H_{\rho_I} \in \mathfrak{a}_I$ as ρ_I annihilates $\mathfrak{a}^I$; also, $H_\rho = H_{\rho^I} + H_{\rho_I}$. □

Corollary 7.13 implies the following known result.

Proposition 7.14. (See Karpelevič [K3, Theorem 15.4.1]) *The operator* L' *defined by eq.* (7.11) *is the direct sum of the invariant operator*

$$\sum_{k=s+1}^{r} H_k^2 - 2H_{\rho_I} \tag{7.15}$$

on A_I *and the Laplace-Beltrami operator*

$$L^I = \sum_{i=1}^{p} Y_i^2 + \sum_{j=1}^{s} H_j^2 - 2H_{\rho^I} \tag{7.16}$$

on X^I.

Computation of the limit functions: reduction

§ **7.17.** To compute the Martin compactification $X \cup \partial X(\lambda_0)$, one must first determine all the λ_0-**limit functions**, i.e., those functions that arise as limits of the Martin kernel $K^{\lambda_0}(x,y)$ when y converges to a point in the Martin boundary $\partial X(\lambda_0)$.

Consider a sequence (y_n) that converges to the limit function h in $X \cup \partial X(\lambda_o)$. In view of the Cartan decomposition, one may write $y_n = k_n a_n \cdot o$ with $k_n \in K$ and unique $a_n \in \overline{A^+}$. To compute the limit h, it suffices to assume that k_n converges to $k \in K$ (by passing to a subsequence if necessary).

The sequence (a_n) determines a subset I of the set $\Delta = \Delta(\mathfrak{g}, \mathfrak{a}^+) = \{\alpha_1, \alpha_2, \ldots, \alpha_r\}$ of simple roots: if $H_n = \log a_n$, set $\alpha_i \in I$ if and only if the sequence $(\alpha_i(H_n))$ is bounded. By passing to a further subsequence if necessary, one may assume that for all $\alpha_i \in I$, the sequence $\alpha_i(H_n)$ converges and $\alpha_i(H_n) \to \infty$ if $\alpha_i \notin I$. The splitting $\mathfrak{a} = \mathfrak{a}^I \oplus \mathfrak{a}_I$, see § 2.7, decomposes the logarithms H_n of a_n into $H_n = H_n^I + H_{n,I}$ with $H_n^I \in \mathfrak{a}^I$ and $H_{n,I} \in \mathfrak{a}_I$. The last assumption made is equivalent to requiring that the sequence (H_n^I) converge to a point $H^I \in \overline{\mathfrak{a}^{I+}}$. Let $a_1 = \exp H^I$ and $a_{n,I} = \exp H_{n,I}$.

This proves the following lemma.

Lemma. *Every sequence (y_n) in X, converging to infinity, has a fundamental subsequence.*

As a result, to compute a limit function when (y_n) converges to a point of the Martin boundary $\partial X(\lambda_0)$, it suffices to assume that (y_n) is fundamental. If $y_n = k_n a_n \cdot o$ and $k_n \to k$, $a_n^I \to a^I$, and $a_{n,I} \overset{C_I}{\to} \infty$, it follows from Proposition 6.4 that (y_n) converges to h if and only if $(a_{n,I} \cdot o)$ converges to a function u. Furthermore, $h = S_{ka_1} u$, where the action of G on functions is the one defined in § 6.3. Consequently, to compute all the limit functions h, it suffices to compute the limit functions that arise from sequences like $(a_{n,I} \cdot o)$. They will be referred to as I-canonical sequences — see Definition 7.19.

Remark. This observation is due to Dynkin [D4] and is also used by Karpelevič [K3, §17.5, p. 182].

THE LIMIT OF A C_I-CANONICAL SEQUENCE

§ **7.18.** Recall (Definition 3.35) that a sequence $(y_n) \subset X$, with $y_n = k_n a_n$ and $a_n = \exp H_n \in \overline{\mathfrak{a}^+}$, is said to be C_I-fundamental if (i) (k_n) converges and (H_n) is C_I-fundamental in the sense of Definition 3.24, which implies that $I \neq \Delta$. When (i) $k_n = e$ for all n and the component H_n^I in $\mathfrak{a}^I$ is always zero, the sequence will be called C_I-canonical or I-canonical. For convenience, this concept is given a formal definition.

7.19. Definition. A sequence $(H_n) \subset \mathfrak{a}$ is said to be C_I **-canonical** or I**-canonical**, where $I \subsetneq \Delta$, if

(1) for all $n, H_n \in \overline{\mathfrak{a}^+}$;
(2) $\alpha_i(H_n) = 0$, if $\alpha_i \in I$; and
(3) $\alpha_i(H_n) \to \infty$ as $n \to \infty$, if $\alpha_i \notin I$.

A sequence $(a_n) \subset A$ is said to be **I-canonical** if (H_n) is I-canonical, where $H_n = \log a_n$.

Remark. Each proper subset I of Δ determines the chamber face $C_I = \{H \in \overline{\mathfrak{a}^+} \mid \alpha_i(H) > 0 \text{ if and only if } \alpha_i \notin I\}$ of $\overline{\mathfrak{a}^+}$ (see Definition 3.3). An I- canonical sequence is a sequence (H_n) in C_I such that $\| H_n \|$ and the distance of H_n to the walls of C_I both converge to infinity.

The subgroup of G that leaves invariant a function that is the limit of an I-canonical sequence is $K^I M A_I N_I$, as shown in Theorem 7.22 . A partial proof of this fact is given by the next result. It will be used to show later that every I-canonical sequence converges in $X \cup \partial X(\lambda_0)$ to one and the same limit. From the explicit formula for the limit function, one sees later that this limit is stabilized by the group $R^I = K^I M A_I N_I$.

7.20. Proposition. (Cf. Karpelevič [K3, §17.7]) *Let (a_m) be an I- canonical sequence in A. Assume that $(a_m \cdot o)$ converges to u. Then $u(x) =$*

$u(n_I a_I \cdot x^I) = u(a_I \cdot x^I)$, for all $x \in X$, where n_I, a_I and x^I are the generalized horocyclic coordinates of x.

In addition, if $k \in K^I M$, then $u(k \cdot x) = u(x)$, for all $x \in X$. Hence, u is N_I- and $K^I M$-invariant (see § 2.7 and § 2.13).

Proof. To simplify notation, let G denote G^{λ_0}. Let $n_1 \in N^I$. Then $G(n_1 a \cdot o, a_m \cdot o) = G(n_1 a a_m^{-1} \cdot o, o)$ as $n_1 a_m = a_m n_1$. Hence,

$$G(n_1 a \cdot o, a_m \cdot o) = G(a_m^{-1} \cdot o, a^{-1} n_1^{-1} \cdot o) = G(s(a^{-1} n_1^{-1} \cdot o), a_m \cdot o),$$

where s is the non-trivial isometry of X that leaves o fixed. This implies that

$$(*) \qquad u(n_1 a \cdot o) = u(s(a^{-1} n_1^{-1} \cdot o)).$$

Let $n_2 \in N_I$ and set $n_m = a_m^{-1} n_2 a_m$. Then $n_m \to e$ as $m \to \infty$: $n_2 = \exp(\sum_{\alpha \in \Sigma_I^+} X_\alpha)$ and so $n_m = \exp(\sum_{\alpha \in \Sigma_I^+} e^{-\alpha(H_m)} X_\alpha)$ (see Helgason [H2, p. 128]).

As $G(na \cdot o, a_m \cdot o) = G(n_2 n_1 a \cdot o, a_m \cdot o) = G(n_m a_m^{-1} n_1 a \cdot o, o) = G(a_m^{-1} \cdot o, a^{-1} n_1^{-1} n_m^{-1} \cdot o) = G(s(a^{-1} n_1^{-1} n_m^{-1} \cdot o), a_m \cdot o)$, it follows from the fact that the convergence is uniform on the compact subsets of X and eq. (*) that $u(x) = u(na \cdot o) = u(n_1 a \cdot o) = u(a_2 n_1 a_1 \cdot o) = u(a_2 \cdot x^I) = u(a_I \cdot x^I)$.

If $k \in K^I$, then $G(k \cdot x, a_m \cdot o) = G(x, k^{-1} a_m \cdot o) = G(x, a_m \cdot o)$. Hence, $u(k \cdot x) = u(x)$ for all $x \in X$ and $k \in K^I$.

Clearly, if $k \in M$, $G(k \cdot x, a_m) = G(x, a_m \cdot o)$ implies that $u(k \cdot x) = u(x)$ for all $x \in X$. □

§ **7.21.** Given any symmetric space X of non-compact type, it is known (see Karpelevič [K3], Guivarc'h [G14] and Theorem 13.23 in Chapter XIII) that the **λ_0-minimal functions**, i.e., the minimal solutions of $Lu + \lambda_0 u = 0$ are the functions

$$\text{(m)} \qquad h_b(g \cdot o) = e^{-\rho(H(g^{-1}k))}, \; b = kM, k \in K.$$

In particular, the Furstenberg boundary $\mathcal{F} = K/M$ may be identified with the set of minimal boundary points for $Lu + \lambda_0 u = 0$. As a result, any positive solution u of $Lu + \lambda_0 u = 0$ has the following unique integral representation, where the positive measure $\mu \in \mathcal{M}(\mathcal{F})$,

$$\text{(i)} \qquad u(g \cdot o) = \int_{\mathcal{F}} e^{-\rho(H(g^{-1}k))} d\mu(b) = \int_{\mathcal{F}} h_b(g \cdot o) d\mu(b).$$

Since the measure μ is unique and $\mu(\mathcal{F}) = u(o)$, it follows that if (i) $u(o) = 1$ and (ii) u is K-invariant, the representing measure μ is the unique K-invariant probability measure m on $\mathcal{F}$. The corresponding function is the unique positive spherical function Φ with $\Phi(o) = 1$ for which $L\Phi + \lambda_0 \Phi = 0$, i.e.,

$$\text{(s)} \qquad \Phi(g \cdot o) = \int_{\mathcal{F}} e^{-\rho(H(g^{-1}k))} dm(b).$$

When $X = X^I$, this probability will be denoted by m^I and the corresponding spherical function will be denoted by Φ^I. The spherical function Φ will be called the **ground state.**

Since every I-canonical sequence contains a subsequence that converges in $X \cup \partial X(\lambda_0)$, the following theorem implies that the I-canonical sequences all converge to one and the same limit function.

7.22. Theorem. *There is a unique positive solution u to the equation $Lu + \lambda_0 u = 0$ such that*

(1) *$u(o) = 1$;*
(2) *u is N_I-invariant; and*
(3) *u is K^I-invariant.*

It is the function

$$u(x) = u(n_I a_I \cdot x^I) = \Phi^I(x^I) e^{\rho(\log a_I)}, \tag{†}$$

where Φ^I is the unique positive K^I- and $K^I M$-invariant solution of the equation $L^I f + \lambda_0(X^I) f = 0$ on X^I for which $f(o) = 1$. (Note that and n_I, a_I, and x^I are the generalized horocyclic coordinates of x.)

This function will be denoted by h_I. It is invariant under the subgroup $R^I = K^I M A_I N_I$ of G, i.e., $S_g h_I = h_I$ if $g \in R^I$.

Furthermore, the probability measure on $\mathcal{F} = K/M$ that represents this function is the unique K^I-invariant probability m_I on the orbit $K^I \cdot \dot{e}$ of the base point $\dot{e} = M \in \mathcal{F}$.

Proof. Since u is N_I-invariant, it may be viewed as a function on $A_I \cdot X^I \subset X$ for which $L'u + \lambda_0 u = 0$, where L' is the operator defined by eq. (7.11). Let f be a minimal solution of this equation. By Harnack's inequality and the fact that L' is left invariant under the action of the abelian group A_I that commutes with G^I, it follows that $f(a_I \cdot x^I) = f_2(a_I) f_1(x^I)$ (see Karpelevič [K3, p. 186], Guivarc'h [G14]). More explicitly, if $a_I, a_2 \in A_I$, and $x^I = g_1 \cdot o$, then $d(a_2 a_I g_1 \cdot o, a_I g_1 \cdot o) = d(a_I g_1 a_2 \cdot o, a_I g_1 \cdot o) = d(a_2 \cdot o, o)$. It follows from Harnack's inequality that, for some constant $C > 0$ that does not depend on $a_I \cdot x^I$, $f(a_2 a_I \cdot x^I) \le C f(a_I \cdot x^I)$, i.e., $f \circ L_{a_2} \le C f$. Since f is minimal, it follows that $f \circ L_{a_2} = C(a_2) f$. Hence, $f(a_I \cdot x^I) = f_2(a_I) f_1(x^I)$, where $f_1(x^I) = f(x^I)$ and $f_2(a_I) = C(a_I) = f(a_I \cdot o), a_I = a_2$.

Since $\| \rho \|^2 = \lambda_0(X) = \lambda_0$ and $\| \rho^I \|^2 = \lambda_0(X^I)$, it follows from eq. (7.15) and eq. (7.16) that

$$\sum_{k=s+1}^{r} H_k^2 f_2 - 2H_{\rho_I} f_2 + ||\rho_I||^2 f_2 = 0, \text{ and} \tag{a}$$

$$L^I f_1 + ||\rho^I||^2 f_1 = 0. \tag{b}$$

Since $f_2(a_I) = e^{\rho_I(\log a_I)}$ is the only positive solution of eq. (a) with $f_2(e) = 1$, it follows that the minimal solutions of equation

$$L'u + ||\rho||^2 u = 0 \tag{c}$$

on $A_I \cdot X^I$ are of the form $f_2(a_I)f_1(x^I)$, where f_1 is a minimal solution of the equation

$$\text{(d)} \qquad L^I f_1 + ||\rho^I||^2 f_1 = 0.$$

Consequently, in terms of n_I, a_I, and x^I, the generalized horocyclic coordinate of $x \in X$, an N_I-invariant positive solution u of $Lu + \lambda_0 u = 0$ with $u(o) = 1$ is of the form $u(x) = e^{\rho_I(\log a_I)} f(x^I)$, where f is a positive solution of eq. (d) with $f(o) = 1$.

Since the group K^I acts on $A_I \cdot X^I$ only on the second variable x^I, it follows that u is K^I-invariant if and only if f is K^I-invariant. In view of eq. (s) in § 7.21, this means that $f = \Phi^I$. Hence, $h_I(x) = \Phi^I(x^I)e^{\rho_I(\log a_I)}$ is the only K^I-invariant solution of eq. (c), i.e., h_I is the only function satisfying (1), (2), and (3). Note that if $\ell \in M$, $\Phi^I(\ell \cdot x^I) = \Phi^I(x^I)$ since, in view of § 2.13, the function $x^I \to \Phi^I(\ell \cdot x^I)$ is K^I-invariant and takes the value 1 at o.

It follows also from eq. (m) in § 7.21 that the minimal solutions of eq. (c) are the functions given by the following formula, where $b_1 = k_1 M \in \mathcal{F}$,

$$\text{(7.23)} \qquad f_{b_1}(g \cdot o) = e^{\rho_I(\log a_I)} e^{-\rho^I(H(g_1^{-1}k_1))} = e^{-\rho(H(g^{-1}k_1))},$$

the last equality following from Corollary 7.13 since $g \cdot o = n_I a_I g_1 \cdot o$ implies that $H(g^{-1}k_1) = -\log(a_I) + H(g_1^{-1}k_1)$.

Since $K^I \cap M = M^I$, the natural map $\mathcal{F}^I = K^I/M^I \to K/M = \mathcal{F}$, that maps $k_1 M^I$ to $k_1 M$, is an injection. Hence, the measure m_I that represents h_I is the image of the unique K^I-invariant probability m^I on $\mathcal{F}^I$ under this map. In other words, m_I is the unique K^I-invariant probability on the orbit $K^I \cdot \dot{e} = K^I M$ of $\dot{e}$, in $\mathcal{F}$, under K^I.

Since Φ^I is M-invariant, it follows that h_I is M-invariant. From the formula for h_I, it follows that $S_a h_I = h_I$ if $a \in A_I$. Hence, h_I is invariant under $K^I \cup M \cup A_I \cup N_I$. It remains to show that $R^I = K^I M A_I N_I$ is a group: by Proposition 2.15(1), $K^I M$ is the centralizer in K of $\mathfrak{a}_I$; since P^I normalizes $\mathfrak{n}_I$, this implies that $K^I M$ normalizes $A_I N_I$ and so R^I is a group. □

7.24. Corollary. *Every I-canonical sequence converges in $X \cup \partial X(\lambda_0)$ to the limit function h_I.*

7.25. Remark. The argument that establishes Theorem 7.22 may be used to give a new inductive proof of the result § 7.21 (m), due to Karpelevič ([K3, 17.10, p. 189]) and Guivarc'h [G14], that a function f is a minimal solution of the equation $Lu + \lambda_0 u = 0$ if and only if $f = S_k h_{\dot{e}} = h_b$, where $h_{\dot{e}} = h_\emptyset$ and $b = kP$. The induction is on the rank of X. For rank one, a direct computation shows that the result is valid (see, for example, Lyons–MacGibbon–Taylor [L3]). Assuming the statement to be true when the rank is less than that of X, it is clear that the argument used to prove Theorem 7.22 applies. Since Corollary 7.24 holds, it follows that if the limit h_I of an I-canonical sequence is minimal, then Theorem 7.22 implies that $I = \emptyset$. Since by § 6.1 minimal functions exist, they are necessarily in the G-orbit of $h_{\dot{e}}$ and so coincide with this orbit, which is the same as the K-orbit of $h_{\dot{e}}$. This last observation follows from the fact that $S_g h_{\dot{e}} = h_{\dot{e}}$, for all $g \in N \cup A$ and so $S_{kan} h_{\dot{e}} = S_k h_{\dot{e}}$.

In view of what has been established, the following result gives a complete description of the set of λ_0-limit functions.

7.26. Proposition. *Let I be a proper subset of the set of simple roots. The G-orbit of h_I consists of λ_0-limit functions and conversely, every λ_0-limit function is in one of these orbits. More specifically, f is a λ_0-limit function if and only if there exists a proper subset I of Δ, $a_1 = a^I \in \overline{A^{I+}}$ and $k \in K$ such that*

$$f = S_{ka_1} h_I.$$

CLASSIFICATION OF LIMIT FUNCTIONS AND THE TOPOLOGY OF $X \cup \partial X(\lambda_0)$

§ **7.27.** The topology of $X \cup \partial X(\lambda_0)$ may be described by making explicit the sequences in X, converging to infinity, that converge to points of the boundary $\partial X(\lambda_0)$, in other words, the set of λ_0-limit functions.

It follows from Definition 3.35, Proposition 6.4, and Corollary 7.24, that every fundamental sequence (y_n) converges to a limit function. If the sequence is C_I-fundamental with $y_n = k_n \exp H_n$, where (i) $k_n \to k$ and (ii) the $\mathfrak{a}^I$ component H_n^I of H_n converges to $H^I = \log(a^I)$, this limit function is $S_{ka^I} h_I$.

The following result describes the sequences in X, converging to infinity, that converge in the Martin compactification $X \cup \partial X(\lambda_0)$.

7.28. Proposition. *Let (y_n) be a sequence in X that converges to infinity, with $y_n = k_n a_n \cdot o, a_n \in \overline{A^+}$. The following are equivalent:*

1. *(y_n) converges in $X \cup \partial X(\lambda_0)$; and*
2. *the limit function to which a fundamental subsequence of (y_n) converges does not depend upon the choice of fundamental subsequence.*

Proof. Assume that (y_n) converges to h. Then every subsequence converges to h, in particular every fundamental subsequence converges to h.

Conversely, consider a sequence (y_n) that satisfies (2) and let h be the common limit of all its fundamental subsequences. Let (y_{n_k}) be a subsequence of (y_n) that converges in the Martin compactification $X \cup \partial X(\lambda_0)$ to a limit function h'. Since a fundamental subsequence of (y_{n_k}) is a fundamental subsequence of (y_n), it follows that $h' = h$. As a result, (y_n) converges in the Martin compactification. □

To have a more precise description of the topology, it is necessary to know exactly when two fundamental sequences converge to the same limit function. Recall that the homogeneous space $\mathcal{F} = K/M = G/P$, where $P = MAN$ is a minimal parabolic subgroup (Definition 2.5). Hence, G acts on K/M. This action may also be understood in terms of the Iwasawa action of G on K (see § 2.1): if $gk = k_1an$, and $b = kM$, then $g \cdot b$ is defined to be $k(gk)M$. Letting $b = kM$ denote a generic point of K/M, under this left action a measure μ on K/M is transported to $g \cdot \mu$ by left translation. In other words, $\int \varphi d(g \cdot \mu) = \int \varphi(g \cdot b) d\mu(b)$.

The action of G on the functions normalized to take the value 1 at o leads to another action of G on positive measures μ that will be referred to as the **twisted action**. The measure $S_g\mu$, obtained by the twisted action of g^{-1} on μ, a left action, is given by

$$\int \varphi d(S_g\mu) = \frac{\int \varphi(g \cdot b)\sigma(g,b)d\mu(b)}{\int \sigma(g,b)d\mu(b)}, \text{ where}$$

$\sigma(g,b) = h_\emptyset(k^{-1}g^{-1} \cdot o) = S_k h_\emptyset(g^{-1} \cdot o) = h_b(g^{-1} \cdot o)$ if $b = kM$.

The twisted action has the characteristic property that if u is a positive solution of $Lu + \lambda_0 u = 0$ that is represented by μ, then $S_g\mu$ represents $S_g u$ (see Proposition 10.6). Notice that the support of $S_g\mu$ coincides with the support of $g \cdot \mu$.

Let (y_n^1) be a C_{I_1}-fundamental sequence and (y_n^2) be a C_{I_2}-fundamental sequence with the same limit h. Then, there exist $k_1, k_2 \in K$ such that

$$S_{k_1 a^{I_1}} h_{I_1} = S_{k_2 a^{I_2}} h_{I_2},$$

where $\log(a^{I_i})$ is the limit of the $\mathfrak{a}^{I_i}$-component of $H_n^i = \log a_n^i$ and $y_n^i = k_n^i a_n^i \cdot o$, $a_n^i \in \overline{A^+ \cdot o}, i = 1, 2$.

The following preliminary result determines the orbits under G of the limit functions h_I.

7.29. Proposition. *Let I_1, I_2 be two proper subsets of Δ and $g_1, g_2 \in G$. Then, the following are equivalent:*

(1) *$S_{g_1} h_{I_1} = S_{g_2} h_{I_2}$, and*

(2) *$I_1 = I_2 = I$, and $g_2^{-1} g_1 \in R^I \stackrel{\text{def}}{=} K^I M A_I N_I \subsetneq P^I$,*

where P^I is the standard parabolic subgroup corresponding to I.

Hence, the G-orbits of h_{I_1} and of h_{I_2} intersect if and only if they agree, in which case $I_1 = I_2$.

Proof. Assume $S_{g_1} h_{I_1} = S_{g_2} h_{I_2}$ and let $g = g_2^{-1} g_1$. The uniqueness of the integral representation for a positive solution of $Lu + \lambda_0 u = 0$ and Theorem 7.22 imply that $S_g m_{I_1} = m_{I_2}$.

If I is a proper subset of Δ, the support of m_I is $K^I \cdot \dot{e} = G^I \cdot \dot{e}$. This orbit is also equal to $P^I \cdot \dot{e}$. The stabilizer in G of this orbit is P^I: if $gP^I \cdot \dot{e} \cap P^I \cdot \dot{e} \neq \emptyset$, then $g \in P^I MAN = P^I$.

The support of $S_g m_I$ is the support of $g \cdot m_I = gP^I \cdot \dot{e}$. Consequently, the support $gP^{I_1} \cdot \dot{e}$ of $S_g m_{I_1}$ equals the support $P^{I_2} \cdot \dot{e}$ of m_{I_2}. The stabilizer in G of this common support is $gP^{I_1}g^{-1} = P^{I_2}$. Proposition 2.18 implies that $I_1 = I_2 = I$ and $g \in P^I$.

The following lemma determines the subgroup of P^I that stabilizes h_I and so completes the proof of the proposition.

7.30. Lemma. *Let $g \in P^I$. Then $S_g h_I = h_I$ if and only if $g \in R^I$.*

Proof. It was proved in Theorem 7.22 that h_I is invariant under R^I. Conversely, let $g = kan = ka_1n_1a_2n_2$ be such that $S_g h_I = h_I$. This implies that $S_{ka_1n_1}h_I = h_I$. Hence, $S_{a_1n_1}h_I = S_{k^{-1}}h_I = h_I$ since $g \in P^I = K^IMAN$ implies that $k \in K^IM$. The formula for h_I implies that $a_1 = n_1 = e$ since, as follows from Lemma 10.8, the stabilizer of Φ^I in G^I is K^I. Hence, $g \in R^I$. □

While this proposition gives the decomposition into disjoint G-orbits of the set of limit functions, it does not classify the limit functions h in terms of their **polar realization** as $h = S_{ka_1}h_I, k \in K, a_1 \in \overline{A^{I,+}}$.

7.31. Proposition. *Let $h_1 = S_{k_1a^{I_1}}h_{I_1}$ and $h_2 = S_{k_2a^{I_2}}h_{I_2}$ be two limit functions with $a_i \in \overline{A^{I_i,+}}$. Then the following are equivalent:*

(1) *$S_{k_1a^{I_1}}h_{I_1} = S_{k_2a^{I_2}}h_{I_2}$, and*
(2) *$I_1 = I_2 = I$, $a^{I_1} = a^{I_2} = a$ and $k_2^{-1}k_1 \in (K^I \cap aK^Ia^{-1})M$.*

Hence, the parameters (k, a, I) of a polar realization $S_{ka}h_I$ of a limit function h are unique modulo $k_0 \in (K^I \cap aK^Ia^{-1})M$.

Proof. Obviously (2) implies (1). Assume (1). Then it follows from Proposition 7.29 that $I_1 = I_2 = I$. Let $a = a^{I_1}$ and $a' = a^{I_2}$. The support of the measure $S_a m_I$ that represents $S_a h_I$, and of the other representing measure $S_{a'}m_I$, coincides with the support $P^I \cdot \dot{e}$ of m_I. Since the twisted action of K on probabilities is the same as the usual one, $k_1P^I \cdot \dot{e} = k_2P^I \cdot \dot{e}$. This implies that $k_0 = k_2^{-1}k_1 \in P^I \cap K$ which, by the Langlands decomposition, equals K^IM.

Let $k_0 = \ell m, \ell \in K^I, m \in M$. Since $m \in R^I$ it follows from Lemma 7.30 that $S_{k_0a}h_I = S_{a'}h_I$ if and only if $S_{\ell a}h_I = S_{a'}h_I$. It follows from the formula for h_I in Theorem 7.22, that $S_{\ell a}h_I = S_{a'}h_I$ if and only if $S_{\ell a}\Phi^I = S_{a'}\Phi^I$. This holds if and only if, in X^I, $\ell a \cdot o = a' \cdot o$. From the uniqueness of the $\overline{A^{I,+}}$ component in the polar decomposition on X^I, it follows that $a = a'$ and, so, ℓ stabilizes $a \cdot o$, i.e., $\ell \in aK^Ia^{-1}$. □

7.32. Corollary. (Convergence criterion) *Let (y_n^1) and (y_n^2) be two fundamental sequences with*

$$y_n^1 = k_n^1 a_n^{I_1} a_n^1 \cdot o \text{ and } y_n^2 = k_n^2 a_n^{I_2} a_n^2 \cdot o,$$

where

$$k_n^i \to k_i, \quad a_n^{I_i} \to a^{I_i} \text{ and } (a_n^i) \text{ is } I_i\text{-canonical.}$$

These two sequences converge to the same limit function if and only if

(1) *$I_1 = I_2 = I$;*

(2) $a^{I_1} = a^{I_2} = a$; *and*
(3) $k_2^{-1}k_1 \in (K^I \cap aK^Ia^{-1})M$.

By Lemma 3.36(3) and Theorem 3.38, this convergence criterion is exactly the same as for the dual cell compactification $X \cup \Delta^*(X)$ of X. It follows from Lemma 3.28 that these two compactifications are isomorphic.

7.33. Theorem. *The Martin compactification $X \cup \partial X(\lambda_0)$ and the dual cell compactification $X \cup \Delta^*(X)$ are isomorphic compactifications of X. In particular, if $(y_n) \subset X$ converges in $X \cup \partial X(\lambda_0)$ to $S_{ka_1}h_I$ with $a_1 \in \overline{A^{I,+}}$, then it converges to $(k \cdot C_I(\infty), ka_1 \cdot o) \in \Delta^*(X)$.*

Consequently, the G-action on $X \cup \Delta^*(X)$ defined in § 3.41 is continuous. As a result, $X \cup \partial X(\lambda_0)$ and the dual cell compactification $X \cup \Delta^*(X)$ are isomorphic G-compactifications of X.

Hence, in addition, the Martin compactification $X \cup \partial X(\lambda_0)$ is also G-isomorphic to the maximal Satake–Furstenberg compactification $\overline{X}^{SF}$.

Proof. Let $(y_n) \subset X$ converge to $S_{ka_1}h_I$. If (y_{n_j}) is a C_J-fundamental subsequence, $y_{n_j} = k_j a_j \cdot o$, it converges to $S_{k'a}h_J$, where $k' = \lim_j k_j$ and $a = \lim_j a_j^J$. It follows from Proposition 7.31 that $I = J, a = a_1$, and $k^{-1}k' \in (K^I \cap a_1 K^I a_1^{-1})M$. As a result, Lemma 3.36(3) implies that the limit of (y_n) in $\Delta^*(X)$ equals $(k \cdot C_I(\infty), ka_1 \cdot o)$.

Proposition 6.4 implies that $(g \cdot y_n)$ converges to $S_g h = S_{gka_1}h_I$. Since $gka_1 = k(gk)s_1 s_2 a_1 = k(gk)s_1 a_1 s_2$, with $s_i \in S_i$, it follows that $S_g h = S_{k(gk)s_1 a_1}h_I$. By the first part of this proof, it follows that $(g \cdot y_n)$ converges in $X \cup \Delta^*(X)$ to $(k(gk).C_I(\infty), k(gk)s_1 a_1 \cdot o) = g \cdot (k \cdot C_I(\infty), ka_1 \cdot o)$.

This proves the continuity of the G-action on $X \cup \Delta^*(X)$ that was defined in § 3.41. It follows automatically that $X \cup \partial X(\lambda_0)$ and $X \cup \Delta^*(X)$ are isomorphic G-compactifications.

The final statement follows immediately from Theorem 4.43. □

7.34. Remark. While the identification of $X \cup \partial X(\lambda_0)$ with the dual cell compactification $X \cup \Delta^*(X)$ gives a simple proof of the continuity of the G-action that was defined in §3.41, it is desirable to have a direct proof of this fact. (Added in proof: recently, such an argumemt has been found by Guivarc'h.)

CHAPTER VIII

THE MARTIN COMPACTIFICATION $X \cup \partial X(\lambda)$

Motivated by the study of Brownian motion, Dynkin [D4] was the first to investigate the positive eigenfunctions and corresponding Martin boundaries for the Laplace–Beltrami operator on a symmetric space of noncompact type. He restricted his attention to the space $\mathrm{SL}(n, \mathbb{C})/\mathrm{SU}(n)$. This space is especially amenable to a study of the Martin compactification because one has an explicit formula for the Green function G^λ that is a consequence of a remarkable formula of Weyl — namely, $\prod_{\alpha>0}(e^{\alpha(H)} - e^{-\alpha(H)}) = \sum_{s \in W}(\det s)e^{s \cdot \rho(H)}$, see Freudenthal [F1, Proposition 47.14] — that is true for complex Lie algebras. Later, Nolde [N] published a note announcing the same results as Dynkin for any semisimple Lie group whose Lie algebra is a complex Lie algebra. This was followed by a note of Olshanetsky [O1] that stated asymptotic formulas for the Green function and deduced, as a result, the results analogous to those of Dynkin and Nolde. While the proofs of Olshanetsky's asymptotic formulas were recently published [O2], they are insufficient, as pointed out in footnote 6 in Chapter I, to deduce the (correct) results given in [O2] about the Martin compactification for $\lambda < \lambda_0$.

The essential difference between the case of λ_0 and that of $\lambda < \lambda_0$ is that in the first case the limit functions can be explicitly described without using any asymptotic information about the Green function — in particular, the fact that there is a unique limit function associated with each Weyl chamber and each Weyl chamber face is the key to understanding the compactification. When $\lambda < \lambda_0$, directions inside the Weyl chamber faces play a crucial role in determining the compactification. This chapter begins with an account of Dynkin's determination of the set of limit functions.

§ **8.1.** The following theorem lists the limit functions in $\partial X(\lambda)$ for any semisimple Lie group G, when $\lambda < \lambda_0$. As in Chapter VII, it suffices to consider those limit functions that arise from I-canonical sequences. However, as will become evident, one needs to refine this notion so as to take into account the limiting directional behavior of the sequence.

8.2. Theorem. *The set of limit functions associated with the Martin compactification $X \cup \partial X(\lambda)$ is the set of functions $S_{ka_1} h_{I,v}$, where $k \in K, I \subset \Delta, a_1 \in \overline{A^{I,+}}$ and $v \in \overline{\mathfrak{a}^*_+}$ is the function $v(H) = cB(L, H)$, L a direction in $\overline{C_I}$, and $c = \sqrt{\lambda_0 - \lambda}$.*

The function $h_{I,v}(x)$, in terms of the generalized horocyclic coordinates n_I, a_I, and x_I is given by

$$h_{I,v}(n_I a_I \cdot x^I) = \Phi^I(x^I) e^{-(\rho_I + v)(H(a_I^{-1}))} = h_I(x) e^{-v(H(a_I^{-1}))}.$$

THE CASE OF $X = \mathrm{SL}(n,\mathbb{C})/\mathrm{SU}(n)$ FOR $\lambda < \lambda_0$

§ **8.3.** In this case one begins, following Dynkin, by determining the asymptotics of the Green function G^λ. Rather than use his potential theoretic arguments, systematic use is made of formulas for the radial parts of operators that are found in Helgason [H3].

A formula for the Green function on X can be obtained by viewing X as $\mathfrak{p}$. Recall that since X is a symmetric space of non-compact type, the map $Y \to \exp Y \cdot o, Y \in \mathfrak{p}$ is a diffeomorphism (Helgason [H2, Theorem 1.1, p. 252]). In the case of $\mathrm{SL}(n,\mathbb{C})/\mathrm{SU}(n)$, this says that every positive definite Hermitian symmetric matrix of determinant one is of the form $\exp Y$ for a unique Hermitian symmetric matrix Y of trace zero.

The Laplace–Beltrami operator L corresponds, under this diffeomorphism, to a differential operator on $\mathfrak{p}$, that has the following explicit description for $Ad(K)$-invariant functions f (see Helgason [H3, Theorem 3.15, p. 273])

$$\begin{aligned} Lf(\exp Y \cdot o) &= \Delta_{\mathfrak{p}}\psi(Y) + (\operatorname{grad}\log J)\psi(Y) \\ &= \frac{\Delta_{\mathfrak{p}}(\sqrt{J}\psi)}{\sqrt{J}}(Y) - \frac{\Delta_{\mathfrak{p}}\sqrt{J}}{\sqrt{J}}(Y)\psi(Y), \end{aligned}$$

where $f(\exp Y \cdot o) = \psi(Y)$, $\Delta_{\mathfrak{p}}$ denotes the Laplacian on $\mathfrak{p}$ with respect to the inner product given by the Killing form B and $J(Y)$ is the Jacobian of the map $Y \to \exp Y \cdot o$. (Recall that $\operatorname{grad}\log J$ is a vector field.) The Jacobian is $Ad(K)$-invariant and for $H \in \mathfrak{a}$, $J(H) = \prod_{\alpha\in\Sigma^+}\left[\frac{\sinh\alpha(H)}{\alpha(H)}\right]^{m_\alpha}$.

Let $L_{\mathfrak{p}}\psi = \Delta_{\mathfrak{p}}\psi + (\operatorname{grad}\log J)\psi$. The Green function $G^\lambda(x,y)$ for $Lf + \lambda f = 0$, with pole at y is determined by the K-invariant function $g^\lambda(x) = G^\lambda(x,o)$. Therefore, it is necessary to consider the radial form of $L_{\mathfrak{p}}$ corresponding to $Ad(K)$ acting on $\mathfrak{p}$ with transversal manifold $\mathfrak{a}^+ \cdot o$. This is the operator

$$(\dagger) \qquad \Delta_{\mathfrak{a}}\psi(H) + (\operatorname{grad}\log\delta)\psi(H) = \frac{\Delta_{\mathfrak{a}}(\sqrt{\delta}\psi)}{\sqrt{\delta}}(H) - \frac{\Delta_{\mathfrak{a}}\sqrt{\delta}}{\sqrt{\delta}}(H)\psi(H),$$

where $\delta(H) = \prod_{\alpha\in\Sigma^+}[\sinh\alpha(H)]^{m_\alpha}$, since, by Helgason [H3, Proposition 3.13, p. 270], the radial part of $\Delta_{\mathfrak{p}}$ is

$$(*) \qquad \Delta_{\mathfrak{a}} + \operatorname{grad}\log\delta_0,$$

where $\delta_0(H) = \prod_{\alpha\in\Sigma^+}[\alpha(H)]^{m_\alpha}$.

In the case of a complex Lie algebra, since each $m_\alpha = 2$, $\sqrt{\delta(H)} = \prod_{\alpha\in\Sigma^+}\sinh\alpha(H)$, and the remarkable formula of Weyl implies that

$$\Delta_{\mathfrak{a}}\sqrt{\delta} = \|\rho\|^2\sqrt{\delta} = \lambda_0\sqrt{\delta}$$

(see Helgason [H3, the proof of Proposition 3.10, p. 268]).

The formula (*) for the radial part of $\Delta_{\mathfrak{p}}$ shows that

$$\Delta_{\mathfrak{p}}\sqrt{J} = \lambda_0\sqrt{J}$$

since $\Delta_{\mathfrak{a}}\sqrt{\delta_0} = 0$ (see Helgason [H3, eq. (64), p. 271]). Consequently, in the case of a complex Lie algebra, for a K-invariant function f on X,

$$Lf(\exp Y \cdot o) + \lambda f(\exp Y \cdot o) = 0$$

if and only if

$$\frac{1}{\sqrt{J(Y)}}\Delta_{\mathfrak{p}}(\sqrt{J}\psi)(Y) = (\lambda_0 - \lambda)\psi(Y).$$

From the previous formula it follows that, if γ is the Green function with pole at 0 for $\Delta_{\mathfrak{p}} + (\lambda - \lambda_0)Id$ on $\mathfrak{p}$, the function $g^\lambda(\exp Y \cdot o) \stackrel{\text{def}}{=} \frac{\gamma}{\sqrt{J}}(Y)$ is the Green function $G^\lambda(\exp Y \cdot o, o)$ on $\mathfrak{p}$ with pole at o for the operator $L + \lambda Id$. It follows from Watson [W2, §6.22(15)] that, up to a constant,

$$\gamma(Y) = \| Y \|^{\frac{3-n^2}{2}} K_{\frac{n^2-3}{2}}(c \| Y \|).$$

Hence, one has the following formula for $G^\lambda(\exp H \cdot o, o)$, when $H \in \overline{\mathfrak{a}^+}$.

8.4. Proposition. *Let $X = \mathrm{SL}(n,\mathbb{C})/\mathrm{SU}(n)$. Let $H \in \mathfrak{a}^+$. Then, if $\lambda < \lambda_0$ and $c = \sqrt{\lambda_0 - \lambda}$, for a suitable constant C,*

$$G^\lambda(\exp H \cdot o, o) = C \prod_{\alpha \in \Sigma^+}\left[\frac{\alpha(H)}{\sinh \alpha(H)}\right]\frac{1}{\| H \|^{\frac{n^2-3}{2}}}K_{\frac{n^2-3}{2}}(c \| H \|).$$

Hence, as

$$\| H \| \to \infty, \quad G^\lambda(\exp H \cdot o, o) \sim \frac{e^{-c\|H\|}}{\| H \|^{\frac{n^2-2}{2}}}\prod_{\alpha\in\Sigma^+}\frac{\alpha(H)}{\sinh \alpha(H)}.$$

Let $I \subset \Delta$. If $H \in C_I$, then

$$G^\lambda(\exp H \cdot o, o) = C \prod_{\alpha \in \Sigma_I^+}\left[\frac{\alpha(H)}{\sinh \alpha(H)}\right]\frac{1}{\| H \|^{\frac{n^2-3}{2}}}K_{\frac{n^2-3}{2}}(c \| H \|).$$

Hence, if $H \in C_I$, as

$$\| H \| \to \infty, \quad G^\lambda(\exp H \cdot o, o) \sim \frac{e^{-c\|H\|}}{\| H \|^{\frac{n^2-2}{2}}}\prod_{\alpha\in\Sigma_I^+}\frac{\alpha(H)}{\sinh \alpha(H)}.$$

Proof. In Watson [W2], it is shown that the Bessel function $K_\nu(t) \sim t^{\frac{1}{2}}e^{-t}$ as $t \to \infty$: see [W2, (12), p. 80], if $\nu = n + \frac{1}{2}$, and [W2, (1), p. 202], for $\nu = n$. This gives the asymptotic formula for $H \in \mathfrak{a}^+$.

When $H \in C_I, I \subset \Delta$, the second set of formulas hold because $H \to G^\lambda(\exp H \cdot o, o)$ is a smooth function on $\mathfrak{a} \setminus \{0\}$ and so $\frac{\alpha(H)}{\sinh \alpha(H)} = 1$ for $\alpha \in \Sigma^{I,+}$. $\square$

§ **8.5.** Now that the asymptotic behavior of the Green function G^λ on the closure of the positive Weyl chamber is understood, one may proceed to compute the limit functions that arise from fundamental sequences in $\overline{\mathfrak{a}^+}$. The remaining limit functions are obtained by the group action.

8.6. Definition. Let I be a subset of the set of simple roots. A sequence (y_n), where $y_n = k_n a_n \cdot o$ with $a_n = \exp H_n, H_n \in \overline{\mathfrak{a}^+}$ is said to be I **-directional** if it is C_I-fundamental and $\frac{H_n}{\|H_n\|}$ converges to a limit L as $n \to \infty$.

Note that the limiting direction is in $\overline{C_I} \subset \mathfrak{a}_I$ as the projection H_n^I of H_n onto $\mathfrak{a}^I$ converges and so the direction of the projection $H_{n,I}$ of H_n onto $\mathfrak{a}_I$ also converges to L. Recall that $H_n^I \in \overline{\mathfrak{a}^{I,+}}$ as this chamber is the projection of $\overline{\mathfrak{a}^+}$ onto $\mathfrak{a}^I$.

8.7. Definition. Let L be a direction in $\overline{\mathfrak{a}^+}$ and let $v \in \overline{\mathfrak{a}^*_+}$ be defined by $v(H) = cB(L,H) = B(V,H)$. Denote by $h_{I,v}$ the function of $x = n_I a_I \cdot x^I$ defined by

$$h_{I,v}(x) = \Phi^I(x^I)e^{-(\rho_I+v)(H(a_I^{-1})} = h_I(x)e^{-(v(H(a_I^{-1}))}.$$

Note that $h_{I,v}$ is a K^I- and N_I-invariant solution of the equation $Lu+\lambda u = 0$.

In order to relate the asymptotics to the actual computation, it is important to have the following formula for the spherical function Φ that was defined in § 7.21. (It is the special case of [H3, Theorem 4.7, p. 423] corresponding to $Z = 0$.)

8.8. Lemma. *Assume that the Lie algebra* $\mathfrak{g}$ *of* G *is complex. If* Φ *is the spherical function in* § 7.21 *that is the unique positive* K*-invariant solution of* $Lu + \lambda_0 u = 0$ *with* $u(o) = 1$*, then*

$$\Phi(g \cdot o) = \prod_{\alpha > 0} \frac{\alpha(H(g^{-1}))}{\sinh \alpha(H(g^{-1}))}.$$

Hence,

$$h_{I,v}(g \cdot o) = \prod_{\alpha \in \Sigma^{I,+}} \frac{\alpha(H(g^{-1}))}{\sinh \alpha(H(g^{-1}))} \; e^{-(\rho_I+v)(H(g^{-1}))}.$$

Proof. As mentioned before, in the complex case, $\Delta_{\mathfrak{a}}\sqrt{\delta_0} = 0$. Since $J = \delta/\delta_0$, it follows from eq. (†) in § 8.3 that

$$\begin{aligned} L_{\mathfrak{p}}\sqrt{J^{-1}} &= (\Delta_{\mathfrak{a}} + \operatorname{grad}\log\delta)\sqrt{J^{-1}} \\ &= \sqrt{\delta^{-1}}\,(\Delta_{\mathfrak{a}}\sqrt{\delta_0}) - \sqrt{\delta^{-1}}\,(\Delta_{\mathfrak{a}}\sqrt{\delta})\sqrt{J^{-1}} \\ &= -\lambda_0\sqrt{J^{-1}}. \end{aligned}$$

Hence, since $\sqrt{J^{-1}}$ is $Ad(K)$-invariant and equals 1 at o, $\Phi(\exp H \cdot o) = \sqrt{J^{-1}}(H)$. □

8.9. Proposition. (Dynkin [D4]) *Let (y_n) be an I-directional sequence, where $y_n = k_n a_n \cdot o$. Let $v(H) = cB(H, L)$, where $L \in \overline{C_I}$ is the limiting direction of the H_n and B is the Killing form. Then*

$$k_n a_n \cdot o \to S_{ka_1} h_{I,v},$$

where a_1 is the limit in A^I of the projection of the sequence (a_n) onto A^I and k is the limit of the sequence $(k_n)_{n \geq 1}$.

Proof. It suffices to prove this when $k_n = e$ for all n and for an I-canonical sequence (a_n). As before, let $c^2 = \lambda_0 - \lambda$.

To begin, suppose that $I = \emptyset$. Assume that $\frac{H_n}{\|H_n\|} \to L$, $L \in \overline{\mathfrak{a}^+}$ with $\| L \| = 1$. Let $a_n = \exp H_n \cdot o$ and $a = \exp H \cdot o$.

For large n, $H_n - H \in \mathfrak{a}^+$, and so

$$K^\lambda(a \cdot o, a_n \cdot o) = \frac{G^\lambda(a \cdot o, a_n \cdot o)}{G^\lambda(o, a_n \cdot o)} = F_1(n) F_2(n) F_3(n),$$

where

$$F_1(n) = \Pi_{\alpha > 0} \frac{\alpha(H_n - H)}{\alpha(H_n)} \times \frac{\sinh \alpha(H_n)}{\sinh \alpha(H_n - H)}$$

$$F_2(n) = \left(\frac{\| H_n \|}{\| H_n - H \|} \right)^{\frac{n^2-3}{2}} \quad \text{and}$$

$$F_3(n) = \frac{K_{\frac{(n^2-3)}{2}}(c \| H_n - H \|)}{K_{\frac{(n^2-3)}{2}}(c \| H_n \|)}.$$

Clearly, $\lim\limits_{n \to \infty} F_2(n) = 1$. The asymptotic behavior of the Bessel function implies that $\lim_{n \to \infty} F_3(n) = \lim_{n \to \infty} e^{-c(\|H_n - H\| - \|H_n\|)}$. The Euclidean Busemann function (see § 3.1) is the map $H \to -B(L, H)$ and so $\lim_{n \to \infty} F_3(n) = e^{cB(L,H)} = e^{v(H)}$.

Since

$$\lim_{n \to \infty} \frac{\alpha(H_n - H)}{\alpha(H_n)} = 1$$

and

$$\lim_{n \to \infty} \frac{\sinh \alpha(H_n)}{\sinh \alpha(H_n - H)} = \lim_{n \to \infty} \frac{1 - e^{-2\alpha(H_n)}}{e^{\alpha(-H)} - e^{-\alpha(2H_n - H)}} = e^{\alpha(H)},$$

it follows that $\lim\limits_{n \to \infty} F_1(n) = e^{\rho(H)}$. Therefore,

$$\lim_{n \to \infty} \frac{G(a \cdot o, a_n \cdot o)}{G(o, a_n \cdot o)} = e^{(\rho + v)(H)}.$$

This proves the result for the case $I = \emptyset$: the reason being that, by taking subsequences if necessary, the limit function $h = \lim_{n\to\infty} K^\lambda(\cdot, a_n \cdot o)$ can be seen to satisfy $h(na \cdot o) = h(a \cdot o), a \in A, n \in N$ as the argument of Proposition 7.20 applies.

Now assume that $I \neq \emptyset$. In the computation, the only modification that occurs involves $F_1(n)$. Let $F_1(n) = F_{11}(n)F_{12}(n)$, where

$$F_{11}(n) = \prod_{\alpha\in\Sigma^{I,+}} \frac{\alpha(H_n - H)}{\alpha(H_n)} \times \frac{\sinh\alpha(H_n)}{\sinh\alpha(H_n - H)},$$

and

$$F_{12}(n) = \prod_{\alpha\in\Sigma_I^+} \frac{\alpha(H_n - H)}{\alpha(H_n)} \times \frac{\sinh\alpha(H_n)}{\sinh\alpha(H_n - H)}.$$

Since $G^\lambda(\exp H \cdot o, o)$ is a smooth function of H, as H_n approaches the wall C_I where all the roots in I vanish, it follows from the fact that $\lim_{t\to 0} \sinh t/t = 1$ that $F_{11}(n)$ tends to $\prod_{\alpha\in\Sigma^{I,+}} \frac{\alpha(H)}{\sinh\alpha(H)}$. Consequently, for $H_n \in \mathfrak{a}_I$,

$$F_{11}(n) = \prod_{\alpha\in\Sigma^{I,+}} \frac{\alpha(H)}{\sinh\alpha(H)}.$$

Also,

$$\lim_{n\to\infty} F_{12}(n) = \prod_{\alpha\in\Sigma_I^+} e^{\alpha(H)} = e^{\rho_I(H)},$$

where ρ_I (see Lemma 7.12) is one half the sum of the positive roots in Σ_I with multiplicities counted — in this case 2.

If $\lim_{n\to\infty} \frac{H_n}{\|H_n\|} = L$ and $v(H) = cB(L, H)$, then

$$\lim_{n\to\infty} F_{12}(n)F_3(n) = e^{(\rho_I + v)(H)}.$$

Hence, by Lemma 8.8,

$$\begin{aligned} h(a\cdot o) &= \lim_{n\to\infty} K(a\cdot o, a_n \cdot o) \\ &= \prod_{\alpha\in\Sigma^{I,+}} \frac{\alpha(H(a^{-1}))}{\sinh\alpha(H(a^{-1}))}\, e^{-(\rho_I+v)(H(a^{-1}))} = h_{I,v}(a\cdot o). \end{aligned}$$

As the limit function h is K^I- and N_I-invariant by Proposition 7.20, it follows from Lemma 8.8 that $h = h_{I,v}$. □

Since, for every direction $L \in \overline{C_I}$ and $H_0 \in C_I$, the sequence $(\exp(nL + H_0))$ is I-directional with limiting direction L, this completes the proof of Theorem 8.2 when $X = \mathrm{SL}(n,\mathbb{C})/\mathrm{SU}(n)$. □

Computation of the limit functions for a general semisimple group

§ **8.10.** The main difference between the case when the Lie algebra of G is complex and the general case has to do with the fact that, to date, there is no satisfactory information about the asymptotic behavior of the Green function. In place of that, one makes use of the following estimates for the Green function G^λ and for the spherical function Φ, due to Anker–Ji [A6] and Anker [A5].

As before, let $c = \sqrt{\lambda_0 - \lambda}$ and, for $H \in \overline{\mathfrak{a}^+}$, let

$$g^\lambda(H) = \| H \|^{\frac{(1-r)}{2} - |\Sigma_0^+|} \prod_{\alpha \in \Sigma_0^+} \{1 + \alpha(H)\}\, e^{-\rho(H) - c\|H\|}, \tag{8.11}$$

where $|\Sigma_0^+|$ is the cardinality of the set Σ_0^+ of indivisible positive roots α, i.e., those positive roots α for which $\frac{1}{2}\alpha$ is not a root, and r is the rank of X. According to Anker–Ji [A6], there is a constant $C > 0$ such that, if $H \in \overline{\mathfrak{a}^+}$,

$$\frac{1}{C} g^\lambda(H) \leq G^\lambda(\exp H \cdot o, o) \leq C g^\lambda(H), \text{ when } \| H \| \text{ is large.} \tag{8.12}$$

Also Anker [A5] showed that there is a constant $C > 0$ such that for $H \in \overline{\mathfrak{a}^+}$

$$\begin{aligned} \frac{1}{C} \prod_{\alpha \in \Sigma_0^+} \{1 + \alpha(H)\}\, e^{-\rho(H)} &\leq \Phi(\exp H \cdot o) \\ &\leq C \prod_{\alpha \in \Sigma_0^+} \{1 + \alpha(H)\}\, e^{-\rho(H)}. \end{aligned} \tag{8.13}$$

The estimate (8.13) will be used in place of the exact information given by Lemma 8.8 in the complex case.

By using the function g^λ in place of G^λ, and by making some use of integral representation, it is possible to prove Theorem 8.2 for any semisimple group G with finite center. As a first step, one computes the analogue of the limit functions for g^λ. Note that $g^\lambda(H' + H)$ will correspond to $G^\lambda(\exp -H \cdot o, \exp H' \cdot o)$ when $H' + H \in \overline{\mathfrak{a}^+}$ and $\| H' + H \|$ is large.

8.14. Proposition. *Let H_n be an I-canonical sequence with limiting direction $L \in \overline{C}_I$ and let $H \in \overline{\mathfrak{a}^+}$. Then*

$$\lim_{n \to \infty} \frac{g^\lambda(H_n + H)}{g^\lambda(H_n)} = \prod_{\alpha \in \Sigma_0^{I+}} \{1 + \alpha(H)\} e^{-(\rho + v)(H)},$$

where $\Sigma_0^{I,+}$ denotes the set of positive indivisible roots of $\mathfrak{a}^I$ and $v(H) = cB(L, H)$, for all $H \in \mathfrak{a}$.

Proof. The term of the quotient involving $\| H_n + H \| \,/\, \| H_n \| \to 1$ as $n \to \infty$. The term involving the roots converges to $\prod_{\alpha \in \Sigma_0^{I,+}} \{1 + \alpha(H)\}$

since $\{1+\alpha(H_n+H)\}/\{1+\alpha(H_n)\} = 1+\alpha(H)$, if $\alpha \in \Sigma^{I,+}$ and tends to 1 if $\alpha \in \Sigma_I^+$.

Let v be the linear functional on $\mathfrak{a}$ corresponding to cL, i.e., $v(H) = cB(L,H)$, for all $H \in \mathfrak{a}$. Since $\| H_n + H \| - \| H_n \|$ converges to the Busemann function $-B(L,-H) = B(L,H)$, it follows that

$$\frac{e^{-\rho(H_n+H)-c\|H_n+H\|}}{e^{-\rho(H_n)-c\|H_n\|}} = e^{-\rho(H)-c\{\|H_n+H\|-\|H_n\|\}} \longrightarrow e^{-(\rho+v)(H)}. \quad \square$$

Karpelevič [K3] (see also Guivarc'h [G14] and Theorem 13.23 in Chapter XIII) showed that the minimal solutions of $Lu + \lambda u = 0$ are the functions given by the formula in the following proposition.

8.15. Proposition. *The minimal solutions of $Lu + \lambda u = 0$ are the functions $S_k h_v$, where*

$$h_v(g\cdot o) = e^{-(\rho+v)(H(g^{-1}))},$$

and the linear functional v is given by a direction L in $\overline{\mathfrak{a}^+}$: $v(H) = cB(L,H) = B(V,H)$.

Remark. Let γ be the unit speed geodesic with $\dot\gamma(0) = L$. Then, as shown by the remarks in § 3.1, the function $d_\gamma(g\cdot o) = B(L, H(g^{-1}))$. As a result, the functional $v(H(g^{-1}))$ is a constant multiple of the Busemann function corresponding to the direction L. It can be expressed in terms of cocycles (see § 10.1). Let $\sigma(g,b) = e^{-\rho(H(gk))}$, where $b = kM \in \mathcal{F}$. The Busemann cocycle δ_L corresponding to L is defined (see Remark 10.5(2)) by setting $\delta_L(g,b) = e^{-B(L,H(gk))}, b = kM$. Then one has the following formula for the minimal function $S_k h_v$:

$$S_k h_v(g\cdot o) = \sigma(g^{-1},b)\{\delta_L(g^{-1},b)\}^{\sqrt{\lambda_0-\lambda}}$$

where $b = kM \in \mathcal{F}$.

The set $\overline{\mathfrak{a}^+} \cap \{V \in \mathfrak{a} \mid \|V\| = c = \sqrt{\lambda_0 - \lambda}\} \times K/M$ parametrizes these functions: to (V,b), $b = kM$, corresponds the minimal function

$$S_k h_v, \text{ where } S_k h_v(g\cdot o) = e^{-(\rho+v)(H(g^{-1}k))}.$$

Let $V \in \overline{C_I} \cap \{V \in \mathfrak{a} \mid \|V\| = c\}$, which is the intersection of the closure of $\{V \in \overline{\mathfrak{a}^+} \mid \alpha_i(V) > 0$ if and only if $\alpha_i \notin I\}$ with $\{\|V\| = c\}$. Define the positive solution $h_{I,v}$ of $Lu + \lambda u = 0, \lambda < \lambda_0$ as follows

$$h_{I,v}(x) = \Phi^I(x^I)e^{-(\rho_I+v)(H(a_I^{-1}))}, \text{ if } x = n_I a_I \cdot x^I.$$

8.16. Lemma. *The representing measure for the function $h_{I,v}$ is $\delta_V \otimes m_I$, where $V = cL$. In other words,*

$$h_{I,v}(g \cdot o) = h_I(g \cdot o) e^{-v(H(g^{-1}))} = \int_{K/M} e^{-(\rho+v)(H(g^{-1}k))} dm_I(b).$$

Consequently, if $a_1 \in A^I$ and $k \in K$, the representing measure for the function $S_{ka_1} h_{I,v}$ is $\delta_V \otimes S_{ka_1} m_I$.

Proof. Since $h_{I,v}(g \cdot o) = h_I(g \cdot o) e^{-v(H(g^{-1}))}$, it follows from Theorem 7.22 that the representing measure for $h_{I,v}$ is represented by the measure $\delta_V \otimes m_I$. This proves the first statement. The second one is automatic as the twisted action on measures corresponds to the action of G on the functions normalized to have the value one at o. □

Consider a limit function h that is the limit of a $\emptyset$-canonical sequence (H_n) with limiting direction $L \in \overline{\mathfrak{a}^+}$. Since the argument of Proposition 7.20 applies to $G = G^\lambda$, it is N-invariant and so is completely determined by its restriction to $A \cdot o$. Since

$$h(\exp(-H) \cdot o) = \lim_{n \to \infty} \frac{G^\lambda(\exp(-H) \cdot o, \exp H_n \cdot o)}{G^\lambda(o, \exp H_n \cdot o)},$$

eq. (8.12), Proposition 8.14, and the observation that, for all $H \in \mathfrak{a}$, $H_n + H \in \mathfrak{a}^+$ for large n, imply that $h(\exp(-H) \cdot o) \leq C^2 h_v(\exp(-H) \cdot o) = C^2 e^{(\rho+v)(-H)}$. Since h_v is minimal and $h(o) = 1, h = h_v$. This completes most of the proof of the next result.

8.17. Proposition. *If (H_n) is a $\emptyset$-canonical sequence with limiting direction L, then it converges to h_v, where $v(H) = cB(L, H)$.*

Proof. By what has been proved, every subsequence that converges to a limit function also converges to h_v. □

The situation is very similar for an I-canonical sequence (H_n) with limiting direction L. If it converges to a limit function h then, by the argument of Proposition 7.20, it is N_I-invariant and K^I-invariant. It follows from Corollary 7.13, Proposition 8.14 and the estimate (8.13) for Φ that

$$0 \leq h(n_I a_I \cdot x^I) = h(a_I \cdot x^I) \leq C^2 \Phi^I(x^I) e^{-(\rho_I+v)(H(a_I^{-1}))} = C^2 h_{I,v}(a_I \cdot x^I),$$

since $H_n - (\log a_I) \in \overline{\mathfrak{a}^+}$ for large n.

This implies that the representing measure for h has a bounded density with respect to the representing measure $\delta_V \oplus m_I$ for $h_{I,v}$. Since h is K^I-invariant, this density is a constant and so $h = h_{I,v}$ as $h(o) = 1$. This proves the following proposition.

8.18. Proposition. *Let (H_n) be an I-canonical sequence that has limiting direction $L \in \overline{C_I}$. Then it converges to $h_{I,v}$, where $v(H) = cB(L, H)$.*

Conclusion of the proof of Theorem 8.2.

It follows from Propositions 8.17 and 8.18 that Theorem 8.2 is valid for any symmetric space of non-compact type. □

Determination of the Martin compactification

§ **8.19.** The Martin compactification for $\lambda < \lambda_0$ involves more than establishing the lists of limit functions. One needs to get hold of the exact parameters for the boundary points. This is equivalent to knowing when to identify limit functions.

8.20. Proposition. *Let I_1, I_2 be two subsets of the set $\Delta = \Delta(\mathfrak{g}, \mathfrak{a}^+)$ and let $L_i \in \overline{C_{I_i}} \cap \{H \in \mathfrak{a} \mid ||H|| = 1\}, i = 1, 2$, correspond to the linear functionals $v_i, v_i(H) = cB(L_i, H)$. If $k_i \in K$ and $H_i \in \mathfrak{a}^{I_i}, i = 1, 2$, then the following are equivalent, where $a_i = \exp H_i$:*

(1) $S_{k_1 a_1} h_{I_1, v_1} = S_{k_2 a_2} h_{I_2, v_2}$; *and*
(2) *(i)* $I_1 = I_2 = I$,
(ii) $H_1 = H_2 = H$ *and* $Ad(k_1)H = Ad(k_2)H$, *and*
(iii) $L_1 = L_2 = L$, *and* $Ad(k_1)L = Ad(k_2)L$.

Proof. It is clear that (2) implies (1). Assume (1) holds. Since the representing measures $\delta_{V_i} \otimes S_{k_i a_i} m_{I_i}, i = 1, 2$, coincide, the marginals $S_{k_1 a_1} m_{I_1}$ and $S_{k_2 a_2} m_{I_2}$ agree and so $S_{k_1 a_1} h_{I_1} = S_{k_2 a_2} h_{I_2}$. By Proposition 7.31, $I_1 = I_2 = I$. Since $H_i \in \overline{\mathfrak{a}^{I+}}$, Proposition 7.31 implies that $H_1 = H_2 = H$ and $k_2^{-1} k_1 = k_0 m, k_0 \in (K^I \cap a K^I a^{-1})$ and $m \in M$, where $a = \exp H$. Again, the equality of the other marginals $\delta_{V_i}, i = 1, 2$, implies that the directions L_1 and L_2 coincide. It remains to show that $Ad(k_1)L = Ad(k_2)L$, where $L = L_1 = L_2$

Now $S_{k_1 a} h_{I,v} = S_{k_2 a} h_{I,v}$ and $k_0 \in K^I$ implies that $S_{k_0 a} h_{I,v} = S_a h_{I,v} = f$. The gradient of f at o is $Z + L$, where Z is the gradient of $S_a \Phi^I$ at o. The gradient of $S_k f$ at o is $Ad(k)(Z + L)$. Since $S_{k_0 a} h_I = S_a h_I$, and $k_0 \in K^I$, it is clear that $S_{k_0 a} \Phi^I = S_a \Phi^I$ and so $Ad(k_0)Z = Z$. This implies that $Ad(k_0)L = L$ and, hence, $Ad(k_1)L = Ad(k_2)L$. □

This proposition also makes it possible to give a more geometric description of the Martin compactification $X \cup \partial X(\lambda)$ for $\lambda < \lambda_0$. The result is stated as the next theorem.

8.21. Theorem. *Let X be a symmetric space of non-compact type. The following compactifications are G-isomorphic:*

(1) *the Martin compactification* $X \cup \partial X(\lambda)$;
(2) $\overline{X}^c \vee \overline{X}^{SF}$, *the smallest compactification that dominates both the conical compactification* $\overline{X}^c$ *and the maximal Satake-Furstenberg compactification* $\overline{X}^{SF}$;
(3) $\overline{X}^c \vee (X \cup \partial X(\lambda_0))$, *the smallest compactification that dominates both the conical compactification* $\overline{X}^c$ *and the Martin compactification* $X \cup \partial X(\lambda_0)$; *and*
(4) $\overline{X}^c \vee (X \cup \Delta^*(X))$, *the smallest compactification that dominates both the conical compactification* $\overline{X}^c$ *and the dual cell compactification* $X \cup \Delta^*(X)$.

Recall from Chapter III (preceding Proposition 3.45) that if $(K_1, i_1) = K_1$, and $(K_2, i_2) = K_2$ are two metrizable compactifications of a locally compact space X, then there is a smallest compactification $K_1 \vee K_2$ that dominates them both (see Definition 3.27). Note that if K_1 and K_2 are G-compactifications, then $K_1 \vee K_2$ is also a G-compactification.

The proof of Theorem 8.21 makes use of the next two elementary lemmas.

8.22. Lemma. *The following are equivalent:*

(1) *a sequence* (y_n) *converges in* K_1 *whenever it converges in* K_2*;*
(2) K_2 *dominates* K_1.

8.23. Lemma. *If a sequence* (y_n) *converges in* K_1 *and* K_2*, then it converges in* $K_1 \vee K_2$.

Proof of Theorem 8.21.

In view of Theorem 7.33, it suffices to prove that $X \cup \partial X(\lambda)$ is G-isomorphic to $\overline{X}^c \vee (X \cup \partial X(\lambda_0))$. Let K_1 denote $X \cup \partial X(\lambda_0)$, the Martin compactification at the bottom of the positive spectrum, equivalently, a maximal Satake-Furstenberg compactification, and let K_2 denote the conical compactification $\overline{X}^c$, which is obtained by adjoining the unit sphere in the tangent space $T_o(X)$ at infinity (see § 3.1).

Assume that a sequence (y_n) converges in the Martin compactification $X \cup \partial X(\lambda)$ of X to a function f. Recall that, by Definition 8.6, a subsequence is said to be I-directional if $y_n = k_n a_n \cdot o$ with $a_n = \exp H_n, H_n = H_n^I + H_{n,I}$, where $(H_n^I) \subset \overline{\mathfrak{a}^{I+}}$ converges, $H_{n,I}$ is I-canonical and has a limiting direction, and (k_n) converges. Consider an I_1-directional subsequence $(y_{n_j(1)})$ and an I_2-directional subsequence $(y_{n_j(2)})$. They both converge in the Martin compactification to $f = S_{k_1 a_1} h_{I_1, v_1} = S_{k_2 a_2} h_{I_2, v_2}$, where $k_i = \lim_j k_{n_j}(i), a_i = \lim_j a_{n_j}^{I_i}(i)$, and v_i corresponds to the limiting direction L_i of $\log a_{n_j}(i)$.

It follows from Proposition 8.20 that $I_1 = I_2 = I, H_1 = H_2 = H$, and $L_1 = L_2 = L$, where H_i is the limit in $\overline{\mathfrak{a}^{I_i^+}}$, L_i is the direction in $\overline{C}_{I_i}$ associated with the sequence $(y_{n_j(i)})$, and $Ad(k_2^{-1} k_1)$ fixes H and L. Clearly, both subsequences converge in the conical compactification K_2 to the same point and it follows from Proposition 7.31 that these subsequences converge in K_1 to the same point. Consequently, both subsequences converge in $K = K_1 \vee K_2$ to the same point and so the original sequence (y_n) converges in K. Hence, by Lemmas 8.22 and 8.23, this Martin compactification dominates K.

Conversely, suppose that a sequence (y_n) converges in K. Then, by Lemma 8.22, it converges in both K_1 and K_2. Consider an I_1-directional subsequence $(y_{n_j(1)})$ and an I_2-directional subsequence $(y_{n_j(2)})$. As they converge in K_2 to the same point, it follows that $Ad(k_1)L_1 = Ad(k_2)L_2$. As they converge in K_1 to the same point, by Proposition 7.31, it follows that $I_1 = I_2, H_1 = H_2 = H$, $Ad(k_2^{-1} k_1)$ fixes H, and in addition $k_2^{-1} k_1 \in (K^I \cap a K^I a^{-1})M$, where $a = \exp H$.

Since the directions L_i are in $\mathfrak{a}_I$, it follows that $Ad(k_2^{-1}k_1)$ fixes both L_1 and L_2,. Hence, $L_1 = L_2 = L$. This implies that the limit functions of the two directional subsequences agree in the Martin compactification $X \cup \partial X(\lambda)$. Hence, (y_n) converges in this Martin compactification. In other words, K dominates this Martin compactification. □

This relation between the conical compactification, the maximal Satake-Furstenberg compactification, and the Martin compactification $X \cup \partial X(\lambda)$ can be understood, in view of Theorem 3.39, in terms of a flat F. One has the following schematic representation that corresponds to the interpretation for F of Theorem 8.21.

8.24. Theorem. *Let F be a maximal flat subspace and $\lambda < \lambda_0$. Then*

$$X(\infty) \cap \overline{F} \overset{blow-up}{\longrightarrow} \partial X(\lambda) \cap \overline{F} \overset{blow-down}{\longrightarrow} \partial X(\lambda_0) \cap \overline{F},$$

where $\overline{F}$ is the closure of F in the corresponding compactification.

For example, in the case when X is the bidisc, the flat with its Weyl chambers is the coordinate plane, i.e., the plane with orthogonal coordinate axes and the resulting four quadrants. It is clear that $X(\infty) \cap \overline{F}$ is a circle. It is "blown up" by replacing the four points at the ends of the coordinate axes on the circle at infinity by lines to get $\partial X(\lambda) \cap \overline{F}, \lambda < \lambda_0$ and leaving the four open arcs unchanged. By now collapsing the closures of these arcs one gets $\partial X(\lambda_0) \cap \overline{F}$ which is the boundary of a square (the polyhedral compactification of the coordinate plane). The following diagram illustrates this.

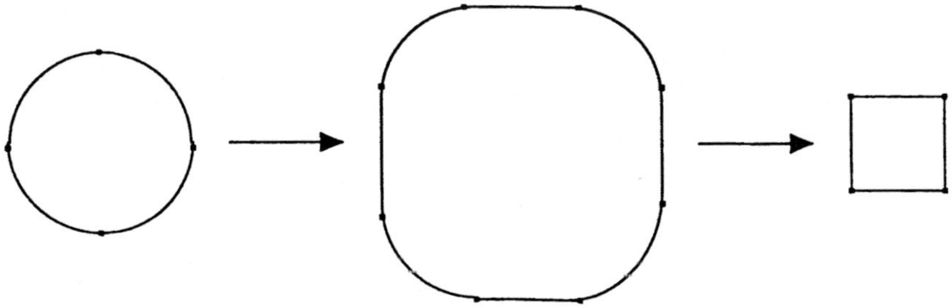

Fig. 4

Theorem 8.21 shows that there are continuous surjections $p_0 : X \cup \partial X(\lambda) \to X \cup \partial X(\lambda_0)$ and $p_c : X \cup \partial X(\lambda) \to \overline{X}^c$. In the first case, the fiber over a point $S_{ka}h_I$ in the boundary $\partial X(\lambda_0)$ consists of all the directions at infinity in $\overline{k \cdot C_I(\infty)}$. In the second case, the fiber over a direction z at infinity is the compactification at the bottom of the spectrum for the symmetric space X_z associated with this direction. When $z \in C_I(\infty)$ this symmetric space is defined to be $X^I = G^I/K^I$ (see § 2.13) that, in terms of the Langlands decomposition of $P_z = P^I = M_I A_I N_I$, equals $M_I/(K \cap M_I)$. When $z \in k \cdot C_I(\infty)$, this symmetric space is $k \cdot X^I = k \cdot G^I \cdot o$. Alternatively, it may be defined in terms of the Langlands decomposition of $P_z = kP^I k^{-1}$ as $(kM_I k^{-1}/(K \cap kM_I k^{-1})$. While this fact about the fiber can easily be deduced from Proposition 3.44, what follows is another proof, using the limit functions rather than the Tits building.

8.25. Proposition. *Let $\lambda > \lambda_0$. The fiber over a point $z \in X(\infty)$ of the natural map $X \cup \partial X(\lambda) \to \overline{X}^c$ is isomorphic to $X_z \cup \partial X_z(\lambda_0)$, where $\lambda_0 = \lambda_0(X_z)$.*

Proof. It suffices to consider the case of $z = L_1(\infty) \in C_I(\infty)$, where L_1 is a unit vector in C_I.

The direction of a point $S_{ka} h_{J,v} \in \partial X(\lambda)$ is $Ad(k)L(\infty)$, where $L \in \overline{C_J}$ is the direction associated with the linear functional v.

Since $L \in \overline{C_J}$, it follows from Lemma 3.19 that P^J stabilizes $L(\infty)$. On the other hand, $Ad(k)L = L_1 \in C_I$ and so, by Proposition 3.9, $k^{-1}P^I k \supset P^J$. Hence, $k^{-1}P^I k$ is a standard parabolic subgroup. As a result, $k \in K^I M$.

It follows that the fiber over $L_1(\infty)$ is $\{S_{ka} h_{J,v} \mid J \subset I, k \in K^I M, a \in \overline{A^{J,+}}\}$. Therefore, each point of the fiber corresponds to a triple of parameters (k, a, J). It follows from Proposition 8.20 that (k_1, a_1, J_1) and (k_2, a_2, J_2) determine the same point in the fiber if and only if $J_1 = J_2 = J$, $a_1 = a_2 = a$, and $k_2^{-1} k_1 \in (K^J \cap aK^J a^{-1})M$.

The proof is completed by comparing this description of the fiber with the description of the compactification of $X_z \cup \partial X_z(\lambda_0(X_z))$ that is implicit in Corollary 7.32. □

With this determination of the fiber over a direction at infinity and the description of the Karpelevič compactification in Chapter V, one is able to determine exactly the relation between the Karpelevič compactification and the Martin compactifications.

8.26. Theorem. *There exists a G-equivariant continuous surjective map $\overline{X}^K \to X \cup \partial X(\lambda)$, $\lambda \le \lambda_0$, that is an isomorphism if and only if the rank of X is one or the rank of X is two and $\lambda < \lambda_0$.*

Proof. First note that, it follows from Corollary 5.28 and Theorem 8.21, and from Proposition 5.27 and Theorem 7.33, that $\overline{X}^K$ dominates $X \cup \partial X(\lambda)$ for any $\lambda \le \lambda_0$. The map involved is G-equivariant. When the rank of X is one these compactifications all coincide with the conic compactification $\overline{X}^c$.

If the rank of X is two, it follows from Proposition 8.25 that the fiber in $X \cup \partial X(\lambda)$ over $z \in X(\infty)$ is the conic compactification $\overline{X_z}^c$. As a result, when the rank of X is two, the Karpelevič compactification is G-isomorphic to $X \cup \partial X(\lambda)$ provided that $\lambda < \lambda_0$. If the rank of X is greater than two, there are boundary symmetric spaces of rank greater than one. It follows from Proposition 8.25 that in this case $\overline{X}^K$ is not isomorphic to any Martin compactification. □

BOUNDED HARMONIC FUNCTIONS ON X

§ **8.27.** Consider the Martin compactification for $\lambda = 0$, i.e., the compactification of X associated with the harmonic functions on X. It is a well-known result due to Furstenberg ([F3, Theorem 4.1]) that the bounded harmonic functions h are exactly the functions that are Poisson integrals of bounded measurable functions on the Furstenberg boundary $\mathcal{F} = K/M$. In other words

$$(*)\qquad h(g\cdot o) = \int P(g,b)f(b)dm(b) = \int e^{-2\rho(H(g^{-1}k))}f(kM)dm(kM).$$

From the point of view of the Martin representation, the minimal functions that are involved are parametrized by the K-orbit, in the Martin boundary, of the minimal function $h_\rho(g\cdot o) = e^{-2\rho(H(g^{-1}))}$. They determine the Poisson kernel $P(g,b) = S_k h_\rho(g)$, where $b = kM, k \in K$. To prove Furstenberg's result it suffices, since the representing measure for a bounded harmonic function is absolutely continuous with respect to μ_1, to identify the representing measure μ_1 for the constant function 1 with the measure m on $\mathcal{F} = K/M$. In [F3], Furstenberg made use of the martingale convergence theorem to show this. It will now be proved by making use of the explicit formula, eq. (9.12), that identifies the Radon-Nikodym derivative $\frac{d(g\cdot m)}{dm}$ with $S_g h_\rho$. A more direct proof, in that it does not use the Martin compactification, is presented in Chapter XII (see Theorem 12.10).

8.28. Proposition. *The measure μ_1 is the measure on the K-orbit of h_ρ that corresponds to m on K/M under the identification of $S_k h_\rho$ with $b = kM \in K/M = \mathcal{F}$.*

Proof. It suffices to show that $\int P(g,b)dm(b) = 1$ for all $g \in G$. Let $F(g) = \int P(g,b)dm(b)$. Then $F(o) = 1$ and so the result follows once it is shown that F is a constant function.

Let $\sigma_1(g^{-1},b) = P(g,b)$. In view of eq. (9.12) and the fact that $\sigma_1 = \sigma^2$ satisfies the cocycle condition (see Lemma 10.3), one has, letting $b_1 = g\cdot b$,

$$\begin{aligned}
F(g^{-1}g_1^{-1}\cdot o) &= \int \sigma_1(g_1 g, b)dm(b)\\
&= \int \sigma_1(g_1, g\cdot b)\sigma_1(g,b)dm(b)\\
&= \int \sigma_1(g_1,b_1)\sigma_1(g,g^{-1}\cdot b_1)d(g\cdot m)(b_1)\\
&= \int \sigma_1(g_1,b_1)\sigma_1(g,g^{-1}\cdot b_1)\sigma_1(g^{-1},b_1)dm(b_1)\\
&= \int \sigma_1(g_1,b_1)dm(b_1)\\
&= F(g_1^{-1}\cdot o).
\end{aligned}$$

Note that the last equality holds because $\sigma_1(g, g^{-1} \cdot b_1)\sigma_1(g^{-1}, b_1) = \sigma_1(e, b_1) = 1$.

8.29. Corollary. *The bounded harmonic functions h on X are the functions of the form*

$$h(g \cdot o) = \int P(g, b) f(b) dm(b),$$

with $f \in L^\infty(m)$.

An application to convergence of Brownian motion

§ **8.30.** Proposition 8.28 and Šur's result on the convergence of a diffusion (see Proposition 6.2) imply that the Brownian motion on X, started from the origin o, almost surely converges to a point in the Martin compactification $X \cup \partial X(0)$ that is in the K-orbit of the minimal function h_ρ. Since the set of irregular points is a subset of a set of at least codimension 2, it is well known that the Brownian motion (X_t) is almost surely never found in the set of irregular points of X (see, for example, Taylor [T2]). This means that, one may assume X_t has unique polar coordinates $k_t M = b_t \in \mathcal{F}$ and $a_t \in A^+$. The convergence to a point of the K-orbit of the minimal function h_ρ means that $\log a_t \xrightarrow{\mathfrak{a}^+} H_\rho(\infty)$ (see Definition 3.24) and the "angular" process (b_t) on $\mathcal{F}$ converges.

This result was first proved by Dynkin [D4] for $X = \mathrm{SL}(n, \mathbb{C})/\mathrm{SU}(n)$. The convergence of $\log a_t$ to $H_\rho(\infty)$ for a general symmetric space of non-compact type was first proved by Orihara [O4]. He used stochastic comparison to show that $\frac{1}{t} \log a_t \to H_\rho$ and, thereby, determined the rate of convergence. The principal contribution of another proof by Malliavin–Malliavin [M2] (see also Taylor [T2]) was to prove the convergence of the process (b_t). Liao in [L1] obtained the results on the convergence of the diffusion by using a result of Prat [P2], enabling him to reduce convergence questions to the corresponding ones for the Markov chain determined by the heat kernel at time 1.

8.31. Remarks. These diffusion results are related to similar results for random walks on a semisimple Lie group G. In [T8] Tutubalin proved that $\frac{1}{n} \log a_n$ converges a.s. if (g_n) is a random walk on $G = \mathrm{SL}(n, \mathbb{R})$ with Cartan decomposition $g_n = k_n a_n k'_n$. He also showed that the random coset $k_n M$ in K/M converges a.s. and related these results to the results of Dynkin in [D4]. It was conjectured by Furstenberg in the sixties that, under very general conditions, the sequence $\frac{1}{n} \log a_n$ converges to an interior point of the Weyl chamber. In [T9], Tutubalin obtained the first result, under strong conditions, and an essential improvement was made by Guivarc'h–Raugi in [G15]. Constructive criteria in terms of Zariski closure were obtained by Goldsheid–Margulis in [G8]. For a recent survey of these and related results, see Kaimanovitch [K2].

CHAPTER IX

AN INTRINSIC APPROACH TO THE BOUNDARIES OF X

It is well-known that any symmetric space X of non-compact type can be realized as the space $\mathcal{S}_0$ of maximal compact subgroups by associating with $g \cdot o$ its isotropy subgroup gKg^{-1}. The space $\mathcal{S}$ of closed subgroups of G is compact in the topology of Hausdorff convergence on the compact subsets of G. As a result, the closure $\overline{\mathcal{S}_0}$ is a compactification of X.

A similar compactification is obtained by identifying each $g \cdot o$ with the group sphere $gKg^{-1} \cdot o$ and thus embedding X into the set of closed subsets of X that contain o, again a compact space with respect to the topology of Hausdorff convergence on the compact subsets of X.

In each case, these compactifications satisfy the criterion of Theorem 3.38 and so are isomorphic to the dual cell compactification, the Martin compactification $X \cup \partial X(\lambda_0)$, and the maximal Satake compactification $\overline{X}^S$ (Theorem 9.18). These compactifications are also defined for semisimple groups defined over a local field $\mathbb{F}$ (see Chapter XV).

In the proof of this result, use was made of the behavior of probabilities on $\mathcal{F} = G/P = K/M$ under the action of G. This compact homogeneous space is a boundary of G in the sense of Furstenberg [F5] and, as shown in [F5], dominates all the other boundaries of G (see Theorem 9.37). These boundaries are the homogeneous spaces G/Q where Q is a standard parabolic subgroup.

A key lemma (Lemma 9.11) in the proof of Theorem 9.18 states that the stabilizer of the unique K-invariant probability m on $\mathcal{F} = G/P$ coincides with K. As a result, the map $g \cdot o \to g \cdot m$ embeds X into the compact space $\mathcal{M}_1(\mathcal{F})$ of probabilities on $\mathcal{F}$ (Theorem 9.42). The closure $\overline{G \cdot m}$ of the orbit $G \cdot m$ is then a compactification of X. It is called the maximal Furstenberg compactification and, as proved by Moore in [M8], it is G-isomorphic to the maximal Satake compactification $\overline{X}^S$ (see Corollary 9.46 for the convergence criterion which gives a new proof of this result).

The orbital structure of $\overline{G \cdot m} = \overline{X}^F$ is determined by showing that the boundary symmetric spaces X^I can be identified with the orbits $G^I \cdot m_I$ and thus embedded in the boundary of $\overline{X}^F$. The closure of the orbit $G^I \cdot m_I$ is isomorphic to the maximal Furstenberg compactification $\overline{X^I}^F$ of X^I. The description of the orbital structure of $\overline{X}^F$ given in Theorem 9.56 and the identification of $\overline{X}^S$ with $\overline{X}^F$ gives a new proof of Satake's characterization of $\overline{X}^S$ that was stated in Proposition 4.42.

Note that in this chapter, from Theorem 9.47 onward, the Lie group G is assumed to have no proper normal compact subgroup, as in Chapter IV.

The space of closed subgroups

§ **9.1.** A symmetric space X of non-compact type may be represented in three ways as a space of compact sets. The first two present X as a subset of $\mathcal{K}(G)$, the set of compact subsets of a semisimple group G. In the first place it may be defined as the set of left K-cosets $g \cdot o = gK, g \in G$. The space X can also be identified with the set $\mathcal{S}_0 \stackrel{\text{def}}{=} \{gKg^{-1} \mid g \in G\}$ of maximal compact subgroups of G : to each point $x = g \cdot o \in X$ one associates its isotropy subgroup gKg^{-1}. This map is an injection since, if $gKg^{-1} = K$ and $g = k_1 a^{-1} k_2$, then $Ka \cdot o = a \cdot o$. By identifying X with $\exp \mathfrak{p}$, this implies that, for any $k \in K$, $Ad(k) \log a = \log a$. It follows from Lemma 3.10 that $\log a = 0$.

The third way that X may be realized as a collection of compact sets is as a space of orbits in X. To each point $x = g \cdot o$ associate the orbit $gKg^{-1} \cdot o$ of o under the isotropy group of x — a so-called **group sphere** (see [K3, p. 101]). This maps X into $\mathcal{K}(X)$. The following lemma shows that it is injective.

9.2. Lemma. *If $g_1Kg_1^{-1} \cdot o = g_2Kg_2^{-1} \cdot o$, then $g_1 \cdot o = g_2 \cdot o$.*

Proof. It suffices to show that the stabilizer S in G of $gKg^{-1} \cdot o$ is gKg^{-1}. Clearly, $S \supset gKg^{-1}$. Since $o \in gKg^{-1} \cdot o$, if $s \in S$, then $s \cdot o = gkg^{-1} \cdot o$ for some $k \in K$. Hence, $s^{-1}gkg^{-1} = \ell \in K$. Since $S \supset gKg^{-1}$, it follows that $\ell \in K \cap S$ and so $S = gKg^{-1}(K \cap S)$. This implies that S is compact and so, by maximality, $S = gKg^{-1}$. $\square$

Remarks. (1) As pointed out by A. Borel, Lemma 9.2 follows from the fact, due to Cartan, that the map $x \to \int_Z d(x,z)dz$ has a unique minimum if Z is a compact submanifold of X and dz is the volume element determined by the induced Riemannian metric on Z. Hence, if K' and K'' are both maximal compact subgroups and $K' \cdot x_0 = K'' \cdot x_0 = Z$, then both K' and K'' are subgroups of the isotropy group of ζ, the minimizing point. This implies that $K' = K''$.

(2) Lemma 9.2 may also be proved using bounded harmonic functions. By a theorem of Godement (see [G6], [F3, Theorem 4.4] and Corollary 12.9), if $g_1Kg_1^{-1} \cdot o = g_2Kg_2^{-1} \cdot o$, then $h(g_1 \cdot o) = h(g_2 \cdot o)$ for all bounded harmonic functions h. It follows from Corollary 8.29 and eq. (9.12) that $g_1 \cdot m = g_2 \cdot m$, where m is the unique K-invariant probability measure on $\mathcal{F} = K/M$. Lemma 9.11 then implies that $g_2^{-1}g_1 \in K$.

While the second realization will be the important one in this chapter, it is useful to know that all three embeddings are topological relative to a canonical topology on $\mathcal{K}(G)$, respectively, $\mathcal{K}(X)$.

Given a left invariant metric on G, there is an associated Hausdorff metric on $\mathcal{K}(G)$: one sets $d(K_1, K_2) < \varepsilon$ if every point of K_2 is within ε distance of a point in K_1 and vice versa. There is a similar metric on $\mathcal{K}(X)$ associated with the left invariant metric on X given by the Riemannian metric.

9.3. Proposition. *Let* $(g_n) \subset G$. *The following are equivalent:*

(1) $g_n \cdot o \to g \cdot o$ *in* X;
(2) $g_n K \to gK$ *in* $\mathcal{K}(G)$;
(3) $g_n K g_n^{-1} \to gKg^{-1}$ *in* $\mathcal{K}(G)$; *and*
(4) $g_n K g_n^{-1} \cdot o \to gKg^{-1} \cdot o$ *in* $\mathcal{K}(X)$.

Proof. Use the Iwasawa decomposition $G = NAK$ to write $g_n = n_n a_n k_n = s_n k_n$ and $g = nak = sk$. Then $g_n \cdot o \to g \cdot o$ if and only if $s_n \to s$.

For any $k \in K$, it is clear that $s_n k \to sk$. By compactness, it follows that $s_n K \to sK$. Hence, (1) implies (2) as $s_n K = g_n K$ and $sK = gK$.

Assume (2). Then, if $k \in K$, it follows that there is a sequence $(k_n) \subset K$ such that $s_n k_n \to sk$. It follows from the smoothness of the Iwasawa decomposition (see Helgason [H2, Theorem 5.1, p. 271]) that $s_n \to s$. This proves (1). Furthermore, since $s_n^{-1} \to s^{-1}$, (3) follows from (2).

Since the map $G \to G/K = X$ is continuous, (3) implies (4).

Assume (4). Then all the orbits $g_n K g_n^{-1} \cdot o$ and $gKg^{-1} \cdot o$ are at a bounded distance from the origin o of X. Let d denote the left invariant distance on the symmetric space X. Then $d(gkg^{-1} \cdot o, g \cdot o) = d(kg^{-1} \cdot o, o) = d(g^{-1} \cdot o, o) = d(o, g \cdot o)$ — in other words, the orbit $gKg^{-1} \cdot o$ is part of a geodesic sphere centered at $g \cdot o$ (recall that this orbit is called a group sphere). It follows that the diameter of the orbit gKg^{-1} is equal to $2d(g \cdot o, o)$. Consequently, $\{g_n \cdot o \mid n \geq 1\}$ and, hence, $\{g_n \mid n \geq 1\}$ are relatively compact.

If (g_{n_k}) is a subsequence that converges to ℓ, then $\ell K \ell^{-1} \cdot o = gK^{-1} g \cdot o$ and so, by the previous lemma, $\ell \cdot o = g \cdot o$. This proves (1) and, hence, the equivalence of all three statements. □

§ **9.4.** The representation of X as the subspace $\mathcal{S}_0$ of $\mathcal{K}(G)$ of maximal compact subgroups of G is intrinsic. The space $\mathcal{K}(G)$ is itself a subspace of the space $\mathcal{C}(G)$ of closed subsets of G, where the topology (the so-called **topology of Hausdorff convergence on compact subsets**, see Bourbaki [B14, Ch.VIII, p. 188]) may be defined as follows (see Appendix A for comments): a compact set C_0 and an open neighborhood U of the identity determine a basic open set $O(C_0, U)$; it consists of all the closed sets F that contain a compact set C such that $C \subset U(C_0)$ and $C_0 \subset U(C)$, where $U(C) = \{x \mid \text{ there exists } y \in C \text{ with } x^{-1}y \subset U\}$. The open set U may always be assumed to be the ε-ball centered at the identity element relative to a left invariant metric on G — in which case $O(C_0, U)$ will also be denoted by $O(C_0, \varepsilon)$ — and so $O(C_0, \varepsilon)$ denotes the set of closed sets F that contain a compact set C whose associated Hausdorff distance from C_0 is less than ε.

Remark. A basic neighborhood of a closed set F_0 is given by $O(C_0, \varepsilon)$, where C_0 is a compact subset of F_0 and $\varepsilon > 0$. As shown in Lemma A.16 of the appendix, this is equivalent to taking, as basic neighborhoods of F_0, the sets $P(K, \varepsilon)(F_0)$, where K is compact and $F \in P(K, \varepsilon)(F_0)$ if and only if every point of $F \cap K$ is within ε of F_0 and every point of $F_0 \cap K$ is within

ε of F. (See the definition given in Bourbaki [B14, p. 188].) Note that the topology satisfies the first axiom of countability and is even metrizable as it is given by a uniform structure with a countable base (see [B14]). Finally, observe that if $F_n \to F$ in this topology and $f_n \in F_n$ converge to f, then $f \in F$: consider any large compact set K that contains all the f_n; since $F_n \in P(K, \varepsilon)(F)$ for large enough n, it follows that $d(f, F) \leq \varepsilon$.

This topological space of closed sets has the property that the set of closed sets that all contain a given point is compact since G is a countable union of compact subsets: this is an easy consequence of the well-known fact that the space of compact subsets of a compact metric space is compact in the topology given by the Hausdorff metric. The set $\mathcal{S}$ of closed subgroups of a Lie group (even of a metrizable locally compact group) is a closed subset of the set of closed sets that contain the identity element and, hence, is compact relative to the topology of Hausdorff convergence on compact subsets. In the case of $G = \mathbb{R}$, for example, if $F_a = a\mathbb{Z}$ then F_a is continuous in a, converges to $\mathbb{R}$ as $a \to 0$, and converges to $\{0\}$ as $a \to \infty$.

Since $\mathcal{S}_0$ is a subspace of $\mathcal{S}$, it follows that the closure $\overline{\mathcal{S}_0}$ of $\mathcal{S}_0$ in $\mathcal{S}$ determines a G-compactification of X as it is clear that the group G itself acts continuously on $\mathcal{S}$ by conjugation. It will be shown that this compactification is G-isomorphic to the compactification $X \cup \Delta^*(X)$ (and, hence, the maximal Furstenberg-Satake compactification and the Martin compactification $X \cup \partial X(\lambda_0)$) and that the closed subgroups D in the closure have orbits that may be interpreted as generalized horocycles. Each such group D will be shown to be a subgroup of a corresponding parabolic subgroup (see Proposition 9.14 and Corollary 14.30) and is in fact exactly the distal part (see Definition 9.6) of the stabilizer of the corresponding Martin limit function under the action $h \to S_g h$.

Consider the case of $G = \mathrm{SU}(1,1)$ and $K = \mathrm{SU}(1)$, for which the symmetric space is the disc model of hyperbolic space (the hyperbolic disc). Suppose that $gKg^{-1} \to D$ as $g \cdot o = z$ converges to a boundary point b along a radius. The orbits $gKg^{-1} \cdot o$ are circles through the origin with geodesic center at $g \cdot o$ and as $z = g \cdot o$ converges to the boundary along a radius, these circles converge in the space of closed subsets of the hyperbolic disc to the horosphere given by the circle, through the origin, tangent to the unit circle at the boundary point b. This is the orbit of o under D. Considering the upper half space model and $\mathrm{SL}(2, \mathbb{R})$, this is equivalent to stating that the circles through i with a diameter on the imaginary axis converge to the line through i parallel to the real axis as the center goes to infinity.

This illustrates the general fact, proved as Theorem 9.19 and mentioned in the introduction, that the maximal Satake–Furstenberg compactification of X is isomorphic to the closure in $\mathcal{K}(X)$ of the set of closed group spheres $gKg^{-1} \cdot o, g \in G$, where the group sphere $gKg^{-1} \cdot o$ is identified with the point $g \cdot o$.

In order to identify the compactification of X given by the closure of $\mathcal{S}_0$

in $\mathcal{S}$, one now proceeds to the identification of the groups that are limits of sequences $(g_n K g_n^{-1}) \subset \mathcal{S}_0$.

LIMIT GROUPS

Assume that F is a limit group, i.e., for some sequence (g_n), $g_n K g_n^{-1} = K_n \to F$. Then, each $g \in F$ is the limit of a sequence $(\ell_n) \subset K_n$. Since the spectrum of an element from a compact linear group is a subset of the unit circle, and since the spectrum of $Ad(g)$ is the limit in $\mathcal{K}(\mathbb{C})$ of the spectra of $Ad(\ell_n)$, it follows that the spectrum of $Ad(g)$ is also a subset of the unit circle. This property of the spectrum of a linear transformation is the key to determining the set $\overline{\mathcal{S}_0} \setminus \mathcal{S}_0$ of limit groups.

In order to do this it is important to characterize it in terms of the orbital action of the transformation on the underlying vector space.

9.5. Proposition. *Let V denote a real finite dimensional vector space and let $g \in GL(V)$. The following properties are equivalent:*

(1) *if $x \in V$ is non-zero then $0 \notin \overline{\{g^n \cdot x \mid n \in \mathbb{Z}\}}$,*
(2) *the spectrum of g is a subset of the unit circle.*

Proof. Assume that, for some $x \neq 0$, $0 \in \overline{\{g^n \cdot x \mid n \in \mathbb{Z}\}}$. Let W denote the subspace of V generated by $\{g^n \cdot x \mid n \in \mathbb{Z}\}$. Let $e_k = g^{n_k}x$ of $W (1 \leq k \leq r)$ denote a basis of W. Define a norm $\|y\|$ in W by setting $\|y\| = \sup_{1 \leq k \leq r} |y_k|$ if $y = \sum_1^r y_k e_k$ and let $M = \sup_{1 \leq k \leq r} \|g^{n_k}\|$.

Since $0 \in \overline{\{g^n \cdot x \mid n \in \mathbb{Z}\}}$, for some $n \in Z$, $\|g^n x\| \leq \frac{1}{2rM}$. This implies that $\|g^n \cdot e_k\| = \|g^{n_k} g^n \cdot x\| \leq M \|g^n \cdot x\| \leq \frac{1}{2r}$ and, hence, $\|g^n \cdot y\| \leq \|y\| \left(\sum_{k=1}^r \|g^n \cdot e_k\|\right) \leq \frac{1}{2}\|y\|$. It follows that $\|g^n\| \leq \frac{1}{2}$. This implies that the spectrum of g^n is contained in $\{z \in \mathbb{C}; |z| \leq \frac{1}{2}\}$. As a result (2) is false.

Let $V_{\mathbb{C}}$ denote the complexification of V and let g also denote the canonical extension of g to $V_{\mathbb{C}}$. Let $v = x + iy \in V_{\mathbb{C}}, x, y \in V$, be an eigenvector of g, i.e., $g \cdot v = zv$ for some $z \in \mathbb{C}$. If $|z| \neq 1$ one can assume, using g^{-1} if necessary, that $|z| < 1$, i.e., $z = \rho e^{i\theta}$ with $0 < \rho < 1$. Since $g \cdot (x + iy) = g \cdot x + ig \cdot y$, it follows that $\overline{g \cdot v} = g \cdot \bar{v}$ and, so, $g \cdot \bar{v} = \bar{z}\bar{v}$. As a result, $g \cdot x = \rho(x \cos\theta - y \sin\theta)$ and $g \cdot y = \rho(x \sin\theta + y \cos\theta)$.

If $v \in V$ then $\|g^n \cdot v\| = \rho^n \|v\| \to 0$ and so (1) is false. Similarly, if $iv \in V$ the same conclusion holds. If neither v nor iv lie in V, then x and y generate a subspace W of V that is invariant under g. Since $0 < \rho < 1$, it follows that $\|g^n \cdot u\| \to 0$ for all $u \in W$, in which case (1) is again false. □

This property of g is called distality and comes from topological dynamics (see [C4]).

9.6. Definition. Let V denote a real finite-dimensional vector space. A subgroup H of GL(V) is said to have a **distal action** on V if the spectrum of each $h \in H$ is a subset of the unit circle. In the case of a Lie group, a closed subgroup H of G is said to be a **distal subgroup** if $Ad(H)$ acts distally on $\mathfrak{g}$.

9.7. Lemma. *Let (F_n) be a sequence of closed subgroups F_n of a locally compact metrizable group H and let (μ_n) be a sequence of probability measures on a locally compact metrizable H-space Y. Assume that*

(1) *each μ_n is F_n-invariant;*
(2) *the sequence (μ_n) converges to μ; and*
(3) *the sequence (F_n) converges to F.*

Then μ is F-invariant.

Similarly, if (f_n) is a sequence of functions on Y that converges uniformly on compact subsets to a function f, then f is F-invariant if each f_n is F_n invariant and the sequence (F_n) converges to F.

Proof. Let $g \in F$. It follows from the remark in § 9.4 that there is a sequence (g_n) with $g_n \in F_n$ and $g = \lim_n g_n$. Let φ be a continuous function on Y that has compact support. Then $\varphi(g_n \cdot x)$ converges uniformly to $\varphi(g \cdot x)$ and, hence, $\int \varphi(g_n \cdot x)\mu_n(dx) \to \int \varphi(g \cdot x)\mu(dx)$. Consequently, $g_n \cdot \mu_n$ converges weakly to $g \cdot \mu$, and so (1) implies the F-invariance of μ.

The argument for functions is similar. If $x \in Y$, the set $C = \{x\} \cup \{g_n \cdot x \mid n \geq 1\} \cup \{g \cdot x\}$ is compact and, since $f_n(x) = f_n(g_n \cdot x) \to f(g \cdot x)$ as $n \to \infty$, it follows that $f(g \cdot x) = f(x)$. □

The first step in determining the limit groups involves the action of I-canonical sequences on the invariant measure m on $\mathcal{F}$.

9.8. Proposition. (Cf. Lemma 3.10) *Let $(a_n) \subset \overline{A^+}$ be an I-canonical sequence. Then $(a_n \cdot m)$ converges to m_I. Hence, the centralizer of $H \in C_I$ in K is $K^I M$.*

In particular, if $I = \emptyset$ and $(a_n) \subset \overline{A^+}$ is a $\emptyset$-canonical sequence, then $(a_n \cdot m)$ converges to $m_\emptyset = \delta_{\dot{e}}$.

Proof. Let $\overline{N}_I$ denote the nilpotent group with Lie algebra $\overline{\mathfrak{n}}_I = \theta(\mathfrak{n}_I) = \sum_{\alpha \in \Sigma_I^+} \theta(\mathfrak{g}_\alpha)$. Let $\alpha \in \Sigma_I^+$ and $X \in \theta(\mathfrak{g}_\alpha) = \mathfrak{g}_{-\alpha}$. Since (a_n) is I-canonical, it follows that $\lim_n Ad(a_n)X = \lim_n e^{-\alpha(\log a_n)}X = 0$. Hence, for any $\overline{n}_2 \in \overline{N}_I$, $\lim_n a_n \overline{n}_2 a_n^{-1} = e$. Note that A acts on $\overline{N}_I$ by conjugation. The Bruhat cellular decomposition (Corollary 2.21 and also Helgason [H2, Corollary 1.9, p. 407]) shows that the orbit of $\dot{e} = P$ in $\mathcal{F} = G/P$ under $\overline{N}$ is open and dense. In addition $m(\overline{N} \cdot \dot{e}) = 1$ (see Helgason [H3, formula in Theorem 5.20, p. 198]).

Also, by Corollary 2.21, the orbit under $\overline{N}_I$ of the origin $\dot{e}_I = P^I$ in G/P^I is also open and dense with $\overline{m}(\overline{N}_I \cdot \dot{e}_I) = 1$ if $\overline{m}$ is the image of m under the quotient map $G/P \to G/P^I$.

The embedding $\overline{n}_2 \to \overline{n}_2 \cdot \dot{e}_I$ is equivariant with respect to the action by conjugation of A on $\overline{N}_I$ and the natural left action of A (as a subgroup of G) on the coset space G/P^I.

Since (a_n) is I-canonical, it follows from what has been proved and the theorem of dominated convergence that the sequence of measures $a_n \cdot \overline{m}$ converges to the Dirac measure $\delta_{\dot{e}_I}$ at the origin $\dot{e}_I$. This means that every cluster value in $\mathcal{M}_1(\mathcal{F})$ of the sequence $(a_n \cdot m)$ has its support contained in

$P^I \cdot \dot{e}$ — the orbit under P^I of $\dot{e} = P$. Furthermore, since each a_n commutes with K^I, every cluster value is a K^I-invariant probability measure on the orbit $P^I \cdot \dot{e}$. The fact that m_I is the unique K^I-invariant probability on $P^I \cdot \dot{e}$ implies the result.

Finally, if $k \in K$ centralizes $H \in C_I$ and $a_n = \exp nH$ is I-canonical, then $ka_n \cdot m = a_n k \cdot m = a_n \cdot m$ and so $k \cdot m_I = m_I$. Hence, k stabilizes the support $K^I \cdot \dot{e}$ of m_I, i.e., if $\ell \in K^I, k\ell = \ell_1 k'$ with $k' \in M, \ell_1 \in K^I$. Thus, $k \in K^I M$. $\square$

9.9. Lemma. *Let $(a_n) \subset \overline{A^+}$ be I-canonical. Assume that $(a_n K a_n^{-1})$ converges to $D \in S$. Then $D \supset K^I M N_I$.*

Proof. Since A_I centralizes $K^I M$, it follows that $a_n K a_n^{-1} \supset K^I M$. Note that, since (a_n) is I-canonical, if $y \in N_I$ then $\lim_n a_n^{-1} y a_n = e$.

Let $a_n^{-1} y a_n = k_n a_n' \overline{y}_n$ with $k_n \in K, a_n' \in A$ and $\overline{y}_n \in \overline{N}$. Since $a_n^{-1} y a_n \to e$, these Iwasawa components (relative to $G = KA\overline{N}$) also converge to e. Now $\overline{y}_n = \overline{y}_{1,n}\overline{y}_{2,n}$, with $\overline{y}_{1,n} \in \overline{N}^I$ and $\overline{y}_{2,n} \in \overline{N}_I$, implies that $a_n \overline{y}_n a_n^{-1} = \overline{y}_{1,n} a_n \overline{y}_{2,n} a_n^{-1} \to e$ as $\overline{y}_{1,n} \to e$. Hence, $y = \lim_n a_n k_n a_n^{-1}$ and so, by the remark in § 9.4, $D \supset N_I$. $\square$

9.10. Lemma. *$K^I M$ is the stabilizer of m_I in $M_I = G^I M$ under the action $g'P \to gg'P \stackrel{\text{def}}{=} g \cdot (g'P)$.*

Recall that in the proof of Theorem 7.22, the Furstenberg boundary $\mathcal{F}^I = K^I / M^I$ of G^I is identified with the orbit of $\dot{e} = M$ in $\mathcal{F}$ under K^I: this amounts to identifying the orbit of P in G/P under P^I with $\mathcal{F}^I$. The measure m_I is the unique K^I-invariant probability measure on this orbit and corresponds, under the identification of $\mathcal{F}^I$ with the orbit, to the unique K^I-invariant probability measure m^I on $\mathcal{F}^I$. As a result, the proof of this lemma reduces to proving the corresponding statement for $I = \Delta$. This is formulated as the following result.

9.11. Lemma. *K is the stabilizer of m in G under the action of G on K/M given by $kM \to k(gk)M \stackrel{\text{def}}{=} g \cdot (kM)$.*

Proof. In Helgason [H3, p. 197], the following formula is given for the Radon-Nikodym derivative of $g \cdot \tilde{m}$ with respect to the normalized Haar measure $\tilde{m}$ on K, where $g \cdot \tilde{m}$ is the image of this Haar measure under the map given by the Iwasawa action $k \to k(gk)$:

$$\int_K F(k(gk)) d\tilde{m}(k) = \int_K F(k) e^{-2\rho(H(g^{-1}k))} d\tilde{m}(k).$$

Consequently, as m is the image of $\tilde{m}$ under the canonical projection of K onto K/M and $H(g^{-1}kx) = H(g^{-1}k)$ for any $x \in M$, it follows that the Radon-Nikodym derivative of $g \cdot m$ with respect to m is

$$\frac{d(g \cdot m)}{dm}(kM) = e^{-2\rho(H(g^{-1}k))}. \tag{9.12}$$

To determine the stabilizer of m amounts to determining those $g \in G$ such that $\rho(H(g^{-1}k)) = 0$ for all $k \in K$. The Cartan decomposition implies that $g = k_1 a k_2$ with $a \in \overline{A^+}$. Then $H(g^{-1}k_1) = -\log a$. If $\rho(\log a) = 0$, then $\log a = 0$ as $\log a \in \overline{\mathfrak{a}^+}$. Hence, $g \in K$. □

9.13. Lemma. *The stabilizer in* G *of the measure* m_I *under the natural action* $g'P \to gg'P$ *is* $R^I = K^I M A_I N_I$. *This group is also the normalizer of the subgroup* $D^I = K^I M N_I$.

Proof. As pointed out in the proof of Proposition 7.29, since the support of m_I is $P^I \cdot \dot{e}$, it follows that if $g \in G$ stabilizes m_I, then $g \in P^I$.

Since, by Theorem 2.8, P^I is the normalizer of $\mathfrak{n}_I$ and G^I centralizes A_I, it follows that P^I normalizes $A_I N_I$. Since $K^I M$ normalizes N_I, it follows that it normalizes $A_I N_I$ and so, as in the proof of Theorem 7.22, R^I is a group. This also implies that D^I is a subgroup.

Since P^I normalizes $A_I N_I$, the subgroup $A_I N_I$ acts trivially on $P^I \cdot \dot{e}$, as $gp \cdot \dot{e} = pp^{-1}gp \cdot \dot{e} = p \cdot \dot{e}$ if $g \in A_I N_I$ and $p \in P^I$. Hence, if L is the stabilizer of m_I in G, it follows that $R^I \subset L \subset P^I$.

It follows from Lemma 9.10 that $L \cap M_I$ is a compact subgroup. By what has been proved, it contains $K^I M$, which is maximal compact in $G^I M = M_I$. Hence, $L \cap M_I = K^I M$. The Langlands decomposition (Corollary 2.16), $P^I = M_I A_I N_I$, implies that $L \cap P^I = R^I$. Since $R^I \subset L \subset P^I$, it follows that $L = R^I$.

Let L denote the normalizer of D^I. Then L also normalizes its nilpotent radical N_I. Hence, $L \subset P^I$. Furthermore, $A_I \subset L$ and so $R^I \subset L$.

The probability measure m_I is the unique probability measure on the orbit $P^I \cdot \dot{e}$ stabilized by the group D^I. Since D^I stabilizes $g \cdot m_I$ if $g \in L$, this implies that $L \subset R^I$. □

Remark. This result is to be compared with Lemma 7.30 that shows that the same group is also the stabilizer of h_I in P^I (and, hence, in G). As will be shown in Corollary 10.7, this amounts to saying that the stabilizer in G of m_I under the twisted action coincides with its stabilizer under the natural action.

9.14. Proposition. *If* (a_n) *is* I*-canonical, then* $(a_n K a_n^{-1})$ *converges to* $D^I = K^I M N_I$.

Proof. Assume that a subsequence $(a_{n_k} K a_{n_k}^{-1})$ converges to a closed group D. Then by Lemma 9.9, $D \supset D^I$. In addition, by Lemma 9.7 and Proposition 9.8, D stabilizes m_I and so $D \subset R^I$.

Proposition 9.5 and Definition 9.6 imply that the group D is distal. This implies that $D \cap A_I = \{e\}$ as no element $a \in A_I$ acts distally other than e: if $0 \neq X \in \mathfrak{g}_\alpha$, then $Ad(a)X = e^{\alpha(\log a)}X$; and for some $\alpha \in \Sigma_I^+$, $\alpha(\log a) \neq 0$ if $e \neq a \in A_I$. Since $R^I = K^I M A_I N_I$, this implies that $D = D^I$. □

To compute the set of limit points in $\mathcal{S}$ of the image $\mathcal{S}_0$ of X under the embedding $g \cdot o \to gKg^{-1}$, it suffices, as in § 7.17, to know the limit of I-canonical sequences.

Then, if (g_n) is a sequence with $(g_n \cdot o)$ a C_I-fundamental sequence — i.e., $g_n \cdot o = k_n a_n \cdot o$ where $k_n \to k$ and $(a_n) \subset \overline{A^+}$ is such that $a_n = a_n^I a_{n,I}$ with $a_n^I \to a^I \in A^I$ and $(a_{n,I})$ an I-canonical sequence — it follows that $g_n K g_n^{-1} \to k a^I D^I a^{I-1} k^{-1}$. Since, as shown in § 7.17, every sequence has a C_I-fundamental subsequence for some I, this determines the set of limit groups.

9.15. Corollary. (Cf. Proposition 7.26) *The ideal boundary $\overline{S_0}\backslash S_0$ of the compactification $\overline{S_0}$ of X is the disjoint union of the G-orbits, under conjugation, of the groups D^I. The orbit of $D^\emptyset = MN$ is the unique compact one. It is of minimum dimension.*

Proof. The last two assertions are left to the reader. It remains to show that if $D^{I_1} = gD^{I_2}g^{-1}$, then $I_1 = I_2 = I$. Note that $D^{I_1} = gD^{I_2}g^{-1}$ implies that the nilpotent radicals N_{I_1} and $gN_{I_2}g^{-1}$ agree. Hence the normalizers of the nilpotent radicals coincide, i.e., $P^{I_1} = gP^{I_2}g^{-1}$. Consequently, by Proposition 2.18, $I_1 = I_2 = I$ and $g \in P^I$. By Lemma 9.13, $g \in R^I$. □

To determine the compactification $\overline{S_0}$, one proceeds, as in Chapter VII, to show how the parameters (k, a^I, I) of the **polar realization of the limit group** $D = ka^I D^I a^{I-1} k^{-1}$ are determined by D.

9.16 Proposition. (Cf. Proposition 7.31) *Let*

$$D = k_1 a_1^{I_1} D^{I_1} (a_1^{I_1})^{-1} k_1^{-1} = k_2 a_2^{I_2} D^{I_2} (a_2^{I_2})^{-1} k_2^{-1},$$

with $a_i^{I_i} \in \overline{\mathfrak{a}^{I_i,+}}$ and $k_i \in K$. Then

$$I_1 = I_2 = I,\ a_1^{I_1} = a_2^{I_2} = a \text{ and } k_2^{-1} k_1 \in (K^I \cap aK^I a^{-1})M.$$

Hence, the parameters $(k, a, I), k \in K, a \in \overline{\mathfrak{a}^{I,+}}$ of a polar realization $kaD^I a^{-1} k^{-1} = D$ of a limit group D are unique modulo $k_0 \in (K^I \cap aK^I a^{-1})M$.

Proof. The nilpotent radical of D is $k_1 N_{I_1} k_1^{-1} = k_2 N_{I_2} k_2^{-1}$ since, for any $a \in A$ and I, $aN_I a^{-1} = N_I$. As a result, $P^{I_1} = kP^{I_2}k^{-1}$, with $k = k_1^{-1}k_2$. By Proposition 2.18 this implies that $I_1 = I_2 = I$ and $k \in K^I M$.

Denote $a_1^{I_1}$ by a_1 and $a_2^{I_2}$ by a_2. Since $a_1 D^I a_1{}^{-1} = ka_2 D^I a_2{}^{-1} k^{-1}$, it follows from Lemma 9.13 that $a_1{}^{-1} k a_2 \in R^I$. Since $a_1^{-1} k a_2 \in G^I M = M_I$, this implies that $a_1^{-1} k a_2 = \ell \in M_I \cap R^I = K^I M$. As a result, $ka_2 \ell^{-1} = a_1$ and so by the uniqueness of the $\overline{A^+}$-component of the Cartan decomposition of a_1, one has $a_1 = a_2$. Hence, $k \in aK^I a^{-1} M$. □

As in Chapter VII, this result has as corollary the following criterion for the convergence of fundamental sequences.

9.17. Corollary. (Convergence criterion) *Let (y_n^1) and (y_n^2) be two fundamental sequences with*

$$y_n^1 = k_n^1 a_n^{I_1} a_n^1 \cdot o \text{ and } y_n^2 = k_n^2 a_n^{I_2} a_n^2 \cdot o,$$

where

$$k_n^i \to k_i, \quad a_n^{I_i} \to a^{I_i} \text{ and } (a_n^i) \text{ is } I_i\text{-canonical.}$$

If $g_n^1 \cdot o = y_n^1$ *and* $g_n^2 \cdot o = y_n^2$*, the two sequences* $(g_n^1 K (g_n^1)^{-1})$ *and* $(g_n^2 K (g_n^2)^{-1})$ *of closed groups converge in* $\mathcal{S}$ *to the same limit group* D *if and only if*

(1) $I_1 = I_2 = I$;
(2) $a^{I_1} = a^{I_2} = a$; *and*
(3) $k_2^{-1} k_1 \in (K^I \cap a K^I a^{-1}) M$.

By Lemma 3.36(3), Theorem 3.38, and Corollary 7.32, the convergence criterion in Corollary 9.17 is exactly the same as for the dual cell compactification $X \cup \Delta^*(X)$ of X and the Martin compactification $X \cup \partial X(\lambda_0)$. This implies the following result.

9.18. Theorem. *The following compactifications are G-isomorphic.*

(1) *The dual cell compactification* $X \cup \Delta^*(X)$;
(2) *the maximal Satake compactification* $\overline{X}^S$;
(3) *the Martin compactification* $X \cup \partial X(\lambda_0)$; *and*
(4) *the compactification* $\overline{\mathcal{S}_0}$ *of* X.

Proof. First observe that a sequence $(g_n K g_n^{-1})$ converges in $\mathcal{S}$ if and only if, for every subsequence (g_{n_k}) such that $(g_{n_k} \cdot o)$ is fundamental, the limit of the sequence $(g_{n_k} K g_{n_k}^{-1})$ — that exists by virtue of Proposition 9.14 — is independent of the subsequence.

It follows from Lemma 3.28 that the compactification $\overline{\mathcal{S}_0}$ is isomorphic to the Martin compactification $X \cup \partial X(\lambda_0)$. Let $\varphi : X \cup \partial X(\lambda_0) \to \overline{\mathcal{S}_0}$ be the homeomorphism such that $\varphi(g \cdot o) = g K g^{-1}$ for all $g \in G$ (i.e., φ is G-equivariant) — its existence is guaranteed by Lemma 3.28.

It remains to show that φ interlaces the action of G. For this it suffices to show that if $h = \lim_n g_n \cdot o \in \partial X(\lambda_0)$ and $D = \lim_n g_n K g_n^{-1} = \varphi(h)$ then $\varphi(S_g h) = g D g^{-1}$: since $g g_n \cdot o \to S_g h$ and $g g_n K g_n^{-1} g^{-1} \to g D g^{-1}$, it follows that $\varphi(S_g h) = g D g^{-1}$.

The identification with the dual cell compactification follows from Theorem 7.33. □

Limits of group spheres

The following theorem shows that $\overline{\mathcal{S}_0}$ can be described in terms of the behavior of group spheres, as mentioned at the end of § 9.4. Recall that a group sphere is the orbit of the point $o = K \in X$ under the isotropy group gKg^{-1} of a point $g \cdot o$ of X, equivalently, it is the orbit of the point $o = K \in X$ under a maximal compact subgroup of G.

9.19. Theorem. *Let* $C(X)$ *denote the set of closed subsets of* X *endowed with the topology of Hausdorff convergence on compact sets. Then the map* π *of* $\overline{\mathcal{S}_0}$ *into* $C(X)$ *given by* $\pi(D) = D \cdot o$ *is an embedding and* π *is a homeomorphism onto its image.*

In particular, the point $g_n \cdot o$ converges to D in $\overline{S}_0$ if and only if the group sphere $g_n K g_n^{-1} \cdot o$ converges to $D \cdot o$ in $\mathcal{C}(X)$.

The following proposition and its corollary, Corollary 9.24 are used to prove this result.

9.20. Proposition. *If $D \in \overline{S}_0$ equals $gD^I g^{-1}$, then $g \cdot m_I$ is the unique probability measure on $\mathcal{F} = K/M$ that is invariant under the action of D.*

9.21. Lemma. (See Abels [A1]) *Assume that V is a finite dimensional real vector space of dimension d, $H \subset \mathrm{SL}(V)$ is an unbounded subgroup of $\mathrm{SL}(V)$, and ν is an H-invariant measure on the projective space $P(V)$. Then ν is supported by the union of two projective proper subspaces of $P(V)$.*

Proof. Suppose $g_n \in H$ is such that $\lim_n \|g_n\| = +\infty$. Let u_n denote $\frac{g_n}{\|g_n\|}$. One can assume that $\lim_n u_n = u \in \operatorname{End} V$ with $\|u\| = 1$. Since $\det u = \lim_n \frac{1}{\|g_n\|^d} \det g_n = 0$,it follows that $\ker u \neq 0$ and $\operatorname{im} u \neq V$. If L, L' are the projective subspaces corresponding, respectively, to $\ker u$ and $\operatorname{im} u$, it follows that $\nu(L \cup L') = 1$. To see this, note that if $x \notin L$ one has $\lim_n g_n \cdot x = u \cdot x \in L'$. Hence, if ν_1, ν_2 denote the restriction of ν to L and its complement, one has, by Lemma 4.39, that $\lim_n g_n \cdot \nu_2 = u \cdot \nu_2$.

On the other hand, one can suppose, by extracting a subsequence if necessary, that $g_n \cdot \nu_1$ converges. The limiting measure is supported by a subspace L_1 of $P(V)$ of dimension equal to dim L. Hence,

$$\lim_n g_n \cdot \nu = \lim_n g_n \cdot (\nu_1 + \nu_2) = u \cdot \nu_2 + \lim_n g_n \cdot \nu_1$$

is supported by $L_2 \cup L_1$. Since $g_n \cdot \nu = \nu$, it follows that $\nu(L_1 \cup L_2) = 1$. □

9.22. Lemma. *Suppose H is a subgroup of $\mathrm{GL}(V)$ such that each element is unipotent (i.e., the sum of the identity and a nilpotent element) and ν is an H-invariant measure on $P(V)$. Let W denote the subspace of H-invariant vectors of V. Then ν is supported by the projective subspace $P(W)$.*

Proof. One proceeds by induction on $\dim V$. If $\dim V = 1, H = \{e\}$ and the lemma is trivial. Suppose $\dim V > 1$ and observe that for every $h \in H$ one has $\det h = 1$ because h is unipotent. If $h \neq \{e\}$, it is clear that h^n is unbounded. Hence, by Lemma 9.21 one knows, that ν is supported on a finite union of projective subspaces. There exists, among such finite unions of subspaces that carry ν, a minimum element because of the finite intersection property. The invariance of ν then implies that this finite union is H-invariant. Hence, the subgroup $H' \subset H$ that leaves each of these subspaces invariant is of finite index in H. The induction hypothesis applies to H' and to each of the previous subspaces: each vector of the corresponding vector subspaces is H'-invariant. One can restrict H to the sum W of these subspaces where H' acts trivially. The unipotency of H then implies that H acts trivially on this subspace. Hence, $\nu(P(W)) = 1$. □

9.23. Lemma. *The unique fixed point of the action of N_I on G/P^I is the coset P^I, i.e., the origin.*

Proof. Assume that $N_I \cdot \dot{g} = \dot{g}$, where $\dot{g} = gP^I \in G/P^I$. This implies that $N_I \subset gP^I g^{-1}$. It follows from Corollary 14.30 that there is a minimal parabolic subgroup hPh^{-1} contained in both P^I and $gP^I g^{-1}$. Since $hPh^{-1} \subset P^I$, it follows from Proposition 2.18 that $h^{-1}P^I h = P^I$ and $h \in P^I$. For the same reason, $hPh^{-1} \subset gP^I g^{-1}$ implies that $g^{-1}h \in P^I$ and so $g \in P^I$. □

Proof of Proposition 9.20.

Since, by Proposition 4.27, there is a representation τ with $P = P_\tau$ and V_τ one-dimensional, one can identify G/P with the G-orbit in $P(V)$ of the direction corresponding to $V_\tau = \mathbb{C}v_\tau$. If ν is a D^I-invariant measure on $G/P \subset P(V)$, it is also N_I-invariant. Hence, from Lemma 9.22 it follows that its support corresponds to N_I-invariant vectors in V. Consequently, ν is supported on the subset of G/P of N_I-invariant points. Project this subset into G/P^I and apply Lemma 9.23. It follows that ν is supported on $P^I/P \subset G/P$. The D^I invariance of ν implies its K^I invariance. Hence, $\nu = m_I$. □

Remark. One of the keys to this proof of Theorem 9.19 is Proposition 9.20. If one wants to make use of more analytic arguments than those given above, then there is a shorter and more direct proof of Proposition 9.20 that makes use of the integral representation of the non-negative solutions of $Lu + \lambda_0 u = 0$.

An alternate proof of Proposition 9.20.

It suffices to prove this result when $g = e$. Recall that, by Theorem 7.22, the measure m_I is the representing measure for the function h_I that is K^I and N_I invariant. Since m_I is K^I-invariant, it suffices to show that it is N_I-invariant.

The formula for h_I is

$$(*) \qquad h_I(g_1 \cdot o) = \int h_b(g_1 \cdot o) m_I(db) = \int e^{-\rho(H(g_1^{-1}k))} m_I(db),$$

where $b = kM$. If $n_2 \in N_I$, the fact that $h_I(n_2 \cdot x) = h_I(x)$ for all $x \in X$ implies that

$$\int e^{-\rho(H(g_1^{-1}k))} m_I(db) = \int e^{-\rho(H(g_1^{-1}n_2^{-1}k))} m_I(db)$$

for all $g_1 \in G$. Applying the cocycle formula eq. (10.4) it follows that $H(g_1^{-1}n_2^{-1}k) = H(g_1^{-1}k(n_2^{-1}k)) + H(n_2^{-1}k)$.

When $k \in K^I$ it follows that $H(n_2^{-1}k) = 0$ as $K^I M$ is a subgroup of P^I and, so, normalizes N_I.

$$\begin{aligned}\int e^{-\rho(H(g_1^{-1}k))} m_I(db) &= \int e^{-\rho(H(g_1^{-1}n_2^{-1}k))} m_I(db)\\ &= \int e^{-\rho(H(g_1^{-1}(n_2^{-1}\cdot k))} m_I(db)\\ &= \int e^{-\rho(H(g_1^{-1}k)}(n_2^{-1}\cdot m_I)(db)\end{aligned}$$

where $n_2^{-1} \cdot k = k(n_2^{-1}k)$ and $b = kM$. It follows from eq. (*) and the uniqueness of the integral representation that m_I is N_I-invariant.

Conversely, if μ is a probability on $\mathcal{F} = K/M$ that is K^I and N_I invariant, the function

$$h(g_1 \cdot o) = \int e^{-\rho(H(g_1^{-1}k))} \mu(db)$$

is K^I-invariant. It is also N_I-invariant since the proof that m_I is N_I-invariant also shows that

$$h(n_2 g_1 \cdot o) = \int e^{-\rho(H(g_1^{-1}n_2^{-1}k))} \mu(db) = \int e^{-\rho(H(g_1^{-1}k))} (n_2^{-1} \cdot \mu)(db).$$

Consequently, it follows from Theorem 7.22 that $h = h_I$. This implies that $\mu = m_I$. □

Lemma 9.24. *Assume that U and H are closed subgroups of $GL(V)$. If U is unipotent and H/U is compact, then H acts distally on V.*

Proof. Since U is unipotent, by appropriate choice of basis, it can be realized in strictly triangular form. Therefore, it acts distally on V.

Choose a norm $\|\cdot\|$ on V. Denote by $C \subset H$ a compact set such that $H \subset CU$. Clearly, there exist two constants $\alpha, \beta > 0$ such that $\alpha\|v\| \leq \|Cv\| \leq \beta\|v\|$, for every non-zero $v \in V$and $c \in C$. Hence, if $v \neq 0$ it follows that, for all $c \in C$, $\|cg \cdot v\| \geq \alpha\|g \cdot v\| \geq \alpha\epsilon > 0$ if $\epsilon = \inf\{\|g \cdot v\| \mid g \in U\}$. It follows that the closure of $CU \cdot v$ does not contain 0 and so, by Proposition 9.5, H acts distally on V. □

9.25. Lemma. *Assume that $H \subset G$ is a closed subgroup that is distal, i.e., acts distally on $\mathfrak{g}$. Then H is unimodular.*

Proof. Let H_0 denote the connected component of e in H. It is a Lie group with Lie algebra $\mathfrak{h}$. The Jacobian of the action of $h \in H$ on H_0 by conjugacy is the determinant of the restriction of $Ad(h)$ to $\mathfrak{h}$. Since h acts distally, this determinant has modulus one. Hence, if η is a left Haar measure on H_0 and $h \in H$, it follows that $\delta_h * \eta * \delta_{h^{-1}} = \eta$. This implies, in particular, that H_0 is unimodular.

If $\eta' = \sum_{g \in H/H_0} \delta_g * \eta$, where g varies over a set of representatives of the elements in the denumerable group H/H_0, then $\eta' * \delta_h = \eta'$ if $h \in H$, since $\eta * \delta_h = \delta_h * \eta$.

On the other hand $\delta_h * \eta' = \sum_{g \in H/H_0} \delta_{hg} * \eta = \sum_{g' \in H/H_0} \delta_{g'} * \eta$, where g' varies over a set of representatives in H/H_0. Hence, $\delta_h * \eta' = \eta'$ if $h \in H$. As a result, η' is a Haar measure on H and so H is unimodular. □

9.26. Corollary. *For every $D \in \overline{S_0}$, the stabilizer of the set $D \cdot o \subset X$ in G is equal to D.*

Proof. Let H denote the subgroup that leaves $D \cdot o$ invariant. Clearly, $D \subset H$ and, as $h \cdot o = d \cdot o$ implies that $d^{-1}h \in K$, it follows that $H = D(H \cap K)$. Consider the adjoint action of H on the Lie algebra $\mathfrak{g}$ of G. Since D acts distally on $\mathfrak{g}$ and H/D is unimodular, it follows from Lemmas 9.24 and 9.25 that H is distal and unimodular. As a result, the compact homogeneous space H/D has a finite H-invariant measure θ.

The measure ν on $\mathcal{F}$ defined by $\nu = \int_{H/D} h \cdot m_I d\theta(h)$ is then well-defined and H-invariant. It is also D-invariant and so, by Proposition 9.20, it follows that $\nu = m_I$. Lemma 9.13 implies that $H \subset gR^I g^{-1}$ if $D = gD^I g^{-1}$. Finally, as $R^I = A_I \ltimes D^I$ and H/D is compact, it follows that $H = D$. □

Proof of Theorem 9.19.

Because the map $g \to g \cdot o$ is proper and D is closed, the set $D \cdot o$ is closed. If $D \cdot o = D' \cdot o$ then $D = D'$. To see this, observe that, by Corollary 9.26, the stabilizer of $D \cdot o$ (respectively, $D' \cdot o$) is D (respectively, D'). Hence $D = D'$. This shows that the map $D \to D \cdot o$ of $\overline{S_0}$ into $\mathcal{C}(X)$ is injective. The continuity of the map $g \to g \cdot o$ implies the continuity of the map $D \to D \cdot o$. Hence, the compactness of $\overline{S_0}$ implies that the map $D \to D \cdot o$ is a homeomorphism onto its image. □

This section concludes with the statement of a theorem (see [G18] for the proof) that gives yet another characterization of $\overline{X}^{SF}$ (not used in this book). It answers a question of Furstenberg about a unified conjugacy theorem valid for maximal compact subgroups and minimal parabolic subgroups (see Moore [M9] for a similar answer). For other results on this topic see [A1] and the references there.

9.27. Theorem. *$\overline{S_0}$ is the space of maximal distal subgroups. These subgroups form 2^r conjugacy classes indexed by the representatives $D^I (I \subset \Delta)$. The normalizer $\hat{D}$ of $D \in \overline{S_0}$ is a maximal amenable subgroup and the map $D \to \hat{D}$ is a G-equivariant injection of $\overline{S_0}$ into the set of maximal amenable subgroups. Every $D \in \overline{S_0}$ leaves invariant a unique probability measure m^D on $\mathcal{F}$ and the map $D \to m^D$ is a G-equivariant isomorphism of $\overline{S_0}$ with $\overline{X}^{SF}$.*

PARABOLIC SUBGROUPS AND BOUNDARY THEORY

9.28. The proof of the convergence criterion (Corollary 9.17) for the compactification $\overline{S_0}$ made use of the behavior of probability measures on the compact homogeneous space $\mathcal{F} = G/P = K/M$ that has already been referred to several times as the Furstenberg boundary. This homogeneous space is a boundary of G in an abstract sense due to Furstenberg (see Definition 9.29).

In order to continue the study of boundaries begun in Chapter IV, it is useful to outline the basic elements of boundary theory as found in [F5] and [M4]. Let H denote a locally compact, metrizable group. If E is a compact metric space with metric d and, if $F \subset E$, let diam $F = \sup_{x,y \in F} d(x,y)$. If (F_n) is a sequence of subsets of E and $x \in E$ it converges to x, denoted by writing $\lim_n F_n = x$, if $\lim_n$ diam $(F_n \cup \{x\}) = 0$, equivalently, if, for any $\epsilon > 0$, the sets F_n are eventually subsets of the ball $B(x;\epsilon)$ of radius ϵ about x.

Let $\mathcal{M}_1(E)$ denote the space of probability measures on E. It is well-known to be a compact metrizable space with respect to the topology of weak convergence, where μ_n converges weakly to μ if, for every continuous function ψ on E, $\int \psi d\mu_n \to \int \psi d\mu$. The subset of Dirac measures (or point masses) on E will be denoted by δ_E. The action of G on E determines a corresponding action of G on $\mathcal{M}_1(E)$: define $g \cdot \mu$ to be the measure such that $\int \psi d(g \cdot \mu) = \int \psi(g \cdot x) d\mu(x)$.

9.29. Definition. (See Furstenberg [F5]) A compact metric H-space (E,d) is called an H-**boundary** if

(1) Every H-orbit is dense in E.

(2) For every probability measure μ on E there is a sequence $g_n \in H$ and $x \in E$ such that $\lim_n g_n \cdot \mu = \delta_x$.

9.30. Remarks. (1) The first condition is clearly satisfied if H acts transitively on E, i.e., if E is a homogeneous space.

(2) By using a refinement of the proof of Proposition 9.8, it was proved in Lemma 4.48 that $\mathcal{F} = G/P$ is a boundary.

(3) While Definition 9.29 does not require that E is a homogeneous space, it reduces to the earlier Definition 4.45 if G is semisimple as a consequence of Theorem 9.37 below). The above definition is useful for discrete groups, for example, hyperbolic groups in the sense of Gromov (see [K2]). The free group Γ with $p > 1$ generators also provides another example. The space of reduced infinite words in the generators and their inverses is then a boundary of Γ (see, for example, Furstenberg [F5]).

(4) Situations that lead to the consideration of boundaries for semisimple groups have already been discussed following Definition 4.45, as well as their projective embeddings.

9.31. Example. Consider the linear group $SL(n,\mathbb{R})$ and its action on the projective space P^{n-1} by projective transformation. More explicitly, if $x \in \mathbb{R}^n \backslash \{0\}$ let $\bar{x} \in P^{n-1}$ denote its image in P^{n-1}, i.e., $\bar{x} = \{\lambda x; \lambda \in \mathbb{R}^*\}$. Then define the projective action by $g \cdot \bar{x} = \overline{g \cdot x}$ if $g \in SL(n,\mathbb{R})$. Given a norm $x \to \|x\|$ on $\mathbb{R}^n$, define a distance d on P^{n-1} by setting $d(\bar{x}, \bar{y}) =$

$\inf\{\|x-y\|; x \in \bar{x}, y \in \bar{y}, \|x\| = \|y\| = 1\}$. (For example, if (e_i), $1 \leq i \leq d$, is the canonical base of $\mathbb{R}^d$, one can take $\|x\| = \sup_{1 \leq i \leq d} |x_i|$.)

If $a = diag(\lambda_1, \lambda_2, \cdots, \lambda_d)$ with $|\lambda_1| > |\lambda_2| > \cdots > |\lambda_d|$, then for every $x \in \mathbb{R}^n \backslash \{0\}$, if $x = \sum_{i=1}^d x_i e_i$, it follows that $\lim_n a^n \cdot \bar{x} = \bar{e}_j$ where $j = \inf\{i \mid x_i \neq 0\}$. In addition, there exists a conjugate b of a such that $\lim_n b^n \cdot \bar{y}_i = \bar{e}_1$, $1 \leq i \leq d$.

If $\bar{y}$ belongs to a neighborhood of $\cup_{i=1}^d \{\bar{e}_i\}$, it is easy to see that, for every $\bar{x} \in P^{n-1}$, one has $\lim_n b^n a^n \cdot \bar{x} = \bar{e}_1$. Hence, if $\mu \in \mathcal{M}_1(E)$ and $\psi \in C(E)$, it follows from dominated convergence that $\lim_n (b^n a^n \cdot \mu)(\psi) = \int \psi(b^n a^n \cdot \bar{x}) d\mu(\bar{x}) = \psi(\bar{e}_1)$ and $\lim_n b^n a^n \cdot \mu = \delta_{\bar{e}_1}$. Hence, P^{n-1} is a $\mathrm{SL}(n, \mathbb{R})$-boundary.

9.32. Definition. A set $F \subset E$ is said to be **contractible** if there exists a sequence $g_n \in H$ such that $\lim_n \mathrm{diam}(g_n \cdot F) = 0$. An H-space E is said to be **H-proximal** if every subset $F = \{x, y\} \subset E$ is contractible.

The arguments given in Example 9.31 show that P^{n-1} is $\mathrm{SL}(n, \mathbb{R})$-proximal and every small ball is contractible. These properties are sufficient to guarantee that P^{n-1} is a $\mathrm{SL}(n, \mathbb{R})$ boundary as shown by the next result, proved by Margulis in [M4, p. 196].

9.33. Proposition. *Let (E, d) be a compact metric H-space such that*

(1) *every H-orbit is dense,*
(2) *E is H-proximal, and*
(3) *every point of E has a neighborhood which is contractible.*

Then E is an H-boundary.

For the proof one needs the following lemma

9.34. Lemma. *Suppose (E, d) is a compact metric H-proximal space. Suppose $F \subset E$ is contractible and $x \in E$. Then $F \cup \{x\}$ is contractible.*

Proof. As in [M4], let (g_n) be a sequence in H with $\lim_n \mathrm{diam}(g_n \cdot F) = 0$. Since E is compact, there exist $y, z \in E$ such that, by passing to a subsequence if necessary, $\lim_n g_n \cdot F = y$ and $\lim_n g_n \cdot x = z$. Let $h_n \in H$ be a sequence with $\lim_n d(h_n y, h_n z) = 0$. Then, since H acts continuously on E, for each $n \in N$, there are neighborhoods Y_n, Z_n of y, z respectively such that $\lim_n \mathrm{diam}\ (h_n \cdot Y_n \cup h_n \cdot Z_n) = 0$. One can assume, by passing to a subsequence if need be, that $g_n \cdot F \subset Y_n$ and $g_n \cdot x \in Z_n$. Hence, $\lim_n \mathrm{diam}\ (h_n g_n \cdot F \cup h_n g_n \cdot x) = 0$. $\square$

Proof of Proposition 9.33.

If $\mu \in \mathcal{M}_1(E)$, let $\alpha(\mu) \stackrel{\text{def}}{=} \sup_{x \in E} \mu(\{x\})$ — the mass of the largest atom of μ — and $\beta(\mu) \stackrel{\text{def}}{=} \sup\{\alpha(\nu), \nu \in H \cdot \mu\}$. Observe that if $\lim_n x_n = x \in E$ and $\lim_n \mu_n = \mu \in \mathcal{M}_1(E)$, then $\limsup_n \mu_n(\{x_n\}) \leq \mu(\{x\})$ and so $\limsup_n \alpha(\mu_n) \leq \alpha(\mu)$. Note that since E is compact, there is a measure $\nu \in \overline{H \cdot \mu}$ such that $\beta(\mu) = \alpha(\nu)$.

If $\mu \in \delta_E$, then $\alpha(\mu) = 1$. However, as is now shown, if $\mu \notin \delta_E$, then $\beta(\mu) > \alpha(\mu)$. By condition (3), there is a finite covering of E by contractible open balls V_i, $1 \leq i \leq p$. As $\mu \notin \delta_E$, there exist $i \leq p$ and $x \in E$ such that $\mu(V_i \cup \{x\}) > \mu(\{x\}) = \alpha(\mu)$. Lemma 9.34 and conditions (2) and (3) imply that $V_i \cup \{x\}$ is contractible. Hence, there exists $g_n \in H$ and $y \in E$ such that $\lim_n g_n \cdot (V_i \cup \{x\}) = \{y\}$. It follows that $g_n \cdot \mu(\{y\}) \geq g_n \cdot \mu(g_n \cdot V_i \cup \{g_n \cdot x\}) = \mu(V_i \cup \{x\}) > \alpha(\mu)$.

Hence, by considering a subsequence if necessary, if $\nu = \lim_n g_n \cdot \mu$, then $\beta(\mu) \geq \alpha(\nu) > \alpha(\mu)$.

Now let $\nu \in \overline{G \cdot \mu}$ denote a measure such that $\alpha(\nu) = \beta(\mu)$. If $\nu \notin \delta_E$, it follows from what has been proved that $\beta(\nu) > \alpha(\nu)$. This is impossible since $\overline{G \cdot \nu} \subset \overline{G \cdot \mu}$ implies that $\beta(\nu) \leq \beta(\mu)$. Hence, $\nu \in \delta_E$ and so $\overline{G \cdot \mu} \cap \delta_E \neq \emptyset$. This, in view of (1), implies that E is an H-boundary. □

9.35. Definition. (See [G9]) A topological group H is defined to be **amenable** if either of the following equivalent conditions is satisfied.

(1) The space $C(H)$ of continuous and bounded functions on H has a left-invariant mean λ, i.e., there is a positive linear form λ on $C(H)$ such that $\lambda(1) = 1$ and $\lambda(f_g) = \lambda(f)$ for every $f \in C(H)$ and $g \in H$, where $f_g(x) = f(g \cdot x)$, or
(2) every continuous affine action of H on a compact convex set of a locally convex topological vector space has a fixed point.

9.36. Examples. (See [G9]) The Markov-Kakutani fixed point theorem implies that a locally compact group A which is abelian or, more generally, solvable is amenable. A compact group C is amenable, as is also a semi-direct product $C \ltimes A$, where A is abelian. Since P^{n-1} is a boundary of $\mathrm{SL}(n, \mathbb{R})$, if $n > 1$ this group is not amenable as, by Example 9.31, its action on the compact convex set $\mathcal{M}_1(P^{n-1})$ has no fixed point.

The following theorem, that sharpens Lemma 4.48 and Proposition 9.8, describes all the boundaries of a semisimple group G.

9.37. Theorem. (See Furstenberg [F3]) *Let G be a semisimple Lie group with finite center. Then $\mathcal{F} = G/P$ is a boundary. Furthermore, E is a boundary of G if and only if there exists a parabolic subgroup $Q \supset P$ such that $E = G/Q$.*

Proof. It is clear from the Iwasawa decomposition of G that $\mathcal{F} = G/P = K/M$ is compact.

Consider the Bruhat decomposition $G = \cup_{w \in W} PwP$,where $P = MAN$. Let $\overline{N} = \theta(N)$ denote the nilpotent group opposite to N, i.e., the nilpotent group with Lie algebra $\overline{\mathfrak{n}} = \theta(\mathfrak{n})$. Then, by Corollary 2.21, the so-called "large cell" $\overline{N} \cdot \dot{e} \subset G/P$ is open and dense and is invariant under the action of $a \in A$ on G/P. If $\overline{N} \cdot \dot{e}$ is identified with $\overline{\mathfrak{n}}$, the action of $a = \exp H$ is the automorphism $Ad(a)$ with eigenvalues $e^{\alpha(H)}$, i.e., $Ad(a)(X) = \sum_{\alpha<0} e^{\alpha(H)} X_\alpha$ if $X = \sum_{\alpha<0} X_\alpha$ (see Helgason [H2, p. 128]). As a result, as is well-known, if $a \in A^+$, it follows that $\lim_n a^n \cdot x = e$ for any $x \in \overline{N}$.

Given $x, y \in G/P$, one can find $g \in G$ such that $g \cdot x, g \cdot y \in \overline{N} \cdot \dot{e}$. To see this, since G/P is homogeneous, one can assume that $x = \dot{e}$. If $y \notin \overline{N} \cdot \dot{e}$,let V denote a small compact neighborhood of $e \in G$ such that $V \cdot \dot{e} \subset \overline{N} \cdot \dot{e}$ and $V \cdot y \cap \overline{N} \cdot \dot{e} \neq \emptyset$. Then, there is a point $g \in V$ with $g \cdot \dot{e} \in \overline{N} \cdot \dot{e}$ and $g \cdot y \in \overline{N} \cdot \dot{e}$. Hence, if $a \in A^+$, $\lim_n a^n g \cdot x = \lim_n a^n g \cdot y = \dot{e}$. As a result, G/P is G-proximal.

By the same argument it follows that every $x \in G/P$ has a contractible neighborhood $V(x)$. By homogeneity one can assume that $x = \dot{e}$. Since $\overline{N} \cdot \dot{e}$ is open there is relatively compact neighborhood $V(\dot{e})$ of $\dot{e}$ whose closure is a subset of $\overline{N} \cdot \dot{e}$. Hence, $\lim_n (a^n \cdot V(\dot{e})) = \{e\}$. As a result, the first part of this theorem follows from Proposition 9.33.

If $Q \supset P$ then $E = G/Q$ is also a boundary. In view of Remark 9.30 (1), it suffices to verify condition (2) of Definition 9.29. To show that it is satisfied, let $\pi : G/P \to G/Q$ denote the canonical G-equivariant projection. It induces an affine, continuous map from $\mathcal{M}_1(G/P)$ onto $\mathcal{M}_1(G/Q)$. Since $g_n \cdot \nu \to x$ implies that $g_n \cdot \pi(\nu) \to \pi(x)$, the result follows.

If E is a G-boundary, consider the action of P on the compact convex set $\mathcal{M}_1(E)$. Since $P = MAN$ is a semi-direct product $M \ltimes S$ of the solvable group $S = AN$ by the compact group M, it follows that P is amenable. It has a fixed point $\nu \in \mathcal{M}_1(E)$ and, so, P has an invariant measure ν on E. The map $g \to g \cdot \nu$ from G to $\mathcal{M}_1(E)$ is continuous and defines a continuous map of G/P into $\mathcal{M}_1(E)$. Its image is the orbit $G \cdot \nu$ which is compact as it is the orbit $K \cdot \nu$. Since E is a G-boundary this implies there is a measure $\mu \in G \cdot \nu \cap \delta_E$. As a result, $\nu \in \delta_E$ and, so, P has a fixed point x_0 on E. Since the orbit $G \cdot x_0 = K \cdot x_0$ is dense in E and compact, it follows that $E = G \cdot x_0 = G/Q$, where $Q \supset P$ is the stabilizer of x_0. □

9.38. Definition. Let E and F be two locally compact H-spaces. F is said to be a **factor space** of E or E is said to **dominate** F if there is an H-equivariant continuous map π of E onto F, i.e., for every x in E and h in H one has $\pi(h \cdot x) = h \cdot \pi(x)$. If π is bijective and bicontinuous then E is said to be isomorphic to F.

9.39. Examples. If G is a classical group, the boundaries can be described in terms of flags (see [B8]). Consider for example $G = SL(n, \mathbb{R})$. A complete flag f in $\mathbb{R}^n$ is a nested sequence of $n-1$ distinct proper subspaces: $f = (V_1, V_2, \cdots, V_{n-1})$, $V_1 \subset V_2 \subset \cdots \subset V_{n-1}, \dim V_i = i$. An incomplete flag is a nested sequence of distinct proper subspaces. The projective space P^{n-1} is the space of lines. Clearly, $\mathcal{F} = G/P$ is the set of complete flags, while the other boundaries of G are spaces of incomplete flags.

Remark. In other words, Theorem 9.37 states that the G-boundaries are, up to G-isomorphism, the factor spaces of G/P, i.e., in each case there is a continuous G-equivariant map from G/P onto E. Every space G/Q with $Q \supset P$ is called a Furstenberg boundary and G/P is the maximal one. Because Q is a standard parabolic subgroup P^I, as observed by Moore [M8], there are only 2^r G-boundaries, where r is the rank of G.

Extension to semisimple algebraic groups over a local field.

§ 9.40. In chapter XV boundaries are considered for the group $H = G^a(\mathbb{F})$ of $\mathbb{F}$-rational points of a semisimple algebraic group G^a defined over a local field $\mathbb{F}$. For such an H, Theorem 9.37 is true, as the proof given also applies when P is defined to be the set of $\mathbb{F}$-rational points of a minimal parabolic subgroup defined over $\mathbb{F}$. As a result, $\mathcal{F} = H/P$ is an H-boundary and every H-boundary is a factor space of $\mathcal{F}$.

In the algebraic context, a parabolic subgroup of G^a is an algebraic subgroup P^a such that G^a/P^a is a projective manifold whose set of $\mathbb{F}$-rational points is $\mathcal{F} = H/P$ [B8]. The Bruhat decomposition holds for $H = G^a(\mathbb{F})$ and P is amenable. Hence, these facts lead to the same results as in the semisimple case (see [M4, p. 87 and p. 151, Lemma 2.2]). Consequently, the projective character of the manifold H/P is a basic fact which is incorporated in the abstract Definition 9.29 of a G-boundary.

THE MAXIMAL FURSTENBERG COMPACTIFICATION

§ 9.41. In Corollary 9.15, it was observed that, corresponding to the distal group $D = MN$, there is a unique compact G-orbit in $\overline{S_0}$ that is of minimal dimension. This orbit is isomorphic to G/P while the other orbits are closely related to the parabolic subgroups P^I and the homogeneous space G/P^I. A similar situation holds for the other realizations of the maximal Satake–Furstenberg compactification. These compactifications, therefore, lead naturally to the examination of the boundaries G/Q (see Chapter IV). In particular, following Furstenberg, one can try to reconstruct the symmetric space and its compactifications from the boundary itself. This point of view is useful, for example, in rigidity problems (see Mostow [M10]). The account that follows completes the earlier investigation begun in Chapter IV.

Given any homogeneous space E of G and a probability μ on E, it is natural to consider the G-orbit $G\cdot\mu \subset \mathcal{M}_1(E)$ of μ. If the measure μ is K-invariant the map $g \to g\cdot\mu$ clearly factors through the symmetric space X. When E is compact this gives a map of X into the compact space $\mathcal{M}_1(E)$. In the case of the Furstenberg boundary $\mathcal{F} = G/P = K/M$ there is a unique probability m on $\mathcal{F}$ that is K-invariant. As a result, it follows from Lemma 9.11 that the resulting map of X onto the orbit $G\cdot m$ is bijective. (As pointed out in § 4.46, this special feature of the Furstenberg boundary is possessed by a boundary G/Q of G if and only if the stabilizer in G of the image m_Q of m under the natural map $\pi : G/P \to G/Q$ coincides with K. For additional details see Appendix B.)

In the case of the Furstenberg boundary $\mathcal{F}$, not only is the natural map of X into $\mathcal{M}(\mathcal{F})$ injective, it is a topological embedding as stated in the following result.

9.42. Theorem. *The map $g \cdot o \to g \cdot m$ embeds X into $\mathcal{M}_1(\mathcal{F})$. Hence, the closure $\overline{G \cdot m}$ of the image $G \cdot m$ of X in $\mathcal{M}_1(\mathcal{F})$ is a compactification of X.*

Proof. It follows from Lemma 9.11 that the map $g \cdot o \to g \cdot m$ is an injection of X into $\mathcal{M}_1(\mathcal{F})$. It remains to show that this map is an embedding.

Since the Iwasawa action of G on K is smooth, it follows that the resulting G action on $\mathcal{F} = K/M$ is continuous. As a result, if $g_n \to g$ then $g_n \cdot m \to g \cdot m$. If $g_n \cdot o \to g \cdot o$ then $g_n = gk_n, k_n \in M$. As a result, for every convergent subsequence $(g_{n_k}), g_{n_k} \cdot m \to g \cdot m$. Hence, $g_n \cdot m \to g \cdot m$.

Conversely, to show that $g_n \cdot m \to g \cdot m$ implies $g_n \cdot o \to g \cdot o$ it suffices to show that the sequence (g_n) lies in a fixed compact set $C \subset G$. Assuming this, if (g_{n_k}) is a convergent subsequence and $g_{n_k} \to g'$, then $g' \cdot o = g \cdot o$ since $g' \cdot m = g \cdot m$. Hence, if $g_n \cdot m \to g \cdot m$ and (g_n) lies in a fixed compact set C, then $g_n \cdot o \to g \cdot o$.

Clearly, (g_n) lies in fixed compact set C if and only if the sequence $(g_n \cdot o)$ is bounded. To verify this observe that, by Lemma 9.11, the stabilizer of $g \cdot m$ in G is gKg^{-1}. If the sequence $(g_n \cdot o)$ is unbounded, it follows that for some subsequence (g_{n_k}) the sequence $(g_{n_k} K g_{n_k}^{-1})$ converges in $\overline{\mathcal{S}_0}$ to a limit group D. It follows from Lemma 9.7 that $D \subset gKg^{-1}$. This is a contradiction as D is not compact. □

9.43. Proposition. *Let $(g_n \cdot o)$ be a C_I-fundamental sequence in X. Then $(g_n \cdot m)$ converges to $ka^I \cdot m_I$, where $k = \lim_n k_n$ and $a^I = \lim_n a_n^I$ if $g_n \cdot o = k_n a_n \cdot o$.*

Proof. This result is an immediate consequence of Proposition 9.8. □

9.44. Corollary. *The closed orbit $\overline{G \cdot m} \subset \mathcal{M}_1(\mathcal{F})$ is equal to the union of the orbits $G \cdot m_I$, where m_I is the unique K^I-invariant probability on $K^I \cdot \dot{e}$ (see Lemma 9.10). The orbit $G \cdot m_I \simeq G/R^I$.*

Proof. The first statement is an immediate consequence of the fact that every sequence in X, converging to infinity, has a fundamental subsequence (see the lemma in § 7.17). The second assertion follows from Proposition 9.13. □

As in the case of the compactification $\overline{\mathcal{S}_0}$, every limit measure in the boundary of the compactification $\overline{G \cdot m}$, as stated in Proposition 9.43, has a **polar realization** $ka_1 \cdot m_I$ with $k \in K$ and $a_1 \in \overline{A^{I,+}}$. It involves the same parameters (k, a, I) as in the case of a limit group D.

9.45. Proposition. *The parameters $(k, a, I), k \in K, a \in \overline{\mathfrak{a}^{I,+}}$ of a polar realization $ka \cdot m_I$ of a limit measure are unique modulo $(K^I \cap aK^I a^{-1})M$. In other words,*

(1) *if $g_1 \cdot m_{I_1} = g_2 \cdot m_{I_2}$, then $I_1 = I_2 = I$ and $g_2^{-1} g_1 \in R^I$, and*
(2) *if $k_1 a_1^{I_1} \cdot m_{I_1} = k_2 a_2^{I_2} \cdot m_{I_2}$, then $I_1 = I_2 = I$, $a_1^{I_1} = a_2^{I_2} = a$ and $k_2 k_1^{-1} \in (K^I \cap aK^I a^{-1})M$.*

Proof. If $g_1 \cdot m_{I_1} = g_2 \cdot m_{I_2}$, it follows from Lemma 9.13 that the stabilizer of this measure is $g_1 R^{I_1} g_1^{-1} = g_2 R^{I_2} g_2^{-1}$. Since their nilpotent radicals $g_1 N_{I_1} g_1^{-1}$ and $g_2 N_{I_2} g_2^{-1}$ are equal, it follows that the normalizers $g_1 P^{I_1} g_1^{-1}$ and $g_2 P^{I_2} g_2^{-1}$ coincide. Hence, $I_1 = I_2 = I$ and $g_2^{-1} g_1 \in P_I$. Since $g_2^{-1} g_1 \cdot m_I = m_I$, it follows from Lemma 9.13 that $g_2^{-1} g_1 \in R^I$.

Now assume that $k_1 a_1 \cdot m_{I_1} = k_2 a_2 \cdot m_{I_2}$ where $a_i = a^{I_i} \in \overline{A^{I_i,+}}$. It follows from what has been proved that $I_1 = I_2 = I$ and $k = k_2^{-1} k_1 \in P^I \cap K = K^I M$. In addition, by Lemma 9.13, $a_2^{-1} k a_1 = \ell \in R^I \cap G^I M = K^I M$. As a result, it follows from the Cartan decomposition that $a_1 = a_2 = a$. The fact that $\ell \in K^I M$ implies that $k \in a K^I a^{-1} M$. □

Remark. If $k \in (K \cap aKa^{-1})$ then $ak = ka$. To see this observe that $k = a^{-1}\ell a, \ell \in K$ implies $ak = \ell a$ and $ka = a\ell$ (use the Cartan involution θ). From this it follows that $a^2 k = a\ell a = a l a^{-1} a^2 = k a^2$. If $a = \exp H$, then $\exp 2Ad(k)H = \exp 2H$. Since $2H$ and $2Ad(k)H$ both belong to $\mathfrak{p}$ and $\exp : \mathfrak{p} \to X$ is a bijection, it follows that $H = Ad(k)H$ and so $ka = ak$.

As in the case of the compactification $\overline{S_0}$, this result has as a corollary the following convergence criterion.

9.46. Corollary. (Convergence criterion) *Let (y_n^1) and (y_n^2) be two fundamental sequences with*

$$y_n^1 = k_n^1 a_n^{I_1} a_n^1 \cdot o \text{ and } y_n^2 = k_n^2 a_n^{I_2} a_n^2 \cdot o,$$

where

$$k_n^i \to k_i, \quad a_n^{I_i} \to a^{I_i} \text{ and } (a_n^i) \text{ is } I_i\text{-canonical.}$$

If $g_n^1 \cdot o = y_n^1$ and $g_n^2 \cdot o = y_n^2$ the two sequences $(g_n^1 \cdot m)$ and $(g_n^2 \cdot m)$ of probabilities converge in $\mathcal{M}_1(\mathcal{F})$ to the same limit measure if and only if

(1) $I_1 = I_2 = I$;
(2) $a^{I_1} = a^{I_2} = a$; *and*
(3) $k_2^{-1} k_1 \in (K^I \cap a K^I a^{-1}) M$.

It follows from the convergence criterion in Corollary 9.46 that $\overline{G \cdot m}$ is another G-compactification isomorphic to any of the G-compactifications listed in Theorem 9.18. In other words, the following result is established.

9.47. Theorem. *The following compactifications of X are G-isomorphic:*

(1) *the dual cell compactification $X \cup \Delta^*(X)$;*
(2) *the maximal Satake compactification $\overline{X}^S$;*
(3) *the Martin compactification $X \cup \partial X(\lambda_0)$;*
(4) *the compactification $\overline{S_0}$ of X; and*
(5) *$\overline{G \cdot m}$, the maximal Furstenberg compactification $\overline{X}^F$.*

In the proof of this result, Satake's characterization of $\overline{X}^S$ (see Proposition 4.42) was used earlier to show that $\overline{X}^S$ and the dual cell compactification are G-isomorphic (see Theorem 4.43).

Note that, from now on in this chapter, G will be assumed to have no proper compact normal subgroups.

The maximal Satake compactification $\overline{X}^S$ was defined to be the closure of $\tau(X)$ in $P(\mathcal{H}_n)$, where τ is an irreducible faithful projective representation of G whose highest weight μ_τ is generic, i.e., it lies in $\mathfrak{a}^*_+$. It will now be shown directly that it can be identified with the maximal Furstenberg compactification $\overline{X}^F = \overline{G \cdot m}$. This identification will be used to give a proof of Satake's characterization of $\overline{X}^S$ by simply transporting the orbit structure of $\overline{G \cdot m}$ to $\overline{X}^S$. The identification of the other Satake compactifications with the compactifications $\overline{G \cdot m_I}$ is given in Appendix B.

First, some preliminary lemmas are established dealing with the action of degenerate projective maps on certain measures on a complex projective space $P(V)$. As in Proposition 4.32, a Hermitian scalar product is fixed and an orthonormal basis chosen.

9.48. Lemma. *Suppose that the sequence $(g_n) \subset \mathrm{GL}(V)$ is such that $\lim_n \frac{g_n}{\|g_n\|} = u \in End\ V$ and ν is a measure on $P(V)$ which gives zero mass to every proper projective subspace. Then the sequence of measures $g_n \cdot \nu$ converges weakly to a measure denoted $u \cdot \nu$.*

Proof. Let ϕ be a continuous function on $P(V)$. Then $g_n \cdot \nu(\phi) = \int \phi(g_n \cdot x) d\nu(x)$. Since $\frac{g_n \cdot x}{\|g_n\|}$ converges to $u \cdot x \neq 0$ for every $x \in V$, provided $x \notin \ker u$, it follows that, for every x outside the projective subspace defined by $\ker u$, the sequence $g_n \cdot x$ converges to a point of $P(V)$ denoted by $u \cdot x$. Hence, by dominated convergence and the non-degeneracy of ν, it follows that $\lim_n g_n \cdot \nu(\phi) = \int \phi(u \cdot x) d\nu(x) = u \cdot \nu(\phi)$. □

9.49. Lemma. *Let τ denote an irreducible representation of G in a complex vector space V for which $\dim V_\tau = 1$. Let $\bar{v}_\tau$ denote the unique point of $P(V)$ corresponding to the highest weight of τ. Denote by m_τ the unique K-invariant measure on the compact orbit $\tau(G) \cdot \bar{v}_\tau$. Then m_τ gives zero measure to every projective subspace.*

Proof. Consider the family Ψ of proper subspaces W of $P(V)$ of minimal dimension for which $m_\tau(W) > 0$. If W and W' belong to Ψ, and $W \neq W'$ then, since $W \cap W'$ has smaller dimension, $m_\tau(W \cap W') = 0$. Because m_τ is a probability measure it follows that, for every $\epsilon > 0$, there exist only a finite number of elements of Ψ of mass greater than ϵ. In particular, there exists a $W_0 \in \Psi$ such that $m_\tau(W_0) = \sup_{W \in \Psi} m_\tau(W)$. Furthermore, the subspaces W_0 which satisfy this equality form a finite family. Consider now a K-bi-invariant probability measure p on G whose support is equal to G. It follows from the K-bi-invariance of p that $\int \tau(g) \cdot m_\tau dp(g) = m_\tau$. In particular, $m_\tau(W_0) = \int \tau(g) \cdot m_\tau(W_0) dp(g) = \int m_\tau(\tau(g^{-1})W_0) dp(g)$. Because $m_\tau(W_0) \geq m_\tau(\tau(g^{-1})W_0)$ and p is a probability measure, it follows that $m_\tau(\tau(g^{-1}W_0)) = m_\tau(W_0)$ for every $g \in G$. As a result, the finite union of the subspaces W_0 is $\tau(G)$-invariant. Since G is connected, each of these subspaces W_0 is $\tau(G)$-invariant. This contradicts the irreducibility of $\tau(G)$. □

9.50. Lemma. *Let $(g_n) \subset G$ be a sequence such that $(g_n \cdot o)$ is a C_I-fundamental sequence on X and $g_n = k_n a_n k_n'$, where $\lim_n k_n' = k'$ exists. Let $a^I = \lim_n a_n^I$ and $k = \lim_n k_n$. Let τ be an irreducible representation of G on V. Then, $\lim_n \frac{\|\tau(g_n\|}{\|\tau(a_{n,I})\|} = \|\tau(a^I)\|$ and $\lim_n \frac{\tau(g_n}{\|\tau(a_{n,I})\|} = \tau(ka^I)\pi_I\tau(k')$ where π_I is the orthogonal projection on the subspace $V_I \subset V$ that is the sum of the weight subspaces V_μ with μ of the form $\mu = \mu_\tau - \sum_{\alpha\in I} c_\alpha \alpha$ $c_\alpha \geq 0$.*

Proof. To simplify notation, G is identified with $\tau(G) \subset SL(V)$, i.e., $g = \tau(g)$ and $\|\tau(g\|$ is denoted by $\|g\|$ if $g \in G$. Then $\|g_n\| = \|a_n^I a_{n,I}\|$ because k_n and k_n' are then viewed as unitary transformations. In addition, $\|a_n^I a_{n,I}\| = \|a_n^I\|\|a_{n,I}\|$ because a_n^I and $a_{n,I}$ are two diagonal matrices in an orthonormal basis given by the weight subspaces of τ. Hence, $\lim_n \frac{\|a_n\|}{\|a_{n,I}\|} = \|a^I\|$. Furthermore, since the weights μ of τ can be written as $\mu = \mu_\tau - \sum_{\alpha\in\Delta} c_\alpha \alpha$ with the $c_\alpha \geq 0$ (see formula (4.9) in Chapter IV), it follows that $\mu(\log a_n) \leq \mu_\tau(\log a_n)$.

Hence, $\|a_{n,I}\| = e^{\mu_\tau(\log a_n)} = e^{\mu_\tau}(a_n)$ and $\frac{a_{n,I}}{\|a_{n,I}\|}$ is a diagonal matrix with dominant eigenvalue 1 of multiplicity $\dim V_\tau = p_\tau$, and the other eigenvalues are of the form $\prod_{\alpha\in\Delta} e^{-c_\alpha \alpha}(a_{n,I})$.

Since $a_{n,I}$ is I-canonical, it follows that $\lim_n e^\alpha(a_{n,I}) = 1$, if $\alpha \in I$, and $\lim_n e^\alpha(a_{n,I}) = 0$, if $\alpha \notin I$. Hence, in the limit, the eigenvalues of $\frac{a_{n,I}}{\|a_{n,I}\|}$ which are not of the form $\prod_{\alpha\in I} e^{-c_\alpha\alpha}(a_{n,I})$ converge to zero. The other eigenvalues converge to 1. As a result, $\lim_n \frac{a_{n,I}}{\|a_{n,I}\|} = \pi_I$. From this it follows that $\lim_n \frac{\tau(g_n)}{\|\tau(a_{n,I})\|} = \tau(kc)\pi_I\tau(k')$. □

9.51. Lemma. *Let $(g_n \cdot o$ be a C_I-fundamental sequence. Then*

$$\lim_n \frac{\tau(g_n)\tau(g_n)^*}{\|\tau(a_{n,I})\|^2} = \tau(k(a^I)^2)\pi_I\tau(k^{-1}).$$

If $\dim V_\tau = 1$ and G/P_τ is identified with $\tau(G) \cdot \bar{v}_\tau$, then

$$\lim_n \tau(g_n) \cdot m_\tau = \tau(ka^I)\pi_I \cdot m_\tau.$$

Proof. Using the Cartan decomposition, it follows that $g_n = k_n a_n^I a_{n,I} k_n'$. Consider a subsequence g_{n_ℓ} such that $\lim_\ell k_{n\ell}' = k'$ exists. It follows from Lemma 9.50 that $\lim_\ell \frac{\tau(g_{n_\ell})}{\|\tau(a_{n_\ell,I})\|} = \tau(ka^I)\pi_I\tau(k')$ and $\lim_\ell \frac{\tau(g_{n_\ell})^*}{\|\tau(a_{n_\ell,I})\|} = \tau(k')^{-1}\pi_I\tau(a^I k^{-1})$. Hence, since $\pi_I = \lim_\ell \frac{\tau(a_{n_\ell,I})}{\|\tau(a_{n_\ell,I})\|}$ commutes with $\tau(a^I)$, $\lim_\ell \frac{\tau(g_{n_\ell})\tau(g_{n_\ell}^*)}{\|\tau(a_{n_\ell,I})\|^2} = \tau(k(a^I)^2)\pi_I\tau(k^{-1})$. Since this limit is independent of the subsequence, the first statement follows.

If $\dim V_\tau = 1$, consider again a subsequence (g_{n_ℓ}) for which $\lim_\ell k_{n_\ell}' = k'$ exists. Then it follows from Lemmas 9.48,9.49, and 9.50 that $\lim_\ell \tau(g_{n_\ell}) \cdot m_\tau = \tau(ka^I)\pi_I\tau(k') \cdot m_\tau = \tau(ka^I)\pi_I \cdot m_\tau$. As before, the limit is independent of the subsequence and, so, the result follows. □

9.52. Theorem. (Convergence criterion) *Let τ be a faithful, irreducible representation of G on a complex vector space V such that the highest weight μ_τ is generic. If (y_n) is a fundamental sequence in X, with $y_n = g_n \cdot o$, then the corresponding sequence $(\tau(g_n)\tau(g_n)^*)$ in $\overline{X}_\tau^S$ converges.*

Furthermore, if (y_n^1) and (y_n^2) are two fundamental sequences with

$$y_n^1 = k_n^1 a_n^{I_1} a_n^1 \cdot o \text{ and } y_n^2 = k_n^2 a_n^{I_2} a_n^2 \cdot o,$$

where

$$k_n^i \to k_i, \quad a_n^{I_i} \to a^{I_i} \text{ and } (a_n^i) \text{ is } I_i\text{-canonical},$$

and if $g_n^1 \cdot o = y_n^1$ and $g_n^2 \cdot o = y_n^2$, the two sequences $(\tau(g_n^1)\tau(g_n^1)^*)$ and $(\tau(g_n^2)\tau(g_n^2)^*)$ converge in $\overline{X}_\tau^S$ to the same point if and only if

(1) $I_1 = I_2 = I$;
(2) $a^{I_1} = a^{I_2} = a$; *and*
(3) $k_2^{-1}k_1 \in (K^I \cap aK^I a^{-1})M$.

Proof. Let ϕ be a hermitian scalar product on V such that, relative to any orthonormal basis, $\tau(\theta(Y) = -{}^t\overline{\tau(Y))}$ (see Proposition 4.32). To simplify notations, let g denote $\tau(g)$. As in Proposition 4.24, let $v_\tau \in \wedge^{p_\tau} V$ be a p_τ-vector that corresponds to the p_τ-dimensional subspace V_τ. If $V^\wedge$ is the subspace generated by $\wedge^{p_\tau}[\tau(G)]v_\tau$, it is an irreducible factor of $\wedge^{p_\tau}\tau$. Let $\tau^\wedge$ denote the corrresponding representation of G on $V^\wedge$ and $\bar{v}_\tau \in P(V^\wedge)$ be the point corresponding to v_τ.

Since $P_\tau = P$, one may identify G/P with $\tau^\wedge(G) \cdot \bar{v}_\tau \subset P(V^\wedge)$ and m with the unique K-invariant measure m_τ on $\tau^\wedge(G) \cdot \bar{v}_\tau$. If $p_\tau = \dim V_\tau$, the weights of $\tau^\wedge$ are of the form $p_\tau\mu_\tau - \sum_{\beta\in\Delta} c_\beta \beta$, where the c_β are positive integers. A basis of weight vectors is given by p-wedge products of weight vectors of τ. If such a wedge product belongs to $V_I^\wedge$, then it belongs to $\wedge^{p_\tau} V_I$. Hence, $V_I^\wedge = \wedge^{p_\tau} V_I \cap V^\wedge$ and $\pi_I^\wedge$ is the restriction of $\wedge^{p_\tau}\pi_I$ to $V^\wedge$.

Let $(g_n \cdot o)$ be a C_I-fundamental sequence in X. Lemma 9.51 implies that the corresponding sequence $(g_n g_n^*)$ in $\overline{X}_\tau^S$ converges with $\lim_n \frac{g_n g_n^*}{\|a_{n,I}\|^2} = k(a^I)^2 \pi_I k^{-1}$. Hence, every fundamental sequence converges in $\overline{X}_\tau^S$ and if two fundamental sequences in X have the same formal limit — i.e., if $k_1 = k_2, I_1 = I_2$ and $a^{I_1} = a^{I_2}$— then their limits in $\overline{X}_\tau^S$ agree.

Conversely, assume that $(g_n^1 \cdot o)$ and $(g_n^2 \cdot o)$ are two fundamental sequences in $\overline{X}_\tau^S$ with the same limit, i.e., $k_1 (a^{I_1})^2 \pi_{I_1} k_1^{-1} = k_2 (a^{I_2})^2 \pi_{I_2} k_2^{-1}$ up to a scalar. Applying $\tau^\wedge$ to this identity and making use of the identification of $\tau^\wedge(G) \cdot v_\tau$ with G/P, it follows that $k_1(a^{I_1})^2 \cdot m_{I_1} = k_2(a^{I_2})^2 \cdot m_{I_2}$. Proposition 9.45 (2) implies that $I_1 = I_2 = I, (a^{I_1})^2 = (a^{I_2})^2 = a^2$, and $k_1^{-1}k_2 \in (K^I \cap a^2 K^I a^{-2})M$. As pointed out in the remark following Proposition 9.45, this implies that $k_1^{-1}k_2 a = a k_1^{-1} k_2$. Hence, $k_1^{-1}k_2 \in (K^I \cap aK^I a^{-1})M$. □

9.53. Corollary. (See Moore [M8]) *Let τ be a faithful, irreducible representation of G on a complex vector space V such that the highest weight μ_τ is generic. Let m denote the unique K-invariant probability measure on G/P. Then the corresponding Satake compactification $\overline{X}^S_\tau$ is G-isomorphic to the closure $\overline{G \cdot m}$ of the orbit $G \cdot m$ in $\mathcal{M}_1(G/P)$.*

9.54. Corollary. *Let (H_n) be a sequence in the closed positive Weyl chamber $\overline{\mathfrak{a}^+} \subset \mathfrak{a}$ such that $\|H_n\| \to \infty$. Let $a_n = \exp H_n$ and let $k_n \in K$. Denote by I the set of simple roots $\alpha \in \Delta$ that are bounded on (H_n). Then $(k_n a_n^2 k_n^{-1})$ converges in $\overline{X}^S$ if and only if for every $\alpha \in \Delta, \lim_n \alpha(H_n) \in [0, \infty]$ exists — in which case $H_n^I \to H^I \in \overline{\mathfrak{a}^{I,+}}$ — and the cluster values of the sequence (k_n) are of the form $k\ell$ with $\ell \in (K^I \cap a^I K^I {a^I}^{-1})M$, where $a^I = \exp H^I$.*

Proof. As in Proposition 7.28, a sequence of the form $(g_n g_n^*)$ in $\overline{X}^S$ that is eventually outside any compact subset of $\tau(X)$ converges if and only if the limits of its fundamental subsequences all agree. Since I is the set of simple roots that are unbounded on (H_n), it follows that every fundamental subsequence of $(k_n a_n \cdot o)$ is a C_I-fundamental subsequence. This implies, therefore, that (H_n^I) converges to a point H^I of $\overline{\mathfrak{a}^{I,+}}$. Finally, the observation about the cluster values of the sequence (k_n) follows immediately from Corollary 7.32, or any other version of the convergence criterion. □

Remark. In the above theorem one has considered a representation of G instead of a projective representation as in Definition 4.4. This is not a restriction because every projective representation of G can be lifted to a representation of a finite cover of G, that can, moreover, be supposed to be faithful.

In the proof of Theorem 7.22 it was shown that the Furstenberg boundary $\mathcal{F}^I$ of the symmetric space $X^I = G^I/K^I$ can be naturally identified with a subset of $\mathcal{F}$, the Furstenberg boundary of $X = G/K$. As observed there, the centralizer M^I of $\mathfrak{a}^I$ in K^I is $M \cap K^I$. As a result, the map $k_1 \to k_1 \cdot \dot{e}$ of K^I onto the orbit of $\dot{e} \in \mathcal{F}$ determines the natural embedding $k_1 M^I \to k_1 \cdot \dot{e}$ of $\mathcal{F}^I = K^I/M^I$ into $\mathcal{F} = K/M$. If m^I denotes the unique K^I-invariant probability on $\mathcal{F}^I$ its image in $\mathcal{M}_1(\mathcal{F})$ is the probability m_I. Since the symmetric space X^I is embedded in $\mathcal{M}_1(\mathcal{F}^I)$ by the map $g_1 \cdot o \to g_1 \cdot m^I$, where m^I is the unique K^I-invariant probability on $\mathcal{F}^I$, it follows that X^I is embedded in $\mathcal{M}_1(\mathcal{F})$ by the map $g_1 \cdot o \to g_1 \cdot m_I$. Its image is the orbit $G^I \cdot m_I$. Note that since $P^I = K^I P$ the orbit $K^I \cdot \dot{e} = P^I \cdot \dot{e}$ when $\mathcal{F}$ is expressed as G/P, i.e., one can view $\mathcal{F}^I$ as P^I/P. In addition, $\mathcal{F}^I = G^I/M^I A^I N^I = G^I/(P \cap G^I)$.

9.55. Proposition. (Cf. Proposition 3.44.) *Let $\iota_I : X^I \to \overline{X}^F$ denote the embedding $g_1 \cdot o \to g_1 \cdot m_I$. The image $\iota_I(X^I) = G^I \cdot m_I$ of X^I is in the boundary of the compactification $\overline{X}^F = \overline{G \cdot m}$. It corresponds to $X^I(C_I(\infty)$ in $\Delta^*(X)$ under the isomorphism of $\overline{X}^F$ with $X \cup \Delta^*(X)$. The closure of* *the orbit $G^I \cdot m_I$ in $\overline{X}^F = \overline{G \cdot m} \subset \mathcal{M}_1(\mathcal{F})$ is isomorphic to the maximal Furstenberg compactification $\overline{X^I}^F$ of the symmetric space X^I.*

Proof. It follows from Proposition 9.8 that m_I is in the boundary of $\overline{X}^F$. As a result, the orbit $G^I \cdot m_I \subset \partial \overline{X}^F$. It also follows from this proposition and the action of G^I that the orbit corresponds to $X^I(C_I(\infty)$ in $\Delta^*(X)$ (see Proposition 3.44).

The statement about the closure of $G^I \cdot m_I$ in $\overline{X}^F$ follows from the fact that $\mathcal{M}_1(K^I \cdot \dot{e})$ is a closed subset of $\mathcal{M}_1(\mathcal{F})$ since the orbit $K^I \cdot \dot{e} = P^I \cdot \dot{e}$ is closed. □

The embeddings of the boundary symmetric spaces X^I into the boundary of $\overline{X}^F$ and the fact that this is a G-compactification of X characterizes this compactification. The next result, is Satake's characterization of his maximal compactification $\overline{X}^S$ (see Proposition 4.42).

9.56. Theorem. *The G-space $Y = \overline{X}^F$ is characterized by the following conditions:*

(1) *Y is a G-compactification of X.*
(2) *For any $I \subset \Delta$, there exists a P^I-equivariant embedding ι_I of X^I into Y. Let $\iota_I(X^I)$ be denoted by X^I_∞ and $o_I = \iota_I(o)$.*
(3) *$Y = \cup_{I \subset \Delta} G \cdot X^I_\infty$ and the condition $g \cdot X^I_\infty \cap X^J_\infty \neq 0$, where $g \in G$, and I and J are subsets of Δ, implies that $I = J$, and $g \in P^I$*
(4) *The closure of $A^+ \cdot o$ in Y is equal to $\cup_{I \subset \Delta} \overline{A^{I,+} \cdot o_I}$.*
(5) *If $I, J \subset \Delta$ then a sequence $(a^I_n \cdot o_I) \subset \overline{A^{I,+} \cdot o_I}$ converges to $a^J \cdot o_J \in A^{J,+} \cdot o_J$ if and only if $J \subset I$ and, for $\alpha \in J, \lim_n \alpha(a^I_n) = \alpha(a^J)$ and, for $\alpha \in I \setminus J, \lim_n \alpha(a^I_n) = +\infty$.*

Proof. One first verifies that $Y = \overline{X}^F = \overline{G \cdot m}$ satisfies these conditions. Note that $X^\Delta = X$ and $\iota_\Delta(x) = x$ for all $x \in X$. In addition $X^\emptyset = \{o\}$. The first two have been established already (see Theorem 9.42 and Proposition 9.55). The third condition is a consequence of Propositions 9.45 and 9.55.

Condition (4) is an immediate consequence of Corollary 9.54. Finally, the convergence condition (5) follows from Proposition 9.55.

The converse was proved earlier in Chapter IV. The second part of the proof of Theorem 4.43 shows that any G-compactification satisfying the five conditions is isomorphic to the dual cell compactification. The result follows from Theorem 9.47. More precisely, it follows from the equivalence of (1) and (5) in this theorem as proved by Corollary 9.46. □

CHAPTER X

COMPACTIFICATION VIA THE GROUND STATE

The K-invariant probability m on $\mathcal{F} = G/P$ represents, by means of the square root of the Poisson kernel, a unique solution of the equation $Lu + \lambda_0 u = 0$ with $u(o) = 1$. It is the spherical function Φ defined by $\Phi(g\cdot o) = \int_K e^{-\rho(H(g^{-1}k))}dk = \int_{\mathcal{F}} P^{1/2}(x,b)dm(b)$, where $x = g\cdot o \in X, b = kP$, and $P(x,b) = e^{-2\rho(H(g^{-1}k))}$ is the Poisson kernel on X (see § 7.21 and § 8.27). It plays a basic role in harmonic analysis on semisimple groups, for example it dominates all the spherical functions associated with the unitary principal series, and is called the Harish-Chandra spherical function (see [G1]). It arose earlier in Chapter VII when determining the limit functions for the Martin compactification at the bottom of the positive spectrum.

Anker in [A5] showed that, for $H \in \overline{\mathfrak{a}^+}$, up to polynomials terms in H, $\Phi(\exp H \cdot o)$ is comparable to $e^{-\rho(H)}$, as stated in formula (8.13). It is of fundamental interest to know its asymptotic behavior.

It turns out that the spherical function Φ takes its maximum value of 1 only at the origin o. This fact has the immediate consequence that the symmetric space can be identified with the orbit of Φ under the action of G on the functions f on X normalized to take the value 1 at o. (Recall that under this action $S_g f(x) = f(g^{-1}\cdot x)/f(g^{-1}\cdot o)$.)

The representing measure m_g for $S_g\Phi$ is obtained from the measure m that represents Φ by a "twisted action" that involves the cocycle $\sigma(g,b) = P^{1/2}(g^{-1},b)$. As a result, one may embed X into $\mathcal{M}_1(\mathcal{F})$ by associating with $g\cdot o$ the measure m_g. It is shown in this chapter, see Theorem 10.9, that the closure of the image of X under this embedding is yet another realization of the maximal Satake–Furstenberg compactification. This means that this compactification encodes some of the qualitative aspects of the asymptotics of Φ.

It follows from the formula for h_I and Theorem 10.9 that the closure of the orbit in $\partial X(\lambda_0)$ of this limit function h_I can be identified with the Martin compactification of the boundary symmetric space X^I at the bottom of the positive spectrum. This fact is closely related to Propositions 3.44 and 9.55.

The chapter starts with a discussion of the twisted action of G on probability measures that is associated with a cocycle.

THE TWISTED ACTION

§ 10.1. Given a left G-space B there is a natural left action of G on functions defined on B given by the formula $f_g(x) = f(g^{-1}\cdot x) \stackrel{\text{def}}{=} L_g f$. This dualizes to a left action on measures $g\cdot\mu \stackrel{\text{def}}{=} L_g\mu$, where $\int(L_g f)d\mu =$

$\int f(g^{-1}\cdot x)\mu(dx) \stackrel{\text{def}}{=} \int f(x)(g^{-1}\cdot\mu)(dx) = \int f d(L_{g^{-1}}\mu)$. (Recall that the dual of a left action is a right action, which accounts for the g^{-1}.) Other left actions are given by a multiplier. A (complex) function σ on $G\times B$ is called a **multiplier** (see Bourbaki [B14, Ch. VIII, p. 132]) if it satisfies the **cocycle condition**

$$\sigma(g_1g_2,x) = \sigma(g_1,g_2\cdot x)\sigma(g_2,x), \text{ for all } g_1,g_2\in G \text{ and } x\in B.$$

If σ is a multiplier, let $\tau(g)f$ be defined by $\tau(g)f(x) = \sigma(g^{-1},x)f(g^{-1}\cdot x)$.

Then $f\to\tau(g)f$ is a linear representation of G on functions on B and, hence, a left action on functions. It too dualizes to a left action on measures μ: let $(\tau(g)\mu)(dx) = \sigma(g,g^{-1}\cdot x)(g\cdot\mu)(dx)$. In other words, if φ is continuous with compact support, $\int(\tau(g)\varphi)d\mu = \int\varphi(g^{-1}\cdot x)\sigma(g^{-1},x)\mu(dx) = \int\varphi(x)\sigma(g^{-1},g\cdot x)(g^{-1}\cdot\mu)(dx) = \int\varphi d(\tau(g^{-1})\mu)$ (one uses the fact that $1=\sigma(g,g^{-1}\cdot x)\sigma(g^{-1},x) = \sigma(g^{-1},g\cdot x)\sigma(g,x)$).

This left action on measures no longer preserves total mass. Hence, one needs to normalize $\tau(g)\mu$ to get a probability measure. This, then, gives a non-linear **twisted** (left) **action** of G on probabilities: $\mu\to S_{g^{-1}}\mu$, where

$$\int\varphi dS_g\mu = \frac{\int\varphi(g\cdot x)\sigma(g,x)d\mu(x)}{\int\sigma(g,x)d\mu(x)}. \tag{10.2}$$

Remark. Note that the twisted action on measures on a compact G-space B is continuous in the weak topology for measures when the cocycle is continuous in b. This will be the case in what follows.

Let $B=\mathcal{F}$. Then there is a natural action of G on $\mathcal{F}=G/P=K/M$. In addition, the **Poisson kernel** $P(g,b) \stackrel{\text{def}}{=} e^{-2\rho(H(g^{-1}k))}$, where $kP = kM = b$, determines the multiplier $\sigma_1(g,b) = P(g^{-1},b)$ on $\mathcal{F}$.

10.3. Lemma. *Let $\sigma(g,b) = P^{1/2}(g^{-1},b) = h_b(g^{-1}\cdot o)$. Then σ and $\sigma_1=\sigma^2$ are multipliers, i.e., they satisfy the cocycle condition.*

Proof. Since $\log\sigma_1(g,b) = -2\rho(H(gk))$, it is enough to show that, for all $g_1,g_2\in G$, and $k\in K$,

$$H(g_1g_2k) = H(g_1k(g_2k)) + H(g_2k). \tag{10.4}$$

If $g_2k=\ell_2a_2n_2$ and $g_1\ell_2=\ell_1a_1n_1$, with $\ell_i\in K$, $a_i\in A$, and $n_i\in N$, then $\ell_2=k(g_2k)$ and $g_1g_2k=\ell_1a_1n_1a_2n_2=\ell_1a_1a_2n_1'n_2$. □

10.5. Remarks. (1) Observe that, by § 7.21, eq. (m), the functions $h_b(g\cdot o) = P^{1/2}(g,b) = \sigma(g^{-1},b), b\in\mathcal{F}$, are the minimal functions associated with $Lu+\lambda_0u=0$. Hence, S_gh_b is a minimal function (since $h\to S_gh$ is an isomorphism of the convex set of positive solutions h of the equation $Lu+\lambda_0u=$ for which $u(o)=1$). In fact, $S_gh_b=h_{g\cdot b}$. This identity is equivalent to the cocycle condition. Let $x=g_0\cdot o$, $g_1=g^{-1}g_0$, and

$b = kM$. Then $h_b(g^{-1} \cdot x) = e^{-\rho(H(g_1^{-1}k))} = e^{-\rho(H(g_0^{-1}k(gk))+H(gk))} = h_{g \cdot b}(x)h_b(g^{-1} \cdot o)$. Conversely, if $h_b(g^{-1} \cdot x) = h_{g\cdot b}(x)h_b(g^{-1} \cdot o)$, then $\sigma(g_0^{-1}g, b) = \sigma(g_0^{-1}, g \cdot b)\sigma(g, b)$.

(2) When $k \in M$, eq. (10.4) implies the following property of the Busemann function $d_z(x)$:

$$(\dagger) \qquad d_z((g_1g_2)^{-1} \cdot o) = d_{g_2 \cdot z}(g_1^{-1} \cdot o) + d_z(g_2^{-1} \cdot o)$$

if $z = [\gamma]$, where γ is a unit speed geodesic with $\gamma(0) = o$ and $\dot\gamma(0) = L \in \overline{\mathfrak{a}^+}$.

To see this, note that, by the remarks on the Busemann function in § 3.1, if $L \in \mathfrak{a}^+$, one has $d_\gamma(g_1g_2)^{-1} \cdot o) = B(L, H(g_1g_2))$ and $d_\gamma(g_2^{-1} \cdot o) = B(L, H(g_2))$. Since $L = \gamma(\dot{0}) \in \mathfrak{a}^+$, it follows from Lemma 13.26(3) that $d_{g_2 \cdot \gamma}(g_1^{-1} \cdot o) = d_{k(g_2)\cdot\gamma}(g_1^{-1} \cdot o) = B(L, H(g_1k(g_2)))$.

When $L \in \overline{\mathfrak{a}^+}$ the same result holds by continuity. It follows that each $L \in \overline{\mathfrak{a}^+}$ determines a cocycle $\delta_L(g, b)$ on $G \times \mathcal{F}$, called the **Busemann cocycle corresponding to** L, defined by setting $\delta_L(g, b) = e^{-B(L,H(gk))}$. Note that $\log \delta_L(g, b) = d_\gamma((gk)^{-1} \cdot o)$ if $\gamma(0) = o, \dot\gamma(o) = L$ and $b = kM$. If $z = k \cdot L(\infty) \in X(\infty)$, then (see § 3.1) $d_z(g^{-1} \cdot o) = d_{L(\infty)}(k^{-1}g^{-1} \cdot o) = \log \delta_L(g, b)$. Hence, $\delta_L(g, b)$ is a function $\delta(g, z)$ of g and z.

Since the G-action on $\mathcal{F}$ is such that $g \cdot b = k(gk) \cdot \dot{e}$ if $b = k \cdot \dot{e} = kM$ (i.e., it is induced by the Iwasawa action), it follows that $\delta(g, z)$ is a cocycle on $G \times X(\infty)$. This is shown in Proposition 13.26 where eq (†) is proved geometrically. (In essence, it amounts to observing that the cocycle $\delta_L(g, b)$ factors through the natural map of G/P onto the orbit $G \cdot L(\infty)$.)

The twisted action of G on probability measures that is associated with σ is the twisted action defined in Chapter VII following Proposition 7.28.

10.6. Proposition. *Let h be a positive solution of $Lu + \lambda_0 u = 0$ and let μ denote its representing measure. Then $S_g\mu$ represents S_gh, i.e.,*

$$S_gh(x) = \frac{\int h_b(x)\sigma(g, b)d\mu(b)}{\int \sigma(g, b)d\mu(b)} = \int h_b(x)S_gd\mu(b).$$

In particular, if Φ is the ground state (see § 7.21 eq. (s)), the representing measure S_gm of $S_g\Phi$ has Radon–Nikodym derivative $\dfrac{dS_gm}{dm}(b) = \dfrac{\sigma(g^{-1}, b)}{\Phi(g^{-1} \cdot o)}$.

Proof. Let $x = g_1^{-1} \cdot o$. Then, since $h(x) = \int h_b(x)d\mu(b)$, it follows that

$$h(g^{-1}g_1^{-1} \cdot o) = \int \sigma(g_1g, b)d\mu(b) = \int \sigma(g_1, g \cdot b)\sigma(g, b)d\mu(b).$$

Hence, by eq. (10.2), letting $g_1 = e$ and noting that $\sigma(e, g \cdot b) = 1$,

$$S_gh(x) = \frac{h(g^{-1} \cdot x)}{h(g^{-1} \cdot o)} = \frac{\int \sigma(g_1, g \cdot b)\sigma(g, b)d\mu(b)}{\int \sigma(g, b)d\mu(b)}$$
$$= \int \sigma(g_1, b)S_gd\mu(b) = \int h_b(x)S_gd\mu(b).$$

When $\mu = m$, by letting $b_1 = g \cdot b$, it follows, from eq. (9.12) for the Radon–Nikodym derivative of $g \cdot m$ and the cocycle condition, that

$$\begin{aligned}\Phi(g^{-1} \cdot x) &= \int h_b(g^{-1} g_1^{-1} \cdot o) dm(b) = \int \sigma(g_1 g, b) dm(b) \\ &= \int \sigma(g_1, g \cdot b) \sigma(g, b) dm(b) \\ &= \int \sigma(g_1, b_1) \sigma(g, g^{-1} \cdot b_1) d(g \cdot m)(b_1) \\ &= \int \sigma(g_1, b_1) \sigma(g, g^{-1} \cdot b_1) \sigma^2(g^{-1}, b_1) dm(b_1) \\ &= \int \sigma(g_1, b_1) \sigma(g^{-1}, b_1) dm(b_1),\end{aligned}$$

where the last equality follows since $\sigma(g, g^{-1} \cdot b_1)\sigma(g^{-1}, b_1) = \sigma(e, b_1) = 1$. This computes the Radon–Nikodym derivative of $S_g m$ with respect to m since $\sigma(g_1, b_1) = h_{b_1}(x)$. □

This result, in view of Proposition 7.22 and Lemma 7.30, has the following immediate corollary.

10.7. Corollary. *The stabilizer in G of m_I under the twisted action is R^I. Hence, this group is the subgroup of elements g of G for which $L_g h_I$ is proportional to h_I.*

Remark. Note that the coefficient of proportionality in the last statement is of course $\chi(g) = h_I(g^{-1} \cdot o)$. This is a positive character on R^I and so, being trivial on $K^I M$ and N_I, factors through the map $K^I M A_I N_I \to A_I$. Its restriction to A_I coincides with $e^{-\rho}$.

Since the normalization of the natural action of G on functions and the twisted action defined by the multiplier σ on measures are interlaced by the map $h \to \mu$, where μ represents h, one could hope for a way of embedding the symmetric space into $\mathcal{M}_1(\mathcal{F})$ that interlaces the natural action on X and the twisted action on measures (note that in the case of the maximal Satake–Furstenberg compactification $\overline{X}^{SF}$, the embedding $g \cdot o \to g \cdot m$ interlaces both natural actions of G). Given such a twisted embedding, it is also reasonable to hope that the limit measures would in fact be the representing measures themselves: more precisely, the resulting compactification would hopefully be isomorphic with the Martin compactification in such a way that a harmonic function and its representing measure correspond under the isomorphism.

It will now be shown that such a twisted embedding exists. The key to it is to make use of the ground state Φ defined by eq. (s) in § 7.21.

The ground state has an orbit under the action $g \to S_g \Phi$ that can be identified with X. This a consequence of the following lemma.

10.8. Lemma. *The ground state Φ has maximum value 1. It takes this maximum value only at o. Consequently, the functions $S_g\Phi, g \in G$ parametrize X, i.e., the map $g \cdot o \to S_g\Phi$ is a bijection. Further, it is a topological embedding of X into the function space of positive solutions of $Lu+\lambda_0 u = 0$ equipped with the topology of uniform convergence on compact subsets of X, i.e., $g_n \cdot o \to g \cdot o$ if and only if $S_{g_n}\Phi \to S_g\Phi$ uniformly on compact sets.*

Proof. $\Phi(x) = \int_{\mathcal{F}} h_b(x)dm(b) \leq \sqrt{\int_{\mathcal{F}} h_b^2(x)dm(b)} \leq 1 = \Phi(o)$, by the Cauchy-Schwarz inequality and the fact that $h_b^2(g \cdot o) = P(g,b)$ is the Poisson kernel. Also, there is equality if and only if the continuous function $b \to h_b(x)$ is the constant function 1. If $x = g \cdot o$ this is equivalent to $H(g^{-1}k) = 0$ for all $k \in K$. If $g = k_1ak'$ with $a \in \overline{A^+}$, by the Cartan decomposition, this implies $a = e$ and so $g \in K$.

It follows that $S_g\Phi$ attains its maximum only at the point $g \cdot o$. Since $S_{gk}\Phi = S_g\Phi$, for all $g \in G, k \in K$, it follows that the map $g \cdot o \to S_g\Phi$ is a well-defined bijection. Assume that $g_n \cdot o \to g \cdot o$. Since $S_{g_n}\Phi(x) \to S_g\Phi(x)$ pointwise and all these functions take the value 1 at o, it follows from Harnack's inequality that the convergence is uniform on compact subsets. Conversely, assume that $S_{g_n}\Phi \to S_g\Phi$ uniformly on compact sets. From the uniqueness of the maximum, it follows that, for any compact neighborhood of $g \cdot o$, the points $g_n \cdot o$ are eventually in it. □

Compactification of X via the Ground State

It follows from Proposition 10.6 and Lemma 10.8, that the map $g \cdot o \to S_g m$ is an embedding of X into the space $\mathcal{M}_1(\mathcal{F})$ of probability measures on $\mathcal{F}$ that interlaces the natural action on X and the twisted action on $\mathcal{M}_1(\mathcal{F})$. Denote by $\overline{X}^T$ the closure in $\mathcal{M}_1(\mathcal{F})$ of the set $\{S_g m \mid g \in G\}$. This defines yet another G-compactification of X that has the desired property, as stated in the next theorem.

Remark. Note that the embedding $g \cdot o \to S_g m$ is not the same as the embedding $g \cdot o \to g \cdot m$ that defines the maximal Satake–Furstenberg compactification since

$$\sigma(g^{-1}, b)\Phi^{-1}(g^{-1} \cdot o) = \frac{dS_g m}{dm}(b) \neq \sigma^2(g^{-1}, b)$$

that, by eq. (9.12), is the Radon–Nikodym derivative $\dfrac{d(g \cdot m)}{dm}(b)$

10.9. Theorem. *The compactification $\overline{X}^T$ is isomorphic to (and distinct from Furstenberg's realization of) the maximal Satake–Furstenberg compactification $\overline{X}^{SF}$ as a G-compactification.*

Proof. This theorem is equivalent to showing that the compactifications $\overline{X}^T$ and $X \cup \partial X(\lambda_0)$ are isomorphic G-compactifications. This is done by showing, in Corollary 10.13, that if (a_n) is I-canonical then $S_{a_n}\Phi \to h_I$. □

The first step in the argument that eventually establishes Corollary 10.13 is the following formal lemma.

10.10. Lemma. *Let h be a limit function and let μ be a probability measure on $\mathcal{F} = K/M$. The following conditions are equivalent:*

(1) *if $(g_n) \subset G$, then $S_{g_n}m \to \mu$ implies $g_n \cdot o \to h$, where h is represented by μ;*
(2) *if $(g_n) \subset G$, then $g_n \cdot o \to h$ implies $S_{g_n}m \to \mu$, where μ is the representing measure of h; and*
(3) *if (a_n) is I-canonical, $S_{a_n}m \to m_I$.*

Proof. In view of Lemma 10.8, it is clear that in (1) and (2) one may assume the sequence is converging to infinity.

To show that (1) implies (2), let (g_{n_k}) be a subsequence such that $S_{g_{n_k}}m \to \nu$. By (1), ν is the representing measure of h. Hence, $S_{g_n}m \to \mu$, the representing measure of h.

Corollary 7.24 and Theorem 7.22 imply that (3) follows from (2).

Assume (3). Since the twisted action on measures is continuous, it follows from (3) that, for any I-fundamental sequence (y_n) with $y_n = k_n a_n \cdot o$, if $y_n = g_n \cdot o$, then $S_{g_n}m \to (m_I)_{ka}$ which, by Proposition 10.6, is the representing measure of $S_{ka}h_I$. (As usual, $k_n \to k$ and the projection of a_n onto A^I converges to a.) As a result, (1) holds whenever the sequence is fundamental.

Assume that (g_{n_k}) is a fundamental subsequence of the given sequence (g_n). It follows that (i) $\mu = \lim_k S_{g_{n_k}}m$ and, hence, (ii) μ is the representing measure of the limit function $\lim_k g_{n_k}$. Consequently, the limit function is independent of the particular fundamental subsequence. By Proposition 7.28, this implies that $(g_n \cdot o)$ converges to this common limit function, whose representing measure has been shown to be μ. Hence, (3) implies (1). □

Note that once part (3) of Lemma 10.10 is established, Lemma 10.10 shows that the compactifications $\overline{X}^T$ and $X \cup \partial X(\lambda_o)$ are isomorphic. They are then also G-isomorphic in view of Proposition 10.6.

For positive solutions of $Lu + \lambda_0 u = 0$, normalized to take value 1 at o, the topology of uniform convergence on compact subsets of X is equivalent to the topology of weak convergence for the representing measures. As a result, part (3) of Lemma 10.10 is equivalent to the following statement.

$$\text{(10.11)} \qquad \text{If } (a_n) \text{ is } I\text{-canonical, then } S_{a_n}\Phi \to h_I.$$

The verification of statement (10.11) involves the study of the orbit of the limit function h_I under the group A^I. To understand this orbit, first note that if $I_1 \subset I$, it induces a decomposition of $N^I A^I$. This is because the roots of $\mathfrak{g}^I$ with respect to $\mathfrak{a}^I$ coincide with the restrictions to $\mathfrak{a}^I$ of the roots of $\mathfrak{g}$ with respect to $\mathfrak{a}$ that are in Σ^I. Consequently, I can be identified with the set $\Delta(\mathfrak{g}^I, \mathfrak{a}^{I,+})$ of simple positive roots and

I_1 is a subset of the set $\Delta(\mathfrak{g}^I, \mathfrak{a}^{I,+})$. Let $\mathfrak{n}^{I,I_1} = \mathfrak{n}^{I_1} = \Sigma_{\alpha\in\Sigma^{I_1,+}}\mathfrak{g}_\alpha$ and $\mathfrak{n}^I_{I_1} = \sum_{\alpha\in\Sigma^{I,+}\setminus\Sigma^{I_1}}\mathfrak{g}_\alpha$. Then N^I is the semidirect product of N^{I_1} and $N^I_{I_1}$ as N^{I_1} conjugates $N^I_{I_1}$. Also $N_{I_1}\supset N_I$ is a semidirect product of $N^I_{I_1}$ and N_I because $N^I_{I_1}$ conjugates N_I. If $n_1\in N^I$, it has a unique decomposition as $n_1 = n_{12}n_{11}, n_{12}\in N^I_{I_1}, n_{11}\in N^{I_1}$. Furthermore, $x^I = n_1a_1\cdot o = n_{12}a_{12}\cdot x^{I_1}$, where $a_1\in A^I$ is written as $a_{12}a_{11}, a_{12}\in A^I_{I_1}$, and $a_{11}\in A^{I_1}$.

Part of the next proposition can be seen as a reflection of the geometry of $\Delta^*(X)$. Namely, the fact that the limit of an I_1-fundamental sequence in the flat $A^I\cdot o\subset X^I$, when viewed at infinity, is the limit of a corresponding I_1-fundamental sequence in the flat $A\cdot o\subset X$. This fact is proved here directly, using analytic techniques.

Proposition 10.12. *Let $a = a_1a_2\in A$, with $a_1\in A^I$ and $a_2\in A_I$. Then, if $x = n_Ia_I\cdot x^I$,*

$$S_ah_I(x) = S_{a_1}h_I(x) = \{S_{a_1}\Phi^I(x^I)\}e^{\rho(\log a_I)} = \frac{\Phi^I(a_1^{-1}\cdot x^I)}{\Phi^I(a_1^{-1}\cdot o)}e^{\rho(\log a_I)}.$$

Furthermore, if I_1 is a subset of I and (a_n) is an I_1-fundamental sequence in A^I, then $\lim_{n\to\infty} S_{a_n}h_I = S_ah_{I_1}$, where a is the limit of the projection onto A^{I_1} of the sequence (a_n).

In particular, if (a_n) is a sequence in A^I such that $\lim_{m\to\infty}\alpha_i(\log a_n) = +\infty$ for all $\alpha_i\in I$, then $\lim_{n\to\infty}S_{a_n}h_I = h_o(= h_\emptyset)$.

Proof. Corollary 10.7 implies that $S_ah_I(x) = S_{a_1}h_I(x)$. The action of a_1 on h_I follows from the formula for h_I.

To prove the second statement it is enough, by the continuity of the action, to prove this when $a = e$, i.e., when (a_n) is I_1-canonical in A^I. Let (b_m) be an I-canonical sequence in A. Then

$$h_I(x) = \lim_{m\to\infty}\frac{G(x, b_m\cdot o)}{G(o, b_m\cdot o)}.$$

Hence, if h_n denotes $S_{a_n}h_I$,

$$h_n(x) = \lim_{m\to\infty} g_{m,n}(x), \text{ where } g_{m,n}(x) \overset{\text{def}}{=} \frac{G(x, a_nb_m\cdot o)}{G(o, a_nb_m\cdot o)}.$$

Fix a point $x\in X$. Let $\varepsilon > 0$. Then $|g_{m,n}(x) - h_n(x)| < \varepsilon, m\geq m(x, n, \varepsilon)$, where one may, for example, assume $m(x, n, \varepsilon)\geq n$.

By choosing a subsequence if necessary, one may assume that $\lim_{n\to\infty}h_n$ exists, uniformly on compact subsets of X. Let f denote this limit function. Let $n_\ell\geq\ell$ be such that $|f(x) - h_n(x)| < \frac{1}{2^\ell}, n\geq n_\ell$.

Set $m_n = m(x, n, \frac{1}{2^n})$ and let $c_n = a_nb_{m_n}$. Then (c_n) is an I_1-canonical sequence in A. Hence, $\lim_{n\to\infty} g_{m_n,n}(x) = h_{I_1}(x)$. As a result,

$$\begin{aligned}|f(x) - h_{I_1}(x)| &\leq \frac{1}{2^\ell} + |h_n(x) - g_{m_n,n}(x)| + |g_{m_n,n}(x) - h_{I_1}(x)|\\ &\leq \frac{1}{2^\ell} + \frac{1}{2^n} + |g_{m_n,n}(x) - h_{I_1}(x)|,\end{aligned}$$

if $n \geq n_\ell$. This implies $f(x) = h_{I_1}(x)$. □

10.13. Corollary. *If (a_n) is an I-canonical sequence, then $S_{a_n}\Phi \to h_I$.*

Proof. Consider the auxiliary space $X \times D$, where D is the hyperbolic disc D.

Let $G_1 = G \times SU(1,1)$. If $\{\alpha_1, \alpha_2, \ldots, \alpha_r\}$ denotes the set $\Delta(\mathfrak{g}, \mathfrak{a}^+)$ of simple roots with respect to $\mathfrak{a}^+$ and β is the unique positive root of $\mathfrak{su}(1,1)$ with respect to $\mathfrak{a}_0^+$, a choice of positive Weyl chamber for $\mathfrak{su}(1,1)$, then $\{\alpha_1, \alpha_2, \ldots, \alpha_r\} \cup \{\beta\}$ is the set $\Delta(\mathfrak{g}_1, \mathfrak{a}^+ \times \mathfrak{a}_0^+)$ of simple roots for $\mathfrak{g}_1$ with respect to the Weyl chamber $\mathfrak{a}^+ \times \mathfrak{a}_0^+$. Denote a generic limit function on $X \times D$ by $\tilde{h}$.

Let $I_0 = \{\alpha_1, \alpha_2, \ldots, \alpha_r\}$ and let $(ka \cdot o, b_\ell \cdot o)$ be an I_0-fundamental sequence in $X \times D$. Since, by Theorem 7.22, $\tilde{h}_{I_0} = \Phi(x)h_1(y)$, where $h_1(y)$ for $y \in D$, is the limit on D of the sequence $(b_\ell \cdot o)$ and Φ is the spherical function on X, it follows that $(ka \cdot o, b_\ell \cdot o)$ converges to $S_{ka}\tilde{h}_{I_0}(x,y) = S_{ka}\Phi(x)h_1(y)$.

Denote by I a proper subset of I_0. Let $(a_n \cdot o)$ be I-fundamental on X. Then, by Proposition 10.12, $S_{a_n}\tilde{h}_{I_0} \to S_a\tilde{h}_I$. Since $S_{a_n}\tilde{h}_{I_0}(x,y) = S_{a_n}\Phi(x)h_1(y)$, the result follows from the fact that $\tilde{h}_I(x,y) = h_I(x)h_1(y)$ if $I \subsetneq I_0$ and $S_a\tilde{h}_I(x,y) = S_a h_I(x)h_1(y)$. □

CHAPTER XI

HARNACK INEQUALITY, MARTIN'S METHOD AND THE POSITIVE SPECTRUM FOR RANDOM WALKS

The study of positive eigenfunctions of the Laplace operator L is closely related to the study of convolution equations defined by probability measures p. With applications to other non-semisimple Lie groups in mind, several results for general convolution equations on a locally compact metrizable group H are established in this chapter.

More specifically, a Harnack inequality is proved in Theorem 11.5 for the positive solutions of convolution equations and the corresponding Martin compactification is determined, in general terms, in Theorem 11.12. As in the case of the Laplace operator, a convolution operator $*p$ has a positive spectrum that contains its infimum. This infimum is determined in Theorem 11.17 and, as shown in Proposition 11.19, is less than or equal to the spectral radius of the convolution operator in $L^2(H)$, with equality when p is symmetric. A fixed-line property is introduced in Definition 11.28 and is used to compute this infimum in Theorem 11.40 and the spectral radius in Proposition 11.43.

The results and methods developed here will play an important role in the following chapters that are devoted to the study of the eigenfunctions of convolution equations and the corresponding Martin compactifications.

BASIC NOTATIONS

Denote by Y a locally compact, metrizable topological space and by $C_c(Y)$ the space of compactly supported continuous functions on Y. Let $R(Y) = C_c(Y)^*$ denote the space of **Radon measures** on Y and $C(Y)$ the space of continuous functions on Y. The sets of positive elements of these spaces will be denoted, respectively, by $C_c^+(Y), R^+(Y)$, and $C^+(Y)$. The space $R(Y)$ will be endowed with the topology of weak convergence. Unless otherwise stated, the space $C(Y)$ will be equipped with the topology of uniform convergence on compact sets. The integral $\int \phi d\mu$ will also be denoted by $\mu(\phi)$.

Let H denote a locally compact metrizable topological group that acts continuously on Y. The formula $(\mu * \nu)(\phi) = \int \phi(x \cdot y) d\mu(x) d\nu(y)$ if $\mu \in R^+(H), \nu \in R^+(Y)$, and $\phi \in C_c^+(Y)$ makes sense and defines the convolution $\mu * \nu$. This is a Radon measure on Y if μ or ν has a compact support. When $Y = H$ this defines the **convolution of positive measures** on H. Let $\check{\mu}$ denote the image of $\mu \in R(H)$ under the symmetry $g \to g^{-1}$. Choose a right invariant Haar measure η on H and, letting dg

denote $d\eta(g)$, consider the scalar product on $L^2(H)$ defined by

$$\langle f, f' \rangle = \eta(f\overline{f'}) = \int_G f\overline{f'}dg.$$

If μ has a density f with respect to the right Haar measure on H, the same is true of $\mu * \nu$ and the density $f * \nu$ is given by $(f * \nu)(x) = \int f(xy^{-1})d\nu(y)$. If ν is a bounded measure, then $\langle f * \nu, f' \rangle = \langle f, f' * \check{\nu} \rangle$ and, moreover, the linear map $f \to f * \nu$ on $L^2(H)$ is bounded and its norm is the total mass of ν. In addition, if $\phi \in C_c(H)$ one has $(\mu * \nu)(\phi) = \mu(\phi * \check{\nu})$.

Let δ denote the **modular function** of H, defined by $\delta_x * \eta = \delta(x)\eta$. If $f \in C_c(H)$ define the function $\check{f}$ to be the density of the measure $(f\eta)\check{}$ with respect to η. Then $\check{f}(x) = \delta(x)\tilde{f}(x)$, where $\tilde{f}(x) \overset{def}{=} f(x^{-1})$. Measures that are absolutely continuous with respect η and their densities will be identified. For example, if $\nu = \varphi\eta$, where φ is a compactly supported Borel function, then $\mu * \varphi$ denotes $\mu * \nu$ and its density f. Note that $(\mu * \varphi)(x) = f(x) = \mu(\delta_x * \check{\nu}) = \mu[\delta\tilde{\varphi}_x] = \int \delta(y)\varphi(y^{-1}x)d\mu(y)$. Also, $f\eta * g\eta = (f * g)\eta$, where $(f * g)(x) = \int f(xy^{-1})g(y)d\eta(y)$.

It is well-known, and often used in what follows, that if ν is a measure with compact support and continuous density, the map $\mu \to \mu * \nu$ from $R(H)$ into $C(H)$ is continuous with respect to the natural topologies.

11.1. Definition. A positive measure p on H will be said to be **well-behaved** if it has a continuous density with compact support S such that $H = \cup_{n>0} S^n$.

11.2. Remark. If H admits a well-behaved measure, it is compactly generated.

11.3. Definition. For a well-behaved measure p on H, denote by $\mathcal{C}$ (respectively, $\tilde{\mathcal{C}}$) the cone of positive Radon measures (respectively, continuous functions) such that

$$\mu * p \leq \mu \text{ (respectively, } f * p \leq f),$$

and denote by $\mathcal{H}$ (respectively, $\tilde{\mathcal{H}}$) the cone of positive measures (respectively, functions) that satisfy the equation

$$\mu * p = \mu \text{ (respectively, } f * p = f).$$

11.4. Remarks. (1) These cones may be trivial, i.e., equal to $\{0\}$. This situation is discussed later.

(2) It is clear that, for a unique $r > 0$, $rp = p_1$ is a probability. Then $f * p = f$ if and only if $f * p_1 = rf$, $r > 0$. This last equation is analogous to the equation $Lf = r'f$, where L is the Laplacian. The role of L is played by the measure $p_1 - \delta_e$ and the role of r' is played by $r - 1$, because, if

$f \geq 0$, f superharmonic (i.e., $Lf \leq 0$) corresponds to f excessive (i.e., $f * p_1 \leq f$).

(3) Free use will be made of the theory of convex cones, extremals, Choquet simplices and integral representations as explained in [C2].

(4) In the following chapters, the equation $f * p = f$ will be considered in detail, mainly for a p that is K-bi-invariant, where K is a compact subgroup. In this chapter, K is taken to be $\{e\}$. The results obtained, when K is not trivial, can easily be interpreted in terms of the homogeneous space H/K (see the discussion preceding Definition 14.23).

(5) While this book is concerned with symmetric spaces, it is useful to realize that a Martin compactification may be defined by a convolution equation in the context of groups. This leads to interesting questions in the case of Lie groups that have no non-trivial compact subgroup K. To date, this type of question is very poorly understood, except for $\mathbb{R}^d$ and the group of rigid motions of $\mathbb{R}^d$. Information about this for some classes of solvable groups can be found in [D1], [B2], and [E2].

The discussion of the relation of the cones in Definition 11.3 with a corresponding Martin compactification and the calculation of this latter object when the group is a semisimple group are the main purpose of the following chapters. The present chapter is mainly concerned with general facts and can be considered as an introduction to the later chapters. Closely related considerations can be found in [F4], [G13], and [G14].

Cones with compact bases and the Harnack inequality

11.5. Theorem. *Consider the cones $\mathcal{C}, \tilde{\mathcal{C}}, \mathcal{H}, \tilde{\mathcal{H}}$ defined in Definition 11.3. The equation $\mu(\check{p}) = 1$ (respectively, $\check{p}(f) = 1$) defines a base $\mathcal{C}_1$ of $\mathcal{C}$ (respectively, a base $\tilde{\mathcal{C}}_1$ of $\tilde{\mathcal{C}}$) that is compact in the weak topology of $\mathcal{C}$ (respectively, $\tilde{\mathcal{C}}$). The cone $\mathcal{H}$ is naturally identified with the cone $\tilde{\mathcal{H}}$ and the equation $f(e) = 1$ defines a compact base $\mathcal{H}_1$ of $\tilde{\mathcal{H}} = \mathcal{H}$. Moreover, the weak topology on $\tilde{\mathcal{H}} = \mathcal{H}$ coincides with the topology of uniform convergence on compact sets. Furthermore, if C is a compact subset of H, there exists a constant $\gamma(C)$ such that for every $f \in \mathcal{H}, u \in C, g \in H$, one has the inequality $f(gu) \leq \gamma(C) f(g)$.*

11.6. Lemma. *Let $p \in C_c^+(H)$. For $\varepsilon < p(1)^{-1}$, the series $\sum_{n=1}^{\infty} \varepsilon^n p^n$ converges uniformly as a series of functions and its sum is a strictly positive continuous function on H.*

Proof. The fact that p is positive implies the inequality

$$\|p^n\|_\infty \leq \|p\|_\infty [p(1)]^{n-1}.$$

Hence, the series $\sum_{n=1}^{\infty} \varepsilon^n p^n$ converges normally if $\varepsilon < [p(1)]^{-1}$. In particular, the sum $s(x) = \sum_{n=1}^{\infty} \varepsilon^n p^n(x)$ is continuous and non-negative. Clearly the equation $\varepsilon s * p + \varepsilon p = s$ implies $\varepsilon s * p \leq s$. Hence, the condition

$s(x) = 0$ implies $\int s(xy^{-1})dp(y) = 0$ and so $s(xy^{-1}) = 0$ as soon as $y \in S$. The condition $H = \cup_{n>0} S^n$ implies that $s = 0$ on H. Hence, s is strictly positive. □

Proof of Theorem 11.5.

Fix a positive function $\phi \in C_c^+(H)$ and consider the sequence of positive functions $\sum_{k=1}^n \varepsilon^k \check{p}^k$. By Lemma 11.6, this sequence converges uniformly to $s \geq \alpha > 0$ on the support of ϕ. Hence, for n large one has

$$(*) \qquad 0 \leq \phi \leq \frac{\|\phi\|_\infty}{\alpha} \sum_{k=1}^{n} \varepsilon^k \check{p}^k.$$

On the other hand, the relation $\mu * p \leq \mu$ implies $\mu(\check{p}^{k+1}) = \mu(\check{p} * \check{p}^k) = \mu * p^k(\check{p}) \leq \mu(\check{p})$. Hence, by inequality (*), if one takes $\varepsilon < 1$, it follows that $\mu(\phi) \leq \frac{\|\phi\|_\infty}{\alpha} \sum_{k=1}^{n} \varepsilon^k(\check{p}) \leq \frac{\|\phi\|_\infty}{\alpha} \frac{\varepsilon \mu(\check{p})}{1-\varepsilon}$.

If, for some $\mu \in \mathcal{C}$, one has $\mu(\check{p}) = 0$, this implies that $\mu(\phi) = 0$ for $\phi \in C_c^+(H)$. Hence, $\mu = 0$. It follows that the subset $\mathcal{C}_1$ of $\mathcal{C}$ defined by $\mu(\check{p}) = 1$ is a base of $\mathcal{C}$, i.e., every half-ray in $\mathcal{C}$ intersects $\mathcal{C}_1$.

As shown, it follows from inequality ($*$) that there exists, for each $\phi \in C_c^+(H)$, a constant $C(\phi)$ such that $\mu(\phi) \leq C(\phi)$ for every $\mu \in \mathcal{C}_1$. From this it follows that $\mathcal{C}_1$ is relatively compact in the weak topology. Since $\mu * p(\phi) = \mu(\phi * \check{p})$ and the support of p is compact, $\mathcal{C}$ is closed. It follows that the base $\mathcal{C}_1$ is closed and, hence, is compact.

If μ satisfies the equation $\mu * p = \mu$, then μ has a density f given by

$$f(x) = \mu(\delta_x * \check{p}) = \int \delta(y) p(y^{-1}x) d\mu(y).$$

This density is continuous and satisfies$f * p = f$. Conversely, if f satisfies $f * p = f$, then the measure μ with density f satisfies $\mu * p = \mu$.

Hence, this defines a map $\mu \longrightarrow f$ that gives a natural identification between the cones $\mathcal{H}$ and $\tilde{\mathcal{H}}$. The map $\mu \longrightarrow \mu * p$ of $\mathcal{H}$ into itself is the identity map and is continuous with respect to the weak topology and the topology of uniform convergence on compacts sets (see Lemma A.19). Observe that $\mathcal{C}_1 \cap \mathcal{H} = \mathcal{H}_1$ and so $\mathcal{H}_1$ is a compact base of $\mathcal{H}$ in the weak topology defined by the equation $\mu(\check{p}) = f(e) = 1$. Because $\mathcal{H}_1$ is compact and the identity map of $\mathcal{H}_1$ is continuous with respect to these topologies, the two topologies agree on $\mathcal{H}_1$ and, hence, on $\mathcal{H}$ itself. Applying the Ascoli theorem to the compact set $\mathcal{H}_1$, it follows that

$$\sup_{\substack{f \in \mathcal{H}_1 \\ u \in C}} f(u) \leq \gamma(C) < +\infty.$$

When $f \in \mathcal{H}$ and $g \in H$, the function f_g defined by $f_g(x) = \frac{f(gx)}{f(g)}$ belongs to $\mathcal{H}_1$. The previous inequality implies that

$$f(gu) \leq \gamma(C) f(g)$$

if $f \in \mathcal{H}, u \in C$, and $g \in H$. This is a uniform Harnack inequality. □

11.7. Remark. Uniformity in the Harnack inequality means that the ratio $\frac{f(x)}{f(y)}$ is bounded when $f \in \mathcal{H}$ and the left-invariant distance of x to y is bounded. The importance of this type of uniformity was pointed out to us by D. Stroock.

11.8. Remark. Since $\mathcal{H}$ is a lattice ordered cone, the compact set $\mathcal{H}_1$ is a **Choquet simplex** [C2], i.e., every element has a unique representation as a barycenter of extremals. Every **extremal element** f of $\mathcal{H}_1$ will also be called a **minimal function**, because it is characterized by the fact that for any $g \in \mathcal{H}$, $g \leq f$ implies that g is proportional to f (cf. § 6.1).

Martin's method for a random walk

Martin's classical procedure for using potential theory to compactify a domain in $\mathbb{R}^n$ has led to a number of variants that together may be referred to as Martin's method. The resulting space is called a Martin compactification.

Assume, as before, that p is a well-behaved probability measure and that $r > 0$ is a real number such that the series $\sum_{n=0}^{\infty} r^{-n} p^n$ converges in the weak topology. The Radon measure defined by this series will be denoted by V^r. It satisfies the equation

$$p * V^r = V^r * p = r(V^r - \delta_e).$$

It is clear that $V^r(x) = \sum_{n=1}^{\infty} r^{-n} p^n(x)$ is a strictly continuous, positive density for V^r on $H/\{e\}$. Consider the H-left invariant kernel $y \to \delta_y * V^r$ and denote its density on $H/\{y\}$ with respect to the right Haar measure by $V^r(x,y)$. This section is concerned with general facts about the asymptotic behavior of the family of functions $K^r(x,y) = \frac{V^r(x,y)}{\delta_y * V^r(\check{p})}$ when y tends to infinity in H. This leads to an integral representation for every element of $\mathcal{H}_1^r = \{f > 0 \mid f * p = rf, f(e) = 1\}$ in terms of extremal or minimal functions (that have geometrical significance).

11.9. Proposition. *Denote by $\mathcal{P}^r$ (respectively, C^r) the set of positive Radon measures μ^r that satisfy the equation $\mu^r * p = r(\mu^r - \delta_e)$ (respectively, $\mu * p \leq r\mu$). Then $\mathcal{P}^r \subset C^r$ and every $\mu^r \in \mathcal{P}^r$ can be decomposed as $\mu^r = \delta_e + g^r$, where $g^r \in C^r$ has a continuous and strictly positive density. Furthermore, g^r satisfies the equation $g^r * p + p = rg^r$ and the equality $g^r * p = rg^r$ holds outside the support of p.*

Proof. The equation $\mu^r * p = r(\mu^r - \delta_e)$ can be written as $\mu^r = \delta_e + \frac{1}{r}\mu^r * p$. Hence, $g^r = \frac{1}{r}\mu^r * p$ has a continuous, positive density. Taking into account the fact that $\mu^r = \delta_e + g^r$, the equation satisfied by μ^r can be written as

$$g^r * p + p = rg^r.$$

Hence, $g^r * p \leq rg^r$ and, by the proof of Lemma 11.6, g^r has a strictly positive density. Furthermore, if g^r is considered as a function, one has

$$g^r * p = rg^r$$

outside the support of p. □

11.10. Theorem. *Let $\mathcal{P}^r$ and C^r be the convex cones defined in Proposition 11.9. Assume that $0 < r_1 < r_2 < \infty$, $r \in [r_1, r_2]$ and that $\mu^r \in \mathcal{P}^r$. Let $\bar{\mu}_y^r$ denote the Radon measure $\bar{\mu}_y^r = \dfrac{\delta_y * \mu^r}{(\delta_y * \mu^r)(\check{p})}$. Then $\bar{\mu}_y^r \in C_1^r \subset C_1^{r_2}$.*

Furthermore, given a compact subset $C \subset H$, the measures $\bar{\mu}_y^r$ ($r \in [r_1, r_2]$, $y \notin C$) have continuous densities on C that form an equicontinuous and bounded family of functions.

Let $\mu = \lim_n \bar{\mu}_{y_n}^{r_n}$, where $r_n \to r$ and $y_n \to \infty$. Then $\mu * p = r\mu$ and, on every compact subset of H, the densities of the measures $\bar{\mu}_{y_n}^{r_n}$ converge uniformly to the density of μ.

Proof. Clearly, $\bar{\mu}_y^r(\check{p}) = 1$. Also, by Proposition 11.9, $\mu^r * p \leq r\mu^r$. Hence, $\bar{\mu}_y^r * p \leq r\bar{\mu}_y^r$ and $\bar{\mu}_r^y \subset C_1^r$. Since $C^r \subset C^{r_2}$, the family of measures $\bar{\mu}_y^r$ ($y \in H, r \in [r_1, r_2]$) is relatively compact with respect to the topology of weak convergence.

Note that the equation $\bar{\mu}_y^r * p = r(\bar{\mu}_y^r - \gamma_y \delta_y)$, where $\gamma_y = 1/\bar{\mu}_y^r(\check{p})$, implies that the densities of $\bar{\mu}_y^r$ and $\bar{\mu}_y^r * p$ on C are equal if $y \notin C$. By Lemma A.19, the map $\mu \to \mu * p$ from C^r into itself is continuous with respect to the weak topology and the topology of uniform convergence on compact sets. Hence, the family $\bar{\mu}_y^r * p$, where $r \in [r_1, r_2]$ and $y \in H$, is relatively compact in the topology of uniform convergence on compact sets. Therefore, the Ascoli theorem implies the equicontinuity and boundedness of the densities of the measures $\bar{\mu}_y^r$ on C, where $r \in [r_1, r_2]$ and $y \notin C$, .

Consider a limit measure μ of the relatively compact set $\{\bar{\mu}_y^r \mid y \in H, r \in [r_1, r_2]\}$. The fact that $\bar{\mu}_y^r * p = r(\bar{\mu}_y^r - \gamma_y \delta_y)$ with $y = y_n, r = r_n$ implies that $\mu * p = r\mu$ because $r_n \gamma_{y_n} \delta_{y_n}$ converges weakly to the zero measure and p is well-behaved.

On every fixed compact subset of H, for large n, the densities of the measures $\bar{\mu}_{y_n}^{r_n}$ are equicontinuous and bounded. Hence, for any convergent subsequence, these densities converge to the density of μ. □

11.11. Definition. Let K_g^r denote the Radon measure $\frac{\delta_g * V^r}{\delta_g * V^r(\check{p})} \in C_1^r$. Let $K^r(x, g)$ denote its density on $H \setminus \{g\}$. The map $g \to K_g^r$ of H into C_1^r is called the (p,r)-**Martin kernel**. The closure, in the weak topology, of $\{K_g^r \mid g \in H\}$ in C_1^r will be denoted by $\tilde{H}(p,r)$. Identifying H with the image $\{K_g^r \mid g \in H\}$ of the Martin kernel, $\tilde{H}(p,r)$ will also be denoted by $H \cup \partial H(p,r)$. This compactification of H will be called the (p,r)-**Martin compactification** of H and its boundary $\partial H(p,r)$ will be called the (p,r)-**Martin boundary** of H. (If (p,r) is determined by the context the reference to (p,r) will usually be omitted.)

The following result justifies this definition. In particular, it states that the Martin kernel embeds H topologically into C_1.

11.12. Theorem. *The map $g \to K_g^r$ of H into C_1^r is a homeomorphism of H onto its image. Every element of $\partial H(p,r)$ belongs to $\mathcal{H}_1^r$. If $\mu = \lim_n K_{g_n}^r \in \partial H(p,r)$, the density of μ, denoted by $K^r(x, \mu)$, is the* *uniform limit on every compact set of the sequence of continuous functions* $K^r(x, g_n) \in \tilde{\mathcal{H}}^r, x \neq g_n$. *Moreover, every extremal element of* $\mathcal{H}_1^r$ *belongs to* $\partial H(p,r)$.

Remarks. (1) By definition, the boundary $\partial H(p,r)$ is a compact and metrizable space on which H acts continuously. It has a topological structure in contrast to the Poisson boundary [F3] that is only a measure theoretical object. However, as the Poisson boundary is a coarser invariant of p, it can be calculated much more easily than $\partial H(p,r)$.

(2) The probability p determines a Markov kernel P on the group H: set $Pf(g) = \int f(gx)p(dx)$. Similarly, the probability $\check{p}$ determines a second kernel $\check{P}$ where $\check{P}h(g) = \int f(gh)\check{p}(dx)$. These two kernels are in duality relative to right Haar measure dg, i.e., $\int (Pf)h dg = \int f(\check{P}h)dg$ for any two non-negative Borel functions f and h. Assuming that the series $\sum_{n=0}^{\infty} p^n$ converges, the hypotheses of §2 in [R3] are satisfied. From what has been proved, the function $\check{p}$ is a reference function and the Martin compactification $\tilde{H}(p,1)$ is the compact space referred to as the Martin space by Revuz in [R3]. Note that the co-excessive functions h in the sense of [R3] satisfy $Ph \leq h$ since the Martin compactification $\tilde{H}(p,1)$ is related to integral representation of non-negative functions f for which $f * p = \check{P}f = f$.

(3) In the theory that is developed here, the continuity of the density p makes it possible to carry over most of the classical potential theory used in the study of the Martin compactification for a second order elliptic differential operator (briefly outlined in Chapter VI). Additional details concerning remarks (2) and (3) may be found in Taylor [T5].

Proof of Theorem 11.12.

The continuity of the map $g \to K_g^r \in C_1^r$ is clear from the formula $K_g^r = \frac{\delta_g * V^r}{\delta_g * V^r(\check{p})}$. Recall that $\gamma_g^{-1} = \delta_g * V^r(\check{p}) \neq 0$. The identity

$$(\dagger) \qquad K_g^r * p = r(K_g^r - \gamma_g \delta_g)$$

implies that $g \to K_g^r$ is injective since $\gamma_g \delta_g = \gamma_{g'} \delta_{g'}$ implies $g = g'$.

In addition, eq. (†) implies that if $K_{g_n}^r$ converges weakly to K_g^r then (g_n) is bounded. (Otherwise, by Theorem 11.10, $K_g^r * p = rK_g^r$.) By considering convergent subsequences of (g_n), the injectivity of $g \to K_g^r$ implies that $g_n \to g$ if $K_{g_n}^r$ converges weakly to K_g^r. Hence, $g \to K_g^r$ embeds H in C_1.

Theorem 11.10 implies that each $\mu \in \partial H(p,r)$ belongs to $\mathcal{H}_1^r$ and that the densities $K^r(x, g_n)$ converge uniformly on compact sets to the density $K^r(x, \mu)$ of μ.

To prove the last assertion, note that, if $x \in H$ and $y \in \tilde{H}(p,r)$, with $x \neq y$, a density $K^r(x,y)$ is defined: for $y \in H$ see Definition 11.11 and for $y = \mu \in \partial H(p,r)$ see Theorem 11.10. It is continuous in x and y by what has already been proved. It is now shown, using Martin's classical argument [M5], that, for every $u \in \mathcal{H}_1^r$, there exists a probability measure

ν on $\partial H(p,r)$ such that

$$u(x)=\int K^r(x,y)d\nu(y).$$

To simplify the notation, let q denote $r^{-1}p$. Suppose α is a positive measure that has a continuous density with compact support. Then the measure $\alpha * V^r$ has a continuous density and satisfies $\alpha * V^r * q \le \alpha * V^r$. Hence, $\alpha * V^r \in \mathcal{C}^r$ has a strictly positive density. Denote by $C_a \subset H$ a compact set containing e and observe that since u and $\alpha * V^r$ are continuous, there is a constant c_a such that $u \le c_a(\alpha * V^r)$ on C_a.

Denote by v_a the infimum of μ and $c_a(\alpha * V^r)$ on H. Clearly $v_a = u$ on $C_a, v_a * q \le v_a$, and $v_a \le c_a(\alpha * V^r)$. Observe that

$$(\dagger)\qquad v_a = w_a * \sum_{k=0}^{n} q^k + v_a * q^{n+1} \text{ with } w_a = v_a - v_a * q \ge 0.$$

Since $v_a * q^{n+1} \le c_a(\alpha * V^r) * q^{n+1} = c_a(\alpha * \sum_{k=n+1}^{\infty} q^k)$, it follows that $v_a * q^{n+1}$ converges uniformly to zero when n goes to infinity. Hence, the decomposition of v_a in eq. ($\dagger$) implies the convergence of the series $w_a * \sum_{k=0}^{\infty} q^k$ to v_a, i.e., $v_a = w_a * V^r$. Therefore, $v_a = \int(\delta_y * V^r)dw_a(y)$. Let N be a neighborhood of e such that $NS^{-1} \subset C_a$ and $N \subset C_a$. If C_a is sufficiently large, then one has $w_a = 0$ on N. This is proved as follows. Since $w_a(x) = v_a(x) - \int v_a(xy^{-1})dq(y)$ for $x \in N, y \in S, xy^{-1} \in C_a$, then $v_a(xy^{-1}) = u(xy^{-1})$ and $v_a(x) = u(x)$ imply $w_a(x) = 0$, as $u * q = u$. Now fix $x \in N$ and suppose C_a is sufficiently large. Then one has, for $x \in N$,

$$v_a(x) = \int K^r(x,y)\big(\delta_y * V^r\big)(\check{p})dw_a(y),$$

because $K^r(x,y)$ is a continuous function on the support of w_a. Hence, $u(x) = v_a(x) = \int K^r(x,y)d\nu_a(y)$, with $d\nu_a(y) = (\delta_y * V^r)(\check{p})dw_a(y)$, and ν_a is a positive measure, with a continuous density concentrated outside N.

The mass of ν_a is calculated as follows:

$$\nu_a(H) = (w_a * V^r)(\check{p}) = v_a(\check{p}) = u(\check{p}) = 1,$$

because $v_a = u$ on $S^{-1} \subset C_a$. Hence, the family ν_a is a family of probability measures on the compact space $\tilde{H}(p,r)$. One can extract a subsequence that converges weakly to a probability measure ν on $\tilde{H}(p,r)$. From the fact that one can take a large neighborhood N_a of e such that $N_aS^{-1} \subset C_a$, it follows that

$$\nu_a(N_a) = 0,\ \nu(N_a) = 0$$

and, hence, ν is concentrated on the boundary $\partial H(p,r)$. When $x \in N$ is fixed $K^r(x,y)$ is a bounded and continuous function on $\tilde{H}(p,r)/N_a$. Hence, it follows that if $x \in N$

$$u(x) = \int K^r(x,y)d\nu(y).$$

Since N is arbitrary, the formula holds for all $x \in H$. If u is minimal then ν is necessarily a Dirac measure. Hence, there exists $y \in \partial H(p,r)$ such that $f(x) = K^r(x,y)$. $\square$

11.13. Remark. In the course of the proof, a concrete integral representation of u has been obtained in terms of the functions $K^r(x,y)$, $y \in \partial H(p,r)$. Since C_1^r is metrizable, $\{y \in \partial H(p,r) \mid K^r(x,y) \text{ is minimal}\}$ is a G^δ-set and, hence, is a Borel set (see [C2]).

11.14. Definition. The set $\partial_e H(p,r)$, also denoted $\partial_e H$, of points y such that $K^r(x,y)$ is minimal is said to be the **minimal (p,r)-boundary** of H.

11.15. Remarks. (1) It follows that every $u \in \mathcal{H}_1$ has an unique integral decomposition

$$u(x) = \int_{\partial_e H} K^r(x,y)d\nu(y).$$

(2) If one considers continuous time instead of discrete time, p^n is replaced by a convolution semi-group $p^t, t > 0$, and the potential kernel V^r by $V^\lambda = \int_0^\infty e^{-\lambda t} p^t dt$. The integral converges if $\lambda > \lambda_0$ and $e^{-\lambda_0}$ is the analogue of the real number $r(p)$ defined in Definition 11.18. If the generator of p^t is an elliptic operator L on a manifold, the equation $Lf = \lambda f$ corresponds to the equation $f * p = rf$, as pointed out in Remark 11.4(2).

THE POSITIVE SPECTRUM OF A RANDOM WALK

11.16. Lemma. *If $r > \overline{\lim}_n p^n(e)^{1/n}$, the series $\sum_{n=0}^{\infty} r^{-n} p^n$ converges in the weak topology.*

Proof. If $\phi \in C_c^+(H)$ is given, as in the proof of inequality $(*)$ in the proof of Theorem 11.5, there exists $C(\phi) > 0$ and $N \geq 1$ such that $0 \leq \phi \leq C(\phi) \sum_{k=1}^{N} \check{p}^k$. Hence,

$$p^n(\phi) \leq C(\phi) \sum_{k=1}^{N} p^n(\check{p}^k) = C(\phi) \sum_{k-1}^{N} p^{n+k}(e).$$

Hence, $\overline{\lim}_n p^n(\phi)^{1/n} \leq \overline{\lim}_n p^n(e)^{1/n}$. The convergence of $\sum_{n=1}^{\infty} r^{-n} p^n(\phi)$ follows if $r > \overline{\lim}_n p^n(e)^{1/n}$. Because ϕ is arbitrary, this implies the weak convergence of the series $\sum_{n=1}^{\infty} r^{-n} p^n$. $\square$

11.17. Theorem. *Suppose $r \in \mathbb{R}^+$ is a positive real number. Then the following conditions are equivalent:*

(1) $r \geq \overline{\lim}_n p^n(e)^{1/n}$.

(2) *There exists a positive function f that satisfies the equation $f * p = rf$.*

Proof. If $f * p = rf$, then f is continuous and strictly positive (see the proof of Lemma 11.6). Consider $\varepsilon > 0$ such that $\varepsilon\phi \leq f$. Then, since $f*p = rf$ implies that $r^n f(e) = p^n(\tilde{f}) \geq \varepsilon p^n(\tilde{p}) = \varepsilon p^{n+1}(e)$, it follows that $r \geq \overline{\lim}_n p^n(e)^{1/n}$. Hence, (2) implies (1).

Assume that $r > \overline{\lim}_n p^n(e)^{1/n}$. Then, from Lemma 11.16, the series $\sum_{n=0}^{\infty} r^{-n}p^n$ converges and so $\mathcal{P}^r \neq \emptyset$. It follows from Theorem 11.10 that the corresponding Martin boundary contains a function f that satisfies $f * p = rf$. Hence, for every $r > \overline{\lim}_n p^n(e)^{1/n} = r_1$ there exists $f_r \in C_1^r$ such that $f_r * p = rf_r, \check{p}(f_r) = 1$. It follows from Theorem 11.10 that there is a sequence $r_n > r_1$ converging to r_1 for which the measures $f_{r_n} \in C_1$ converge in the weak topology to $\mu \in C_1^r$. By continuity, the equation $f_{r_n} * p = r_n f_{r_n}$ implies $\mu * p = r_1\mu$. The density f of μ satisfies $f * p = r_1 f$. This shows that (1) implies (2). □

11.18. Definition. Let $r(p) \stackrel{\text{def}}{=} \overline{\lim}_n p^n(e)^{1/n}$. The interval $[r(p), \infty]$ is called the **positive spectrum** of the convolution operator $*p$.

11.19. Proposition. *Denote by $r_0(p)$ the* **spectral radius** *of the convolution operator $*p$ in $L^2(H)$. Then $r(p) \leq r_0(p)$. If $p = \check{p}$, then $r(p) = r_0(p)$.*

Proof. To show that $r(p) \leq r_0(p)$, note that by the Cauchy-Schwarz inequality, one has

$$|\langle\psi, \phi * p^n\rangle| \leq \|\psi\|_2 \|\phi * p^n\|_2.$$

Hence, if $\psi = p$,

$$p^{n+1}(\phi) = p(\phi * \check{p}^n) = \langle p, \phi * \check{p}^n\rangle \leq \|p\|_2 \|\phi * \check{p}^n\|_2, \text{ and}$$

$$\lim_n [p^{n+1}(\phi)]^{1/n} \leq \overline{\lim}_n \|\phi * \check{p}^n\|_2^{1/n} \leq r_0(\check{p}) = r_0(p).$$

If $\phi = \check{p}$, the left hand side has the value $\overline{\lim}_n p^{n+2}(e)^{1/n} = r(p)$. Hence, $r(p) \leq r_0(p)$.

To obtain the reverse inequality if $p = \check{p}$, it suffices to show that

$$r_0(p) = \sup\{\overline{\lim_n} |\langle\phi, \phi * p^n\rangle|^{1/n} \mid \phi \in C_c(H)\}.$$

This formula follows from the spectral theory of the self-adjoint operator Q on $L^2(H)$ defined by $Q\phi = \phi * p$. If ν_ϕ is the spectral measure of Q associated with ϕ, it follows that

$$\langle Q^n\phi, \phi\rangle = \int \lambda^n d\nu_\phi(\lambda).$$

Clearly, ν_ϕ is concentrated on the interval $[-p(1), p(1)]$ because the norm of Q is bounded by $p(1)$. If S_ϕ denotes the support of ν_ϕ, it follows that $\overline{\lim}_n |\langle Q^n\phi, \phi\rangle|^{1/n} = \sup\{|\lambda| \mid \lambda \in S_\phi\}$. Hence, if Σ denotes the spectrum of $Q, \Sigma = \overline{\bigcup_{\phi \in L^2(H)} S_\phi}$. Since the map $\phi \to \nu_\phi$ is continuous with respect to the strong topologies, it follows that $\Sigma = \overline{\bigcup_{\phi \in C_c(H)} S_\phi}$ and, hence,

$$r_0(p) = \sup\{\overline{\lim}_n |\langle \phi, \phi * p^n \rangle|^{1/n} \mid \phi \in C_c(H)\}.$$

The definition of $r(p)$ implies that

$$\overline{\lim}_n |\langle \phi, \phi * p^n \rangle| = |p^n(\check{\phi} * \phi)|^{1/n} \leq r(p),$$

because $\check{\phi} * \phi \in C_c(H)$. Hence, $r_0(p) \leq r(p)$. □

11.20. Remarks. (1) If H is non-amenable (see [G9]), then it is known (see [D2]) that $r_0(p) < 1$. Hence, $r(p) < 1$ in this case. This is so, for example, if H is semisimple.

(2) In [G13] a way of calculating $r(p)$ is given when H is a connected Lie group. If $H = \mathbb{R}^d$, then $r(p)$ is nothing but the infimum of the Laplace transform $\hat{p}$ of p, where $\hat{p}(s) = \int e^{\langle s,x \rangle} dp(x)$ if $s \in \mathbb{R}^d$.

(3) Note that $r(p) = r_0(p)$ in the case of a semisimple Lie group G. (See Corollary 11.44, Theorem 13.17, and Proposition 13.19.)

To discuss the question of convergence of the series $\sum_{n=0}^{\infty} r^{-n} p^n$ at $r = r(p)$ the concept of recurrence is useful.

11.21. Definition. A locally compact group H is said to be **recurrent** if there exists a probability measure p, whose support generates H, such that the series $\sum_{n=0}^{\infty} p^n$ diverges weakly, i.e., $\sum_{n=0}^{\infty} p^n(\psi)$ diverges for all $\psi \in C_c^+(H)$.

11.22. Remarks. (1) It is easy to show that if H is recurrent, then it is amenable and unimodular [G11].

(2) The structure of recurrent groups is determined by the results in [V], [L2], and [G11]. Namely, up to compact extensions, the only possibilities are $\mathbb{Z}$ and $\mathbb{Z}^2$.

11.23. Proposition. *Suppose that the series $\sum_{n=0}^{\infty} r^{-n} p^n$ diverges at $r = r(p)$. Then the group H is recurrent.*

Proof. (The following argument was outlined in [G13].) It follows, from Theorem 11.17, that there is a strictly positive function f such that $f * p = r(p) f$. Let $Q = Q(x, dy)$ denote the kernel on $C^+(H)$ defined by $Q\phi = r^{-1} \phi * p$, where $r = r(p)$.

One first shows that the inequality $Q\phi \leq \phi$ implies $Q\phi = \phi$ if $\phi \in C^+(H)$. Since $\sum_{k=0}^{N} Q^k(\phi - Q\phi) = -Q^{N+1}\phi + \phi \leq \phi$, the series $\sum_{k=0}^{\infty} Q^k \psi$, with $\psi = \phi - Q\phi \geq 0$, converges. Unless $\psi = 0$, this contradicts the hypothesis that the series $\sum_{n=0}^{\infty} r^{-n} p^n$ diverges.

The function f is shown to be unique, up to a scalar, as follows. Suppose $h \in C^+(H)$ satisfies $Qh = h$. Let f' denote the infimum of f and kh, where $k > 0$. Then $Qf' \leq f'$. Hence, from what has been proved, $Qf' = f'$. It follows that, given k one has, for every $x \in H$, either $f \geq kh$ or $f \leq kh$ on the support of the measure $Q(x,\cdot)$. Replacing Q by $Q' = \sum_{k=1}^{\infty} \frac{1}{2^k} Q^k$ one has the same result,with the difference that the support of $Q'(x,\cdot)$ is equal to H because p is well-behaved. Hence, for every $k > 0$, one has either $f \geq kh$ or $f \leq kh$. This clearly implies that, for some $k, f = kh$.

Consider now the equation $f * p = r(p)f$. It implies that, for all $g \in H$,

$$\delta_g * f * p = r(p)\delta_g * f.$$

The uniqueness of f implies the existence of a positive number $\ell(g)$ such that $\delta_g * f = \ell(g)f$. It follows that $\ell(g)$ is a multiplicative homomorphism of H into $\mathbb{R}^+$. Hence, $f = \check{\ell}$. The new measure $p_\ell = \frac{1}{r(p)}\ell p$ is a probability measure because $\check{\ell} * p = r(p)\check{\ell}$. Clearly, it satisfies the equation

$$\sum_{n=0}^{\infty} r^{-n} p_\ell^n = \left\{ \sum_{n=0}^{\infty} r^{-n} r(p)^{-n} p^n \right\} \ell.$$

When $r = 1$, the series $\sum_{n=0}^{\infty} p_\ell^n$ diverges. Since the supports of p and p_ℓ are the same, H is recurrent, as p is well-behaved. □

11.24. Corollary. *If H is non-amenable then the series $\sum_{n=0}^{\infty} r^{-n}p^n$ converges at $r = r(p)$.*

Proof. From Remark 11.22.(1) H is non-recurrent. The corollary then follows from Proposition 11.23. □

11.25. Remark. If the group is semisimple, the corollary implies the convergence of $\sum_{n=0}^{\infty} r(p)^{-n}p^n$. This result will be used in Chapter XIV.

The fixed line property

This property was introduced in [F4] and [M3] in similar contexts. A detailed study of the fixed line property can be found in [C5]. It will be used systematically in the next chapter. In the next section it will be used to obtain a simple formula for the number $r(p)$ defined in Definition 11.18.

11.26. Definition. Denote by H a locally compact metrizable group. An **exponential** ℓ on H is a continuous homomorphism of H into the multiplicative group of positive numbers, i.e., $\ell(xy) = \ell(x)\ell(y)$, for all $x, y \in H$,

The set of exponentials on H will be denoted by H^*. It will be given the topology of uniform convergence on compact sets. The trivial exponential $\ell(x) = 1$ for all $x \in H$ will be denoted by 1.

11.27. Remarks. (1) If H is compactly generated, then H^* is isomorphic to a finite dimensional vector space. This follows since, if H' is the commutator subgroup of H, then H/H' is a compactly generated abelian group $\mathbb{R}^n \times \mathbb{Z}^n \times F$, with F a compact abelian group. Hence, $H^* = (H/H')^*$ is isomorphic to $\mathbb{R}^{m+n}$.

(2) If G is a semisimple connected Lie group with finite center and Iwasawa decomposition KAN, any exponential on $S = AN$ is trivial on N since the commutator subgroup of S equals N. This observation is a key to calculating the value of λ_0 for the Laplacian on X (see § 6.1 and Example 11.33). Note that all commutators belong to the subgroup N and that every element of N is a limit of commutators since, as $k \to \infty$, $a^{-k}na^kn^{-1}$ converges to n^{-1}, if $a \in A^+$ (see the proof of Proposition 7.20). A similar argument shows that the commutator subgroup of $R^I = K^I M A_I N_I$ contains N_I (use the positive chamber of $\mathfrak{a}_I$ given by the roots in Σ_I^+). As a result, any exponential on R^I is trivial on $D^I = K^I M N_I$, since it is trivial on $K^I M$, as this group is compact.

11.28. Definition. The topological group H is said to have the **fixed line property** if, for every convex cone $\mathcal{C}$ with compact base $\mathcal{C}_1$ in a locally convex vector space V, and every continuous representation θ of H in V that preserves $\mathcal{C}$, there exists an element $v \in \mathcal{C}$ and an exponential ℓ on H such that $\theta(h)v = \ell(h)v$, for all $h \in H$.

11.29. Remarks. (1) The (linear) representation θ defines a **projective action** of H on the base $\mathcal{C}_1$ as follows: if $x \in \mathcal{C}_1$, define $h \cdot x$ to be the unique $y \in \mathcal{C}_1$ for which $\theta(h)x = \lambda y, \lambda > 0$.

(2) Abelian groups and compact groups have the fixed line property. For $H = \mathbb{Z}$, this is a consequence of the Schauder–Tychonoff fixed point property applied to the projective action of $\mathbb{Z}$ on the compact convex base of $\mathcal{C}$. For a compact group K it follows that, if $x \in \mathcal{C}$, then $v = \int \theta(k)x dk$ is a non-trivial fixed point of $\theta(k)$ in $\mathcal{C}$. Hence, K has the fixed line property. It is clear that this property implies amenability because an affine action on a compact convex set is also a projective action and, consequently, has a fixed point.

The following lemma will be used repeatedly in what follows.

11.30. Lemma. *Consider a continuous representation θ of H into a locally convex vector space such that $\theta(H)$ preserves a convex cone $\mathcal{C}$ with compact base $\mathcal{C}_1$. Then the set of exponentials $\ell \in H^*$, for which there exist $v \in \mathcal{C}$ with $\theta(h)v = \ell(h)v$, for all $h \in H$, is compact in H^*.*

Proof. One can suppose $v \in \mathcal{C}_1$. Clearly the set V_1 of $v \in \mathcal{C}_1$ that satisfies the condition of the lemma is the set of fixed points of the projective action of H on $\mathcal{C}_1$. Hence, $V_1 \subset \mathcal{C}_1$ is compact. Denote by ℓ_v the exponential on H such that $\theta(h)v = \ell_v(h)v$ if $v \in V_1$. Clearly, the map $v \to \ell_v$ from V_1 into H^* is continuous. Its range is compact because V_1 is compact. Hence, the lemma follows. □

11.31. Theorem. *Denote by N a closed normal subgroup of the locally compact group H. If $\ell \in N^*$ and $g \in H$, let $\ell^g(n) \stackrel{\text{def}}{=} \ell(g^{-1}ng)$. This defines a natural action of H on N^*.*

Assume in addition that N and H/N have the fixed line property. Then H has the fixed line property in either of the following situations.

(1) *In the natural action of H on N^*, the only relatively compact orbit is that of the trivial exponential 1.*
(2) *There exists a subgroup $A \subset H$ with the fixed line property such that $H = AN$ and, in the natural action of A on N^*, every relatively compact orbit is reduced to a point.*

Proof. The first statement is clear. Now let θ be a (linear) representation of H on a convex cone $\mathcal{C}$ with compact base $\mathcal{C}_1$. Suppose $v \in \mathcal{C}$ defines a line fixed by N in $\mathcal{C}$, i.e., for all $n \in N$, $\theta(n)v = \ell(n)v$ where $\ell \in N^*$. Then $\theta(g)v$ also determines a line fixed by N, since $\theta(n)\theta(g)v = \theta(g)\theta(g^{-1}ng)v = \ell(g^{-1}ng)\theta(g)v$. Moreover, the associated exponential is ℓ^g where $\ell^g(n) = \ell(g^{-1}ng)$. Lemma 11.30 implies that the H-orbit of ℓ under the natural action is relatively compact.

Now consider the situations (1) and (2). In case (1), by hypothesis, $\ell = 1$. Hence, $\theta(n)v = v$ for all $n \in N$ and so v is N-fixed. The set of N-fixed points is an H-invariant closed convex subcone $\mathcal{C}'$ of $\mathcal{C}$ on which H/N acts. Since, by assumption, H/N has the fixed line property, there exist $v' \in \mathcal{C}', \ell' \in H^*$ such that, for all $g \in H$, $\theta(g)v' = \ell'(g)v'$. Hence, H has the fixed line property.

In case (2), since the A-orbit of ℓ is relatively compact, by assumption, $\ell^a = \ell$ for $a \in A$. Hence, $\theta(n)\theta(a)v = \ell(n)\theta(a)v$. Consider the subcone $\mathcal{C}'$ of $\mathcal{C}$ of vectors w such that, for all $n \in N$, $\theta(n)w = \ell(n)w$. From what has been proved, it follows that $\mathcal{C}'$ is closed, convex, and A-invariant. Because A has the fixed line property and acts on $\mathcal{C}'$, there exist $v' \in \mathcal{C}'$ and $\ell' \in A^*$ such that $\theta(a)v' = \ell'(a)v'$, for all $a \in A$. Hence, for every $a \in A$ and $n \in N$, $\theta(an)v' = \theta(a)\theta(n)v' = \ell'(a)\ell(n)v'$. This shows that the expression $\ell'(a)\ell(n)$, if $g = an$, defines a function ℓ'' on H. It is an exponential as $\ell''(gg_1) = \ell''(aa_1a_1^{-1}na_1n_1) = \ell'(aa_1)\ell^{a_1}(n)\ell(n_1) = \ell'(a)\ell'(a_1)\ell(n)\ell(n_1)$ if $g = an$ and $g_1 = a_1n_1$. Hence, H has the fixed line property. □

The following corollary will play an essential role in the next chapters, where P will be a minimal parabolic subgroup of a semisimple group G.

11.32. Corollary. *Let P be a locally compact group that is a semidirect product $H \ltimes N$, where N is a nilpotent normal subgroup. Assume that M is a compact subgroup of H, A is central in H, and $H = MA$. If the only relatively compact H-orbits in N^* are given by fixed points, then P has the fixed line property.*

Proof. If $H = \{e\}, P = N$ is nilpotent and one can apply case (1) of Theorem 11.31 to the center N_1 of N and the factor group N/N_1. By induction on the length of the ascending central series, one may assume that the fixed line property is valid for N/N_1 and that N acts trivially on N_1 and, hence, on N_1^*. Because N_1 is abelian, the fixed line property for N follows. On the other hand, case (2) of Theorem 11.31 implies that

$H = MA$ has the fixed line property. Then case (2) applies again to the semi-direct product $H \ltimes N$ because of the condition on the action of H on N^*. Hence, the result follows. □

11.33. Example. Let KAN be an Iwasawa decomposition of a semisimple Lie group G. If M is the centralizer of A in K, then the group $P = MAN$ is a minimal parabolic subgroup of G (see Definition 2.5). The action of A on N and N^* is such that there are no non-trivial compact orbits of A, or of MA in N^*. Hence, AN and P have the fixed line property. (The same is true for the local field analogues of G.) Since $S = AN$ has the fixed line property, it follows that if there is a positive solution of the equation $Lu + \lambda u = 0$ on $X = G/K$, then there is one that is N-invariant. This is because S acts on the cone of positive solutions by $u(x) \to u(s \cdot x) = (L_{s^{-1}}u)(x), s \in S$. The fixed line property of S is equivalent to having a solution u with $S_{s^{-1}}u = u$ for all $s \in S$, i.e., $u(s \cdot x) = u(s \cdot o)u(x)$ for all $s \in S, x \in X$. Now $\ell(s) = u(s \cdot o)$ is an exponential on S. It is trivial on N by Remark 11.27(2). Hence, u is an N-invariant solution.

FORMULAS FOR $r(p), r_0(p)$

Assume that $H = KT$, where T is a closed subgroup with the fixed line property and K is a compact subgroup, a typical example being any semisimple Lie group with T a minimal parabolic subgroup, e.g, $T = P = MAN$. Denote by p a K-bi-invariant, well-behaved positive measure. In this section $r(p)$, and $r_0(p)$ will be determined in terms of some special functions on T^*. Moreover, for simplicity H is supposed to be unimodular. Connected semisimple Lie groups with finite center and reductive groups belong to this class.

11.34. Definition. If $g \in H$ equals kt, let $k(g)$ and $t(g)$ denote k and t, respectively. Let $t(g,k) = t(gk)$. If $\ell \in T^*$ let $\sigma_\ell(g,k)$ denote $\ell[t(gk)] = \ell[t(g,k)]$. If $b \in H/T = K/K \cap T$, define $\sigma_\ell(g,b)$ to be $\sigma_\ell(g,k)$ for $b = k(K \cap T)$.

11.35. Remarks. (1) $t(g)$ is defined up to an element of the compact subgroup $K \cap T$. The value $\ell[t(g)]$ is well-defined, since ℓ is trivial on $K \cap T$.

(2) For the same reason as in (1), for fixed g, $\sigma_\ell(g,k)$ depends only on $k(K \cap T) \in K/K \cap T = H/T = B$. Hence, $\sigma_\ell(g,b)$ is well-defined and satisfies the cocycle identity

$$\sigma_\ell(gg', b) = \sigma_\ell(g, g' \cdot b)\sigma_\ell(g', b)$$

for $g, g' \in H, b \in B$, where $g' \cdot b$ denotes the obvious action of g' on $b \in H/T$. Hence, $\sigma_\ell(g,b)$ is a multiplier in the sense of § 10.1.

Let $d\tilde{m} = dk$ denote the normalized Haar measure on K. Because the modular function of H is trivial on the compact subgroup K, the right Haar measure η on H is uniquely defined, up to a coefficient, by requiring it to be right invariant under T and left invariant under K. Hence, there is a unique right Haar measure γ on T such that $\eta = \tilde{m} * \gamma$. Since H is assumed to be unimodular in this section, the Haar measure of H is equal to $\tilde{m} * \gamma$. Given an exponential ℓ on T, the corresponding measure $\ell\gamma$ on T (with density ℓ relative to γ) will also be denoted by ℓ. It satisfies $\ell * \delta_t = \ell(t^{-1})\ell$ and $\check{\ell} = \tilde{\ell\gamma} = \ell^{-1}\check{\gamma}$, with $\ell^{-1}(t) = \ell(t^{-1})$.

11.36. Proposition. *Given $\ell \in T^*$, denote by Γ^ℓ the convex cone of positive measures μ on H for which $\delta_t * \mu = \ell(t)\mu$. Then, up to a scalar, there exists a unique $\mu \in \Gamma^\ell$ such that $\mu * p$ is proportional to μ. This measure is equal to $\check{\ell} * \tilde{m}$ and has density on H equal to $\ell[t(g^{-1})] = \sigma_\ell(g^{-1}, b)$. The constant of proportionality is $\int \ell[t(g)]dp(g)$.*

Proof. Because p is K-right invariant, the condition $\mu * p = r\mu$ implies that μ has a strictly positive, continuous density f that is K-right invariant. Furthermore, since $\mu \in \Gamma^\ell$, the fact that $f(t^{-1}x)$ is the density of $\delta_t * \mu$ implies that $f(t^{-1}x) = \ell(t)f(x)$ for all $x \in H$.

Let $g = kt$. Then $f(g^{-1}) = f(t^{-1}k^{-1}) = f(t^{-1}) = \ell(t)f(e)$. This proves that μ is unique.

The density f is given by $f(g) = \check{f}(g^{-1}) = \ell[t(g^{-1})] = \sigma_\ell(g^{-1}, b)$. Hence, $f * p = rf$ and $f(e) = 1$ implies that $r = \int \ell[t(g)]dp(g)$. □

11.37. Definition. The constant r in Proposition 11.36 for which $\mu * p = r\mu$, where $\mu \in \Gamma^\ell$, will be denoted by $\hat{p}(\ell)$ and called the value at ℓ of the **Laplace transform** of p, i.e., $\hat{p}(\ell) = \int \ell[t(g)]dp(g)$.

Let Φ^ℓ denote the measure $\tilde{m} * \check{\ell} * \tilde{m}$.

11.38. Lemma. *The function equal to $\Phi^\ell(g) = \int \ell[t(g^{-1}k)]dk$ is the density of Φ^ℓ. Furthermore, $\hat{p}(\ell) = \int \Phi^\ell(g^{-1})dp(g)$.*

Proof. The formula for $\Phi^\ell(g)$ follows from the formulas $f(g) = \ell[t(g^{-1})]$, $\Phi^\ell = \tilde{m} * f$, and $\Phi^\ell(e) = 1$. Since $\Phi^\ell * p = \hat{p}(\ell)\Phi^\ell$ and $\Phi^\ell(e) = 1$, it follows that $\hat{p}(\ell) = \int \Phi^\ell(g^{-1})dp(g)$. □

11.39. Remark. Note the fact that $\Phi^\ell(e) = 1$ is independent of the choice of γ.

11.40. Theorem. *The logarithm of the Laplace transform of p is strictly convex on T^*. Moreover, $\hat{p}(\ell)$ reaches its minimum at a unique point $\ell'_0 \in T^*$ and $r(p) = \inf_{\ell \in T^*} \hat{p}(\ell) = \hat{p}(\ell'_0)$.*

Proof. Let $\alpha_1, \alpha_2 \in [0,1]$, with $\alpha_1 + \alpha_2 = 1$. If $\ell_1, \ell_2 \in T^*$, let $\ell = \ell_1^{\alpha_1}\ell_2^{\alpha_2}$ (a convex combination). Then, since $\hat{p}(\ell) = \int \ell[t(g)]dp(g)$, it follows from Hölder's inequality that

$$\hat{p}(\ell) = \int \ell_1^{\alpha}[t(g)]\ell_2{}^{\alpha_2}[t(g)]dp(g) \leq \Big[\int \ell_1[t(g)]dp(g)\Big]^{\alpha_1}\Big[\int \ell_2[t(g)]dp(g)\Big]^{\alpha_2}.$$

Hence, $\log \hat{p}(\ell) \leq \alpha_1 \log \hat{p}(\ell_1) + \alpha_2 \log \hat{p}(\ell_2)$. Equality in this inequality is only possible if $\ell_1[t(g)] = \ell_2[t(g)]$ p-a.e.

Since $p * \delta_k = p$ for every $k \in K$, and $\ell_1[t(g)], \ell_2[t(g)]$ are continuous, the equality $\ell_1[t(g)] = \ell_2[t(g)]$, for almost every g, implies that $\ell_1[t(gk)] = \ell_2[t(gk)]$, for every $k \in K$ and every g in the support of p. It follows that the equality $\log \hat{p}(\ell) = \alpha_1 \log \hat{p}(\ell_1) + \alpha_2 \log \hat{p}(\ell_2)$ implies that $\sigma_{\ell_1}(g, b) = \sigma_{\ell_2}(g, b)$, for every $b \in H/T$ and every g in the support S of p. Because $H = \cup_{u>0} S^n$ and because of the cocycle relation, it follows that $\sigma_{\ell_1} = \sigma_{\ell_2}$ and so $\ell_1 = \ell_2$.

Since H is compactly generated and H/T is compact, T is also compactly generated and T^* is a finite dimensional vector space. Suppose that a sequence ℓ_n tends to infinity in T^*. Then the form of the exponentials shows that there is a compact set C with $\overset{\circ}{C} \neq \emptyset$ such that $\ell_n(s)$ converges uniformly to $+\infty$ on C, i.e., $\inf_{s \in C} \ell_n(s) = u_n$ with $\lim_n u_n = +\infty$. Because p is well-behaved one can find $\varepsilon > 0$ and N such that KC is contained in the support of $q = \sum_{k=1}^N \varepsilon^k p^k$. Hence, $q(KC) > 0$.

Therefore,

$$\hat{q}(\ell_n) = \sum_{k=1}^{N} \varepsilon^k [\hat{p}(\ell_n)]^k \leq \frac{\varepsilon^{N+1} \hat{p}(\ell_n)^{N+1} - 1}{\varepsilon \hat{p}(\ell_n) - 1}.$$

Clearly,

$$\hat{q}(\ell_n) = \int \ell_n[t(g)] dq(g) \geq \int_{KC} \ell_n[t(g)] dq(g) \geq u_n q(KC).$$

If ℓ_n tends to infinity, the fact that $q(KC) > 0$ and $\lim_n u_n = +\infty$ implies that $\lim_n \hat{q}(\ell_n) = +\infty$. The relation between $\hat{p}(\ell_n)$ and $\hat{q}(\ell_n)$ implies that $\lim_n \hat{p}(\ell_n) = +\infty$. It follows that the map $\ell \to \log \hat{p}(\ell)$ is proper. Since it is continuous and T^* is isomorphic to $\mathbb{R}^d$, there exists ℓ_0' in T^* with $\inf_{\ell \in T^*} \hat{p}(\ell) = \hat{p}(\ell_0')$. The strict convexity of $\log \hat{p}$ implies that ℓ_0 is unique.

From the definition of the positive spectrum and of $\hat{p}(\ell)$, it follows that $r(p) \leq \inf_{\ell \in T^*} \hat{p}(\ell)$. Consider the convex cone $\mathcal{H}$ of functions $f \in C(H)$ such that $f * p = r(p) f$. It has a compact base and the fixed line property is valid for T acting on the left on $\mathcal{H}$. In other words, there exist $f \in \mathcal{H}$ and $\ell \in T^*$ such that $\delta_t * f = \ell(t) f$. The condition $f * p = r(p) f$ implies $r(p) = \hat{p}(\ell)$. Hence, $r(p) \geq \inf_{\ell \in T^*} \hat{p}(\ell) = \hat{p}(\ell_0')$. □

Consider now the spectral radius of the convolution operator $*p$ on $L^2(H)$. As explained earlier, there is a unique right invariant Haar measure γ on T such that $\eta = \tilde{m} * \gamma$. One denotes by $\bar{m}$ the image of $\tilde{m}$ on $K/K \cap T = H/T$ under the map $k \to b = kT \in H/T$.

11.41. Lemma. *Let $\delta = \delta_T$ denote the modular function of T defined by $\delta_t * \gamma = \delta(t)\gamma$. Then, for $g \in H, k \in K$,*

$$\frac{dg \cdot \overline{m}}{d\overline{m}}(b) = \delta[t(g^{-1}k)], \quad \textit{where} \quad b = k \cdot e.$$

Proof. Let $gk = k(gk)t(gk) = k_1 t(gk)$. Then $k = g^{-1}k_1 t(gk)$. Hence, $\delta[t(g^{-1}k_1)]\delta[t(gk)] = 1$. Now $\delta_g * (\tilde{m} * \gamma) = \int(\delta_{g\cdot k} * \delta_{t(g,k)} * \gamma)d\tilde{m}(k)$. Since $\int(\delta_{g\cdot k} * \delta_{t(g,k)} * \gamma)d\tilde{m}(k) = \int \delta[t(g,k)](\delta_{g\cdot k} * \gamma)d\tilde{m}(k)$ and $\delta[t(g,k)] = \delta[t^{-1}(g^{-1}, g\cdot k)]$, it follows that

$$(*) \qquad \delta_g * (\tilde{m} * \gamma) = \int \delta[t^{-1}(g^{-1}, g\cdot k)](\delta_{g\cdot k} * \gamma)d\tilde{m}(k).$$

Since H is assumed to be unimodular, $\delta_g * (\tilde{m} * \gamma) = (\tilde{m} * \gamma)$. The result follows from eq. $(*)$. □

In terms of the expression for the Radon–Nikodym derivatives given in Lemma 11.41, one defines the regular representation ρ of H in $L^2(H/T)$ by the formula $\rho(g)[\phi](b) = \left(\frac{dg\cdot\overline{m}}{d\overline{m}}\right)^{1/2}(b)\phi(g^{-1}\cdot b)$, for $\phi \in L^2(H/T)$ and $b \in H/T$.

11.42. Proposition. *The convolution operators on $L^2(H)$ and $L^2(H/T)$ corresponding to p, defined by the regular representations, have the same spectral radius.*

Before giving the proof, some facts are presented about unitary representations that have to do with weak containment and the Fell topology (see Zimmer [Z] for details).

Given two unitary representations ρ and ρ' of H, one says that ρ is weakly contained in ρ', denoted by writing $\rho \ll \rho'$, if every coefficient $\langle\rho(g)x, y\rangle$ is a uniform limit on compact subsets of coefficients $\langle\rho'(g)x_n, y_n\rangle$ of ρ'.

If ρ is a representation of H and p is a bounded measure on H one defines the operator $\rho(p)$ by $\rho(p) = \int \rho(g)dp(g)$. It is known that the spectral radius of $\rho(p)$ is less than or equal to that of $\rho'(p)$ if $\rho \ll \rho'$.

Another useful fact is the so-called continuity of the inducing process. This means that, if L is a closed subgroup of the locally compact group H, and if ρ and ρ' are two unitary representations of L, the condition $\rho \ll \rho'$ implies that $Ind_L^H \rho \ll Ind_L^H \rho'$.

Now, on the other hand, it follows from Herz's majoration principle (see Eymard-Lohoue [E3]) that, for any $u, v \in L^2(H)$, there exist $\bar{u}, \bar{v} \in L^2(H/L)$ such that $\|u\| = \|\bar{u}\|, \|v\| = \|\bar{v}\|$ and $\langle\gamma(g)\bar{u}, \bar{v}\rangle \geq |\langle\delta_g * u, v\rangle|$, where γ is the regular representation of H on $L^2(H/L)$. Hence, the norm of a convolution operator increases when one replaces $L^2(H)$ by $L^2(H/L)$.

Proof of Proposition 11.42.

From what has been said, it is clear that $r_0(p)$ is less than or equal to the spectral radius of $\rho(p)$. On the other hand, as T is amenable, $L^2(T)$ contains the identity representation of T weakly [G9]. By definition, ρ is the induced representation from T to H of the identity representation of T. The regular representation of H into $L^2(H)$ is induced from the regular representation of T into $L^2(T)$. Hence, the continuity of the inducing process implies that ρ is weakly contained in the regular representation of H into $L^2(H)$. Consequently, $r_0(p)$ dominates the spectral radius of $\rho(p)$.

Proposition 11.42 may be reformulated, using these concepts, as follows.

11.43. Proposition. $r_0(p) = \hat{p}(\delta^{1/2})$.

Proof. Consider the operator $\rho(p)$ in $L^2(H/T)$. Since p is K-bi-invariant, $\rho(p)1$ is a constant function because

$$\begin{aligned}
[\rho(p)1](b) &= \int \Big[\frac{dg\cdot\overline{m}}{d\overline{m}(b)}\Big]^{1/2} dp(g) \\
&= \int \Big[\frac{dg\cdot\overline{m}}{d\overline{m}}(b)\Big]^{1/2} d(m*p)(g) = \int \Big[\frac{dkg\cdot\overline{m}}{d\overline{m}}(b)\Big]^{1/2} dm(k)dp(g) \\
&= \int \Big[\frac{dg\cdot\overline{m}}{d\overline{m}}(k^{-1}b)\Big]^{1/2} dm(k)dp(g) \\
&= \int \sigma_{\delta^{1/2}}(g^{-1}, k^{-1}\cdot b)dm(k)dp(g) = \int \Phi^{\delta^{1/2}}(g)dp(g).
\end{aligned}$$

The same calculation implies that

$$\rho(\check{p})1 = \int \Phi^{\delta^{1/2}}(g^{-1})dp(g) = \hat{p}(\delta^{1/2}).$$

The operator $\rho(\check{p})$ is a compact, positive operator on $L^2(H/T)$. Hence, its spectral radius is equal to its dominant eigenvalue, which is equal to $\hat{p}(\delta^{1/2})$. On the other hand, the spectral radii of the operators $\rho(p)$ and $\rho(\check{p})$ in $L^2(H/T)$ are equal since they are adjoint. Hence, Proposition 11.42 implies that $r_0(p) = \hat{p}(\delta^{1/2})$. □

11.44. Corollary. *If* $\Phi^{\delta^{1/2}} \le \Phi^{\ell}$ *for every* $\ell \in T^*$ *then* $r(p) = r_0(p)$.

Proof. The condition $\Phi^{\delta^{1/2}} \le \Phi^{\ell}$ implies that $\hat{p}(\delta^{1/2}) \le \hat{p}(\ell)$ because $\hat{p}(\delta^{1/2}) = p(\Phi^{\delta^{1/2}})$, $\hat{p}(\ell) = p(\Phi^{\ell})$, and p is positive. Hence, by Proposition 11.43, $r_0(p) = \hat{p}(\delta^{1/2}) = \inf_{\ell\in T^*} \hat{p}(\ell) = r(p)$. □

11.45. Remark. It will be shown in the next chapter that for H a semisimple group with finite center, K a maximal compact subgroup and T a minimal parabolic subgroup, the conditions of the corollary are satisfied. Hence, in this case, $r(p) = r_0(p)$.

OUTLINE OF THE FOLLOWING CHAPTERS

Assume, as was the case earlier, that $H = KT$ with K compact, where T has the fixed line property, and p is a well-behaved K-bi-invariant positive measure. In Proposition 11.36 a class of solutions of the equation $f*p = rf$ was determined. Namely, if $r = \hat{p}(\ell)$, where $\ell \in T^*$, then $f(g) = \sigma_\ell(g^{-1}, b)$ is such a solution.

On the other hand, the limits of the Martin kernel $K^r(g,y)$ give all the minimal solutions of the equation $f * p = rf$ and the Martin compactification of H is a well-defined object with the non-minimal points of the Martin boundary integrals of the minimal solutions. Hence, the calculation of the Martin compactification contains the description of the integral representations of all the solutions and, in particular, of the non-minimal boundary points.

In this general situation one can prove the following result.

11.46. Theorem. (See Theorems 13.12 and 13.17) *Suppose that the convolution algebra of K-bi-invariant continuous functions on H with compact support is commutative and let B denote H/T. Then, for $r \geq r(p)$, every minimal solution of the equation $f * p = rf$ is of the form $f(g) = \sigma_\ell(g^{-1}, b)$ for some $b \in B$ and $\ell \in T^*$, with $\hat{p}(\ell) = r$.*

If $r = r(p)$, there exists a unique $\ell'_o \in T^*$ such that $\hat{p}(\ell'_o) = r(p)$, and every minimal solution of the equation $f * p = r(p)f$ is of the form $f(g) = \sigma_{\ell'_o}(g^{-1}, b)$ for some $b \in B$.

In addition, the following result is proved in Chapter XII. (Refer to Theorem 11.46 for the notations.)

11.47. Theorem. (See Theorem 12.13) *Let $\overline{m}$ denote the unique K-invariant normalized measure on B. Then there exists an H-equivariant linear and positive isometry of $L^\infty(H)$ onto $L^\infty(B) : f \to \hat{f}$ such that every bounded solution of the equation $f * p = f$ is given by the Poisson formula, namely,*

$$f(g) = \int \hat{f}(b) \frac{dg \cdot \overline{m}}{d\overline{m}}(b) d\overline{m}(b) = g \cdot \overline{m}(\hat{f})$$

with $\|\hat{f}\|_\infty = \|f\|_\infty$.

11.48. Corollary. *$\Phi^{\ell'_o}$ is the unique positive function F such that $p * F = F * p = rp$ and $F(e) = 1$. Furthermore, if $p(e) > 0$, then $\lim_n \frac{p^{n+1}(e)}{p^n(e)} = r$ and the sequence of functions $\frac{1}{p^n(e)} p^n$ converges uniformly on compact sets to F.*

11.49. Remarks. The last statement of this corollary is not proved in this book. A more general result, that shows the significance of F as a probabilistic ground state, can be found in [G12].

To go further it is necessary to have a precise description of those $\ell \in T^*$ for which the corresponding cocycle is a minimal solution of the equation $f * p = rf$. This result of Karpelevič [K3], used in Chapter VIII, is proved in Chapter XIII for a semisimple Lie group G. Note that in Theorem 8.2 none of the non-minimal limit functions determined by Martin's method is of the form $\sigma_\ell(g^{-1}, s)$.

The result is also valid for the group $H = G(\mathbb{F})$ of $\mathbb{F}$-rational points of a semisimple algebraic group G^a defined over a local field $\mathbb{F}$, provided G^a is simply connected and K is a "good" maximal subgroup (see [B16], [B17]).

In Chapter XIV it is shown that the boundary of the Martin compactification $\tilde{G}(p, r_0(p))$ for a semisimple Lie group G is the boundary of the maximal Satake–Furstenberg compactification of $X = G/K$. It is natural to conjecture, since it occurs in the case of the Laplacian, that, if $r > r_0(p)$, the Martin compactifications should be the same up to simple modifications in the form of the limiting functions. This conjecture can be proved for $r \geq r_0(p)$ in the ultrametric situation. Some additional details are given in this case in Chapter XV. An important role is played in this study by the description of the maximal Satake–Furstenberg compactification in terms of the space of closed subgroups of G given in Chapter IX.

In the general case of a Lie group, one needs to know first of all the minimal solutions. Presumably, Martin's method is also useful for this restricted problem. However, very little information is available for a general Lie group. The convergence of the Martin kernel in "general directions" for some classes of Lie groups has been considered in [B1], [B2], [D1] and [E2], and the minimal eigenfunctions have been investigated in [C5] and, more recently, in [R2].

CHAPTER XII

THE FURSTENBERG BOUNDARY AND BOUNDED HARMONIC FUNCTIONS

Let L denote the Laplace–Beltrami operator on $X = G/K$. The main purpose of this chapter is to give another, elementary, and self-contained proof of the so-called Poisson formula (see Theorem 12.10) for the integral representation of the bounded harmonic functions, i.e., solutions of the equation $Lf = 0$ [F3]. This was proved earlier (see Corollary 8.29), using the Martin boundary of X for $\lambda = 0$. The key to the proof, presented here, is the fact that (G, K) is a Gelfand pair. As a result it follows, see Corollary 12.9, that a bounded C^2-function is harmonic if and only if it satisfies the mean-value property. This is not so easily proved as in Euclidean space because, if the rank of X is greater than one, K is not transitive on the geodesic spheres centered at o.

This new proof is easily adapted to prove the corresponding result for a random walk on a suitable locally compact metrizable group H, where a function f is said to be $\check{p}$-harmonic if $f * p = f$. The group $H = KP$, where K is a compact subgroup and P is amenable. In addition, it is assumed that H/P is a boundary and (H, K) is a Gelfand pair. The point of using convolution equations is that it extends the scope of the proof of this Poisson formula to other classes of topological groups, including the local field analogues of G (see Theorem 12.13).

Basic notations

As in § 2.1, given a Cartan decomposition $\mathfrak{k}\oplus\mathfrak{p}$ of $\mathfrak{g}$ and a maximal abelian subalgebra $\mathfrak{a}$ of $\mathfrak{p}$, an Iwasawa decomposition KAN of G is determined by a choice of a positive Weyl chamber $\mathfrak{a}^+$. Let $a(g) = \exp H(g)$ denote the A-part of g in the Iwasawa decomposition $g = kan$.

Let M denote the centralizer of A in K and P the minimal parabolic subgroup MAN. Denote by m the unique K-invariant probability measure on $\mathcal{F} = G/P$. Note that the choice of a minimal parabolic subgroup P and of K determines the Weyl chamber $\mathfrak{a}^+$ since $\mathfrak{a}^+(\infty)$ is the set of points in $X(\infty)$ whose stabilizer is P.

If ρ denotes half the sum of positive roots, then $e^{2\rho}$ is an exponential on A and

$$P(g,b) = \frac{dg \cdot m}{dm}(b) = e^{-2\rho(H(g^{-1}k))} = e^{-2\rho}[a(g^{-1}k)],$$

where $k \in K, b = kP = k \cdot \dot{e} \in \mathcal{F}$. This formula for the Radon–Nikodym derivative was stated earlier as eq. (9.12).

THE MEAN-VALUE PROPERTY

12.1. Definition. A bounded measurable function f on G has the **mean-value** property if, for all $g \in G$, its average over any orbit of gKg^{-1} equals its value at g. Equivalently, for all $g, g_1 \in G$,

$$f(g) = \int_K f(gkg_1)dk.$$

A bounded measurable function f on X has the **mean-value** property if, for all $g \in G$, its average over any orbit of gKg^{-1} equals its value at $g \cdot o$. Equivalently, for any $g \in G$ and $x \in X$,

$$f(g \cdot o) = \int_K f(gk \cdot x)dk.$$

Remark. A K-right invariant, bounded measurable function f on G has the mean-value property if and only if $f = F \circ \pi$, where $\pi(g) = g \cdot o$, and F is bounded and has the mean-value property on X.

12.2. Proposition. *Let ν be a K-bi-invariant probability measure on G. Assume that g_1^{-1} belongs to the support of ν. If f is a bounded continuous function on G, and if $f = f * \nu$, then $f(g) = \int f(gkg_1)dk$ for all $g \in G$.*

Proof. When $\nu = \tilde{m} * \delta_{g_1^{-1}} * \tilde{m}$, the result is trivial. If $\nu \neq \tilde{m} * \delta_{g_1^{-1}} * \tilde{m}$, let V be a K-bi-invariant neighborhood of $Kg_1^{-1}K$ for which $0 < \nu(V) < 1$. Clearly, V exists as ν is not concentrated on $Kg_1^{-1}K$. Decompose the probability measure ν into its normalized restrictions ν_V and ν'_V to V and its complement. Then $\nu = \alpha\nu_V + (1-\alpha)\nu'_V$, with $0 < \alpha < 1$. Since ν_V and ν'_V are K-bi-invariant, they commute under convolution as (G,K) is a Gelfand pair. Furthermore, as they are probability measures and commute with ν, they define contractions $f \to f * \nu_V$ and $f \to f * \nu'_V$ on the space $C_b(G)$ of bounded continuous functions on G. Moreover, $f = f * [\alpha\nu_V + (1-\alpha)\nu'_V]$. Proposition 12.3 implies that $f = f * \nu_V = f * \nu'_V$. Let V_n be a decreasing sequence of K-bi-invariant neighborhoods with $\cap_n V_n = Kg_1^{-1}K$. Then, for any bounded continuous function ϕ on G, $\lim_n \frac{1}{\nu(V_n)} \int_{V_n} \phi(h)d\nu(h) = \int_K \int_K \phi(\ell^{-1}g_1^{-1}k^{-1})d\ell dk$. If $\phi(h) = f(gh^{-1})$ this implies, since f is K-right invariant, that $f(g) = \int_K \int_K f(gkg_1\ell)dkd\ell = \int_K f(gkg_1)dk$ for all $g \in G$. □

12.3. Proposition. (see [R3, Lemma 1.1, p. 133]) *Let E be a Banach space and let P and Q be two contractions of E that commute. Then, every fixed point of $\alpha P + (1-\alpha)Q$, for $0 < \alpha < 1$, is also a fixed point of P and Q.*

The proof of this result depends on the following lemma, due to Guivarc'h.

12.4. Lemma. *Let P and Q be as in Proposition 12.3. Then*

$$\lim_{n\to+\infty} \|[\alpha P + (1-\alpha)Q]^n (P - Q)\| = 0.$$

Proof. Since $PQ = QP$ and $\|P\|, \|Q\| \leq 1$, it follows from the binomial formula that

$$[\alpha P + (1-\alpha)Q]^n = \sum_{k=0}^{n-1} \gamma_n^k P^k Q^{n-k},$$

with $\gamma_n^k = C_n^k \alpha^k (1-\alpha)^{n-k}$. Hence,

$$[\alpha P + (1-\alpha)Q]^n (P - Q) = \sum_{k=0}^{n-1} (\gamma_n^k - \gamma_n^{k+1}) P^{k+1} Q^{n-k} + \gamma_n^n P^n - \gamma_n^0 Q^n$$

and, therefore,

$$\| [\alpha P + (1-\alpha)Q]^n (P - Q)\| \leq \sum_{k=0}^{n-1} |\gamma_n^k - \gamma_n^{k+1}| + \gamma_n^n + \gamma_n^0.$$

It is well-known that the binomial coefficients γ_n^k increase to a maxmum at $k = n\alpha_n$, where $|\alpha_n - \alpha| < \frac{1}{n}$, and then decrease. As a result

$$\sum_{k=0}^{n-1} |\gamma_n^k - \gamma_n^{k+1}| + \gamma_n^n + \gamma_n^0 \leq 2\gamma_n^{n\alpha_n}.$$

Since, by a standard use of Stirling's formula, $\gamma_n^{n\alpha_n} \leq \frac{c}{\sqrt{n}}$ the result follows. □

12.5. Corollary. *Let f be a bounded measurable function on G and denote by p a probability on G with compact support S such that $G = \cup_{n>0} S^n$. The following conditions are equivalent.*

(1) *f has the mean-value property;*

(2) *$f * p = f$ for every such K-bi-invariant probability p ;*

(3) *$f * p = f$ for at least one such K-bi-invariant probability p.*

In particular, f has the mean-value property if and only if $f = f * p$ for some well-behaved and K-bi-invariant probability p. As a result, a function with the mean-value property is continuous.

Proof. Assume that f is bounded measurable and has the mean-value property, i.e., for all $g, g_1 \in G$, $f(g) = \int_K f(gkg_1)dk$. Let $\tilde{m}$ denote normalized Haar measure on K. Then, $f * (p * \tilde{m})(g) = \int\int f(g(hk)^{-1})dp(h)d\tilde{m}(k) = \int\left[\int f(gk^{-1}h^{-1})d\tilde{m}(k)\right]dp(h)$. It follows from the mean-value property that $\int f(gk^{-1}h^{-1})d\tilde{m}(k) = f(g)$. Hence, $f * (p * \tilde{m})(g) = \int f(g)dp(h) = f(g)$. If p is any K-bi-invariant probability on G, then $p * \tilde{m} = p$. Hence, $f * p = f$ and, so, (1) implies (2).

Trivially, (2) implies (3). Assume (3). Since $f * p = f$ it follows that $f * p^n = f$ for all $n \geq 1$. Let $g_1^{-1} \in G$ be in the support of p^n. It follows from Proposition 12.2 that, for all $g, g_1 \in G$, $f(g) = \int_K f(gkg_1)dk$. Since $G = \cup_{n>0}^{\infty} S^n$, this proves (1). □

12.6. Proposition. *The formula $f(g) = g \cdot m(\phi) = \int \phi d(g \cdot m)$, where $\phi \in L^\infty(\mathcal{F})$, defines a G-equivariant isometry between $L^\infty(\mathcal{F})$ and the space $\mathcal{H}_m$ of bounded measurable functions on G that satisfy the mean-value property.*

To prove this result, one makes use of the following lemma.

12.7. Lemma. *There exists a positive G-equivariant positive contraction $\overline{\beta}$ of $L^\infty(G)$ into $L^\infty(\mathcal{F})$ such that $\overline{\beta}(1) = 1$. If $\bar{\beta}(f)$ is considered as an element of $L^\infty(G)$, then, for every bounded measure μ on G, $\bar{\beta}[\mu * f] = \mu * \bar{\beta}(f)$.*

Proof. Denote by β a positive linear form on $L^\infty(G)$ that is invariant under left translation by MAN and satisfies $\beta(1) = 1$. (The existence of such a form follows from the amenability of MAN [G9].) Let $f_x(g)$ denote $f(xg)$ and suppose initially that $f \in L^\infty(G)$ is left uniformly continuous. Then the formula $\hat{f}(x) = \beta[f_x]$ defines a continuous function on G that is right invariant under MAN and, hence, an element of $C(G/MAN) = C(\mathcal{F})$.

When f is arbitrary in $L^\infty(G)$ one can define $\beta(\bar{f})$ by duality as follows. For every $\alpha \in L^1(G)$, set $\hat{f}(\alpha) = \beta[\check{\alpha} * f]$. Then $\alpha \to \hat{f}(\alpha)$ is a linear form on $L^1(G)$ that is bounded by $\|f\|_\infty$, because β is positive, $\beta(1) = 1$, and $\|\check{\alpha} * f\|_\infty \leq \|\check{\alpha}\|_1 \|f\|_\infty$. Hence $\hat{f}$ defines an element of $L^\infty(G)$. Clearly, $\hat{f}(\alpha * \delta_t) = \beta[\delta_{t^{-1}} * \check{\alpha} * f] = \hat{f}(\alpha)$, if $t \in MAN$. It follows that $\hat{f}$, as an element of $L^\infty(G)$, is right-invariant under MAN. As a result, it defines an element of $L^\infty(\mathcal{F})$, again denoted by $\hat{f}$. Let $\overline{\beta}(f)$ denote $\hat{f}$. The required properties of $\overline{\beta}$ follow from the formula $\hat{f}(\alpha) = \beta[\check{\alpha} * f]$. In particular the last formula follows since

$$\bar{\beta}[\mu * f](\alpha) = \beta[\check{\alpha} * \mu * f] = \hat{f}(\check{\mu} * \alpha) = (\mu * \hat{f})(\alpha).$$

Hence, from the definition of $\bar{\beta}(f)$ as an element of $L^\infty(G)$, it follows that

$$\bar{\beta}[\mu * f] = \mu * \bar{\beta}(f). \quad \square$$

Proof of Proposition 12.6.

Let $\phi \in L^\infty(\mathcal{F})$ and ν be a K-bi-invariant probability measure. Define f on G by setting $f(g) = g \cdot m(\phi)$. Then, since $\check{\nu} * \tilde{m} = \tilde{m}$, it follows that

$$(f * \nu)(g) = \int f(gh^{-1}) d\nu(h) = \delta_g * \check{\nu} * \tilde{m}(\phi) = g \cdot \tilde{m}(\phi) = f(g).$$

Hence, by Corollary 12.5, f satisfies the mean-value property. In particular, $f(g) = \int f(gkg_1) dk$ for all $g_1 \in G$.

Conversely, suppose that f satisfies this formula. Then f is continuous and the formula may be written as $f(g) = \tilde{m} * \delta_{g^{-1}} * f$, where $f(g)$ is considered as a constant function and the expression $\tilde{m} * \delta_{g^{-1}} * f$ is to be evaluated at $g_1 \in G$.

Applying $\bar{\beta}$ to both sides of the equation, it follows that

$$f(g) = \bar{\beta}[\tilde{m} * \delta_{g^{-1}} * f] = \tilde{m} * \delta_{g^{-1}} * \bar{\beta}(f),$$

where the last equality follows from Lemma 12.7. The last term can be written as $(g \cdot m)(\hat{f})$ if $\hat{f}$ is considered as an element of $L^\infty(\mathcal{F})$. Clearly, this formula implies that $\|f\|_\infty \leq \|\hat{f}\|_\infty$. On the other hand, since $\hat{f} = \overline{\beta}[f]$, the properties of $\overline{\beta}$ imply that $\|\hat{f}\|_\infty \leq \|f\|_\infty$. Hence $f \to \hat{f} = \overline{\beta}(f)$ is a G-equivariant isometry of $\mathcal{H}_m$ onto $L^\infty(\mathcal{F})$, as is its inverse $\hat{f} \to f$, where $f(g) = g \cdot m(\hat{f})$. □

Harmonic functions and the mean-value property

As pointed out in § 8.3, it is well-known that $X = G/K$ is diffeomorphic to $\mathbb{R}^n$ and that L can be viewed as an elliptic operator on X with smooth coefficients. Hence, as is well-known (see, for example, [G3]), the classical Dirichlet problem for L on the geodesic unit ball B_r is solvable. This means that any continuous function ϕ on ∂B_r has a continuous extension as a harmonic function $h = h_\phi$ in B_r. Furthermore, by the maximum principle (see [G3]), this extension is unique.

12.8. Proposition. *Let $r > 0$. There exists a K-invariant probability ν_r, with support the boundary ∂B_r of the geodesic ball B_r in G/K with radius r and center the origin o, such that, for every harmonic function f,*

$$f(g \cdot o) = \int_{\partial B_r} f(g \cdot s) d\nu_r(s).$$

The measure ν_r is the harmonic measure of o, relative to B_r.

Proof. The harmonic measure ν_x on ∂B_r is the unique Radon measure for which $h_\phi(x) = \int \phi d\nu_x$ if $x \in B_r$, where h_ϕ is the unique harmonic extension of the boundary value ϕ to B_r. This unicity implies that the support of ν_x is ∂B_r. Let $\nu_r = \nu_o$. Since B_r is K-invariant and L commutes with the left action of K, it follows that ν_r is K-invariant.

Since $f_{|B_r} = h_\phi$ if $\phi = f_{|\partial B_r}$, when f is harmonic on X, it follows that $f(o) = \int f d\nu_r = \nu_r(f)$ for every harmonic function f. □

12.9. Corollary. *A bounded measurable function on X is harmonic if and only if it has the mean-value property.*

Proof. First, assume h is bounded and harmonic. Let $r > 0$. Since ν_r is K-invariant, one may define the Markov kernel N_r by $N_r F(g \cdot o) = \int F(g \cdot x) d\nu_r(x)$. It maps the Banach space of bounded measurable functions into itself and has norm equal to one. Note that $N_r h = h$.

Let $x_0 \in \partial B_r$ and let V be a neighborhood of x_0 with $\nu_r(V) < 1$. Let kernels N_V and $N_{V'}$ be the Markov kernels defined by the K-invariant probabilities that are the respective restrictions of ν to the K-orbits $K \cdot V$ and $K \cdot \complement V$, normalized so as to be probabilities.

These Markov kernels commute. It suffices to lift them to G as convolution operators given by K-bi-invariant probabilities: if ν is K-invariant on X, let $\mu = \tilde{\nu}$ be defined by $\mu(A) = \int [\int 1_A(gk) \tilde{d}m(k)] d\nu(g \cdot o)$ (see § 14.5 for further details). If $g_1 \in G$ then, for some $r > 0$, the support of μ_r contains g_1. Since $N_r F(g \cdot o) = f * \tilde{\mu}_r(g)$, if $f = F \circ \pi$, it follows from Propositions 12.2 and 12.8 that $h \circ \pi$ has the mean-value property on G if h is bounded, harmonic on X. As a result, by the remark following Definition 12.1, h has the mean-value property on X.

To prove the converse, it is shown in Appendix A (Lemma A.21) that if $\phi \in C^2(X)$ then, for $\gamma = \frac{1}{2(d+2)}$, where d is the dimension of X,

$$\gamma L\phi(x) = \lim_{r \to 0} r^{-2}[\mu_x^r(\phi) - \phi(x)],$$

where μ_x^r is the normalized uniform measure on the ball of center $x \in X$ and radius r.

If $f \in \mathcal{H}_m$, then $\mu_x^r(f) = f(x)$. Let p be any K-bi-invariant probability measure on G with a C^2-density. Since $f = f * p$, f is C^2 and, hence, it follows that $\gamma L f(x) = \lim_{r \to 0} r^{-2}[\mu_x^r(f) - f(x)] = 0$. This completes the proof. □

The main result of this chapter follows from Corollary 12.9 and Proposition 12.6.

12.10. Theorem. *Denote by $\mathcal{H}_b$ the space of bounded harmonic functions on X. Then there exists a G-equivariant isometry of $\mathcal{H}_b$ onto $L^\infty(\mathcal{F}) : f \to \hat{f}$ such that, for every $f \in \mathcal{H}_b$, the* **Poisson formula** *holds, i.e.,*

$$f(g \cdot o) = g \cdot m(\hat{f}) = \int \hat{f}(b) P(g, b) dm(b).$$

Remark. It will follow from the next chapter that, as stated in Chapter VIII, the functions $h_b(x)$ are minimal harmonic functions. Hence, the Poisson formula appears as a special case of the integral representation of f in terms of the minimal functions. This is how it was established in Corollary 8.29.

Convergence theorems for harmonic functions

It follows from Theorem 9.45 that the maximal Satake–Furstenberg compactification may be realized as the closure of the orbit $G \cdot m$ in the compact space $\mathcal{M}_1(\mathcal{F})$ of probability measures on $\mathcal{F}$.

Recall that it was shown in Lemma 4.48 and in Theorem 9.37 that G/P is a boundary in the sense that there exist sequences $g_n \in G$ such that $\lim_n g_n \cdot m = \delta_b$ for every $b \in \mathcal{F}$.

For example, if $g_n = a^n$ and $\log a \in \mathfrak{a}^+$, then $\lim_n a^n \cdot m = \delta_{\dot{e}}$.

12.11. Proposition. *Let f be a bounded function on G with the mean-value property. If $g_n \in G$ is a sequence in G such that $\lim_n g_n \cdot m = \delta_b$, the sequence of functions $f(kg_n)$ on K converges weakly to $\phi(k \cdot b)$ in $L^2(K)$.*

Proof. Since f has the mean-value property, Proposition 12.6 implies that f is of the form $f(g) = g \cdot m(\phi)$ with $\phi \in L^\infty(\mathcal{F})$. Consider the measure μ_n on K defined by $d\mu_n(k) = f(kg_n)d\tilde{m}(k)$.

Let $\psi \in L^2(K)$. Then, $\mu_n(\psi) = \int f(kg_n)\psi(k)dk = \int kg_n \cdot \tilde{m}(\phi)\psi(k)dk = [(\psi\tilde{m}) * (g_n \cdot \tilde{m})](\phi) = (g_n \cdot \tilde{m})[(\check{\psi}\tilde{m}) * \phi]$. The function $(\check{\psi}\tilde{m}) * \phi$ on $\mathcal{F}$ is continuous because $\phi \in L^\infty(\mathcal{F})$ and $\psi \in L^2(K)$. The hypothesis $\lim_n g_n \cdot \tilde{m} = \delta_b$ implies that $\lim_n \mu_n(\psi) = \int \phi(k \cdot b)\psi(k)dk$. This proves the weak convergence stated in the proposition. □

12.12. Corollary. *Suppose f is a bounded harmonic function $f(g \cdot o) = (g \cdot m)(\hat{f})$ with $\hat{f} \in L^\infty(\mathcal{F})$ and consider a sequence $g_n \in G$ such that $\lim_n g_n \cdot m = \mu \in \overline{X}^{SF}$. Then the sequence of functions $\phi_n(k) = f(kg_n \cdot o)$ converges weakly in $L^2(K)$ to the function $\int \hat{f}(k \cdot b)d\mu(b)$.*

In particular, if $g_n \in G$ is a sequence such that $\lim_n g_n \cdot m = \delta_b$, then the sequence of functions f_n on K given by $f_n(k) = f(kg_n \cdot o)$ converges weakly in $L^2(K)$ to $\hat{f}(k \cdot b)$.

If $\hat{f}$ is continuous, then the convergence is uniform.

Proof. For every $b \in \mathcal{F}$, $\hat{f}(k \cdot b)$ is well defined as an element of $L^\infty(K)$, as is the barycenter $\int \hat{f}(k \cdot b)d\mu(b)$ with respect to μ.

Denote by ψ an element of $L^\infty(K)$ and consider the scalar product $\langle \phi_n, \psi \rangle$ given by

$$\begin{aligned}\langle \phi_n, \psi \rangle &= \int f(kg_n \cdot o)\psi(k)dk = \int (kg_n \cdot m)(\hat{f})\psi(k)dk \\ &= (\psi\tilde{m} * \delta_{g_n} * m)(\hat{f}) = (g_n \cdot m)[\check{\psi}\tilde{m} * \hat{f}].\end{aligned}$$

Because $\psi \in L^2(K)$ and $\hat{f} \in L^\infty(\mathcal{F})$, the function $(\check{\psi}\tilde{m}) * \hat{f}$ is continuous on $\mathcal{F}$ and the definition of μ implies that

$$\lim_n \langle \phi_n, \psi \rangle = \mu[\check{\psi}\tilde{m} * \hat{f}] = \int \hat{f}(k \cdot b)\psi(k)dk d\mu(b) = \langle \phi, \psi \rangle,$$

with $\phi(k) = \int \hat{f}(k \cdot b)d\mu(b)$.

If $\hat{f}$ is continuous, $f(g \cdot o) = (g \cdot m)(\hat{f})$ is left uniformly continuous and the sequence $\phi_n(k)$ is uniformly equicontinuous. Extracting subsequences and identifying the limit as $\int \hat{f}(k \cdot b)d\mu(b)$, shows that ϕ_n converges uniformly. □

The Poisson formula for random walks

Let H denote a locally compact metrizable group H and p denote a K-bi-invariant probability measure on H. Denotes its support by S and assume that $H = \cup_{n>0} S^n$. Consider the space $\mathcal{H}_b$ of bounded measurable functions on H such that $f * p = f$. When is there a "Poisson formula" for these functions?

For the proof of Theorem 12.10 to apply, it suffices that the following conditions are satisfied.

(1) P is a closed amenable subgroup of H, and K is a compact subgroup such that $H = KP$.
(2) The convolution algebra $C_c(H, K)$ of K-bi-invariant functions with compact support is commutative, i.e., (H, K) is a Gelfand pair.
(3) H/P is a boundary.

As before, let m denote the unique K-invariant probability measure on $\mathcal{F} = H/P$.

12.13. Theorem. *Assume that H and p satisfy conditions (1) and (2). Then there exists an H-equivariant, positive isometry $f \to \hat{f}$ of $\mathcal{H}_b$ onto $L^\infty(H/P)$ such that one has, for all $g \in H$,*

$$f(g) = g \cdot m(\hat{f}), \quad \text{if } f \in \mathcal{H}_b.$$

Assume condition (3) holds. Then, in addition, if $f \in \mathcal{H}_b$ is given and (g_n) is a sequence in H such that $\lim_n g_n \cdot m = \delta_b$, the sequence of functions on K defined by $\phi_n(k) = f(kg_n)$ converges weakly in $L^\infty(K)$ to $\hat{f}(k \cdot b)$. If f is left uniformly continuous, then the convergence is uniform.

Proof. Note that the mean-value property makes sense in this larger context. Also, a function $f \in \mathcal{H}_b$ satisfies the mean-value property since the relevant part of the proof of Corollary 12.5 applies to H. As a result, the first part of this theorem is a consequence of Proposition 12.6. and the last statement follows from Proposition 12.11. □

12.14. Examples. (1) If G is a semisimple Lie group, p is K-bi-invariant with support S, and $G = \cup_{n>0} S^n$, then the conditions of Theorem 12.13 are satisfied. Again K is a maximal compact subgroup, P is a minimal parabolic and $G/P = \mathcal{F}$ is the Furstenberg boundary. (2) Assume that G^a is a simply connected, semisimple algebraic group defined over a local field $\mathbb{F}$. Let $H = G(\mathbb{F})$ denote the set of its $\mathbb{F}$-rational points. Then one can find a maximal subgroup K that is transitive on H/P and such that the algebra $C_c(H, K)$ is commutative. Here P is the group of $\mathbb{F}$-rational points of a minimal parabolic subgroup of G^a [M1]. In this case conditions (1), (2), (3) are satisfied (it is indicated in § 9.40 that H/P is a boundary). Hence, if p satisfies $H = \cup_{n>0} S^n$, the Poisson formula remains valid and characterizes the bounded functions that satisfy the mean-value property. The explicit formula for the Poisson kernel is the same. Some additional details will be given in Chapter XV.

(3) Denote by A a locally compact and metrizable abelian group. Assume that H satisfies the conditions (1), (2), and (3) of Theorem 12.11. Then the direct product $H \times A$ also satisfies these conditions. Here P

is replaced by $P \times A$ and $H \times A = K(P \times A)$, with $P \times A$ amenable. Clearly, $C_c(H \times A, K) = C_c(H, K) \otimes C_c(A)$ is commutative. Moreover, $H \times A/P \times A = H/P$ is a boundary.

(4) In particular, another class of examples is obtained as the products $G \times A$ with G semisimple and A abelian. This class includes the reductive groups.

12.15. Remark. If K-bi-invariance of p is not assumed but p has a density and the support of p generates H, the statement of Theorem 12.10 remains essentially valid for the bounded solutions of the equation $f * p = f$ (see [F3]). Presumably, the same is true of Theorem 12.13. A natural extension of Theorem 12.10 to the class of connected Lie groups has been given in [R1].

CHAPTER XIII

INTEGRAL REPRESENTATION OF POSITIVE EIGENFUNCTIONS OF CONVOLUTION OPERATORS

For the integral representation of λ-eigenfunctions of the Laplacian, it is important to have an explicit description of $\partial_e X(\lambda)$, the set of minimal eigenfunctions. When X is a general symmetric space of non-compact type, these eigenfunctions were first determined by Karpelevič [K3]. In this chapter they are determined by using convolution equations (see Theorems 13.1, 13.23, and 13.28), a method first used by Furstenberg for semisimple Lie groups. This method is to used prove analogous results for convolution equations on a general class of groups that includes local field analogues of G as well as reductive Lie groups.

THE MAIN RESULT OF THIS CHAPTER

As elsewhere in this book, G is a semisimple Lie group with finite center, $G = KAN$ is an Iwasawa decomposition, $P = MAN$ is a minimal parabolic subgroup, and L is the Laplacian on $G/K = X$. The Haar measure on K will be denoted $d\tilde{m}(k)$ and also by dk. In this chapter the solvable group $S = AN$ is denoted by T. If $\ell \in A^* = T^* = P^*$, the multiplier $\sigma_\ell(g,b) = \ell[a(gk)]$, where $b = k \cdot \dot{e} \in \mathcal{F}$, determines the K-right invariant function h_b^ℓ on G (equivalently, on G/K) defined by $h_b^\ell(g) = \sigma_\ell(g^{-1}, b)$. The natural class of functions $h_b^\ell, \ell \in T^*, b \in \mathcal{F}$ plays a fundamental role in this book because of the following theorem due to Karpelevič [K3]. (See Guivarc'h [G14] for a short proof and Furstenberg [F4] for a much more general approach.)

13.1. Theorem. *If $\ell \in P^*$, then $Lh_b^\ell + \lambda(\ell)h_b^\ell = 0$ for some $\lambda(\ell) \in \mathbb{R}$. Every minimal solution of the equation $Lf + \lambda f = 0$ is of the form $f = h_b^\ell$, for some $b \in G/P$ and $\ell \in P^*$, with $\lambda = \lambda(\ell)$.*

13.2. Remarks. (1) Not all these functions are minimal. The exact restrictions on ℓ will be given in Lemmas 13.24 and 13.25.

(2) Note that no connection is given between $\| \ell \|$ and $\lambda(\ell)$ (cf. Proposition 8.15). In this form the theorem is valid for convolution equations (see Theorem 13.33).

Theorem 13.1 is a consequence of the following propositions, the first of which shows that the functions h_b^ℓ are eigenfunctions.

13.3. Proposition. *The functions h_b^ℓ are eigenfunctions of the Laplacian, as well as for all positive G-invariant kernels on G/K. The eigenvalue depends only on ℓ.*

Proof. It will suffice to prove this for a G-invariant kernel. Such a kernel Q is defined by a K-bi- invariant positive measure q on G by setting $Qf = f*q$. Clearly, if θ is a positive measure on $T = AN$, the K-bi-invariance of q implies that

$$\theta * q * \tilde{m} = (\theta * \tilde{m}) * (q * \tilde{m}).$$

This identity remains valid if q is replaced by its projection $\overline{q}$ on T relative to the decomposition $G = TK$, i.e.,

$$(\dagger) \qquad (\theta * \overline{q}) * \tilde{m} = (\theta * \tilde{m}) * (\overline{q} * \tilde{m}).$$

The function $h^{\ell}_{\dot{e}}$ on G/K can be identified with the exponential $\check{\ell}(t) = \ell(t^{-1})$. Because $\check{\ell}$ is an eigenfunction for every convolution operator on T, eq. (†) implies that $h^{\ell}_{\dot{e}}(g \cdot 0)$ and $h^{\ell}_{\dot{b}}(g \cdot 0)$ are also eigenfunctions under right convolution by q. These remarks are valid for distributions as well as for measures. Hence, $h^{\ell}_{\dot{b}}$ is also an eigenfunction of L. (This also follows directly from Proposition 7.4.)

Since $h^{\ell}_{\dot{b}}(g) = h^{\ell}_{\dot{e}}(k^{-1}g)$, when $b = kM$, and right convolution by q commutes with left translation by $k \in K$, the result follows. □

Let $\mathcal{H}^{\lambda}$ denote the convex cone of positive functions f on G/K such that $Lf + \lambda f = 0$, where $\lambda \leq \lambda_0$. As pointed out in Chapter VI, Choquet's theory of integral representation is valid for this cone. More explicitly, one has the following result.

13.4. Proposition. *The cone $\mathcal{H}^{\lambda}$ is closed for the topology of uniform convergence on compact sets of X and the condition $f(o) = 1$ defines a compact base of this cone. The elements of $\mathcal{H}^{\lambda}$ satisfy a uniform Harnack condition: for every $f \in \mathcal{H}^{\lambda}$, the ratio $f(x)/f(y)$ is bounded by a constant depending only on the distance of x to y.*

Proof. As mentioned before, the fact that the map $Y \to \exp Y \cdot o$ is a diffeomorphism of $\mathfrak{p}$ with X (see Helgason [H2, Theorem 1.1, p. 252]) implies that the differential operator L on X can be considered as a strictly elliptic operator D on $\mathbb{R}^n, n = \dim \mathfrak{p}$, with smooth coefficients such that $D1 = 0$. Hence, the theory of second order elliptic partial differential equations on $\mathbb{R}^n$ can be applied to L (see [S3]). In particular, at each point of $\mathbb{R}^n$, Harnack's inequality holds ([S3]). As a result, each function $f \neq 0 \in \mathcal{H}^{\lambda}$ is strictly positive. Since, as observed in Proposition 12.8 for $\lambda = 0$, f is an eigenfunction if and only if it is determined by harmonic measures, it is clear that $\mathcal{H}^{\lambda}$ is closed under uniform convergence on compact sets. The Schauder estimates show that the normalized eigenfunctions are locally equicontinuous. Hence, the normalizing condition $f(o) = 1$ defines a compact base of $\mathcal{H}^{\lambda}$.

Considering the situation on X, it follows from the left G-invariance that, if $u(x) \leq Cu(o)$, for all $u \in \mathcal{H}^{\lambda}$ when $d(x, o) < R$, then, for all $g \in G$, one has $f(g \cdot x) \leq Cf(g \cdot o)$, since $u(x) = f(g \cdot x) \in \mathcal{H}^{\lambda}$. This implies that, for all non-zero eigenfunctions $f \in \mathcal{H}^\lambda$, the ratio $f(x)/f(y)$ is bounded by a constant depending only on the geodesic distance. □

The following key lemma means, in essence, that the action of G on an extremal h of $\mathcal{H}^\lambda$ reduces to that of K. In particular the G-orbits of h (in the twisted action) are K-orbits as will be evident later.

Recall that $C_c(G,K)$ is the subset of K-bi-invariant elements of $C_c(G)$.

13.5. Lemma. *Let h be an extremal function in $\mathcal{H}^\lambda$ and let $\alpha \in C_c(G,K)$. Then there exists a positive number $r(\alpha)$ such that*

$$h * \alpha = r(\alpha)h, \quad \alpha * h = r(\alpha)(\tilde{m} * h).$$

Proof. By definition, $(h*\alpha)(g\cdot o) = \int h(gx^{-1}\cdot 0)d\alpha(x)$. Since the distance of $gx^{-1}\cdot o$ to $g\cdot o$ is bounded when x belongs to the support of α, the uniform Harnack inequality implies that

$$h(gx^{-1}\cdot o) \leq ch(g\cdot o).$$

Hence, for all $y \in X$, $h*\alpha(y) \leq ch(y)$. Because L commutes with right convolution by elements of $C_c(G,K)$ [H3], $h*\alpha \in \mathcal{H}^\lambda$. The minimality of h implies that $h*\alpha = r(\alpha)h$.

The fact that $\phi * \varphi = \varphi * \phi$, if both functions are K-bi-invariant, implies that

$$\begin{aligned} \alpha * h &= (\alpha * \tilde{m}) * (h * \tilde{m}) = \alpha * (\tilde{m} * h * \tilde{m}) \\ &= (\tilde{m} * h * \tilde{m}) * \alpha = \tilde{m} * h * \alpha = r(\alpha)\tilde{m} * h. \end{aligned} \tag{†}$$
□

13.6. Lemma. *Let h be an extremal of $\mathcal{H}^\lambda$. The function $\phi = \tilde{m} * h$ is K-bi-invariant and is a common eigenfunction of right convolution by elements from $C_c(G,K)$. Furthermore, one has that, for all $g \in G$,*

$$\delta_g * \phi \leq d(g)\phi,$$

and for all $g_1 \in G$,

$$\int h(gkg_1)dk = \phi(g_1)h(g). \tag{*}$$

It follows that ϕ is a spherical function.

Proof. It follows from Lemma 13.5 that $\phi = \tilde{m} * h$ satisfies the equations $\phi * \tilde{m} = \phi$ and $\phi * \alpha = r(\alpha)\phi$. Further, eq. (†) in the proof of this lemma implies that

$$r(\alpha)(\delta_g * \phi) = \delta_g * \alpha * h \leq \beta * h = r(\beta)(\tilde{m} * h) = r(\beta)\phi,$$

with a certain function $\beta \in C_c(G,K)$ that dominates $\delta_g * \alpha$. This establishes the first inequality with $d(g) = r(\beta)/r(\alpha)$.

Let α_n be a sequence in $C_c(G,K)$ that is identified with a sequence of measures which converges weakly to $\tilde{m} * \delta_{g_1^{-1}} * \tilde{m}$. Since $h * \alpha_n = r(\alpha_n)h$, it follows that $h * \tilde{m} * \delta_{g_1^{-1}} * \tilde{m} = \lim_n (h * \alpha_n) = \lim_n r(\alpha_n)h$. The fact that $h * \tilde{m} = h$ implies that $h * \delta_{g_1^{-1}} * \tilde{m} = r(g_1)h$. In other words,

$$\int h(gkg_1)dk = r(g_1)h(g).$$

If $g = e$ and $h(e) = 1$, it follows that

$$r(g_1) = \int h(kg_1)dk = \tilde{m} * h(g_1) = \phi(g_1).$$

This proves eq. $(*)$.

Let $g = \ell g'$ in eq $(*)$ and average over K. Then

$$\phi(g_1)\phi(g') = \int \Big[\int h(\ell g' k g_1) d\ell\Big] dk = \int \phi(g' k g_1) dk.$$

Hence, ϕ is a spherical function. □

13.7. Definition. Let ϕ be a spherical function on G. A function $f \in C(G)$ is said to satisfy the **ϕ-mean-value property** if, for all g and $g_1 \in G$,

$$\phi(g_1)f(g) = \int f(gkg_1)dk.$$

13.8. Definition. Let $\ell \in P^* = A^*$. The **ℓ-twisted action** ρ_ℓ of G on $G/P = \mathcal{F}$ is defined by

$$\rho_\ell(g)\psi(b) = \sigma_\ell(g^{-1}, b)\psi(g^{-1} \cdot b), \text{ and}$$
$$\rho_\ell(g)\nu = \int \sigma_\ell(g,b)\delta_{g\cdot b} d\nu(b),$$

where $b \in G/P, \psi \in C(G/P)$, $\nu \in R(G/P)$ is a measure on G/P, and $\int \varphi d(\rho_\ell(g)\nu) = \int \varphi(g \cdot b)\sigma_\ell(g,b)d\nu(b)$.

Since σ_ℓ is a cocycle on $G \times K$, these actions are linear and coincide with those considered in § 10.1. If μ is a compactly supported measure on G, the operators $\rho_\ell(\mu)$ on the spaces $C(G/P)$ and $R(G/P)$ are defined by

$$\rho_\ell(\mu) = \int \rho_\ell(g)d\mu(g).$$

One needs an analogue, for the l-twisted action, of the Poisson formula of Chapter XII. To determine this formula, consider the space of continuous functions f satisfying the continuity condition $\lim_{g\to e} \|\frac{\delta_g * f - f}{\phi}\|_\infty = 0$, with f/ϕ bounded. This space is normed by $\|f\|_\phi = \|f/\phi\|_\infty$ and will be denoted by $C_\phi(G)$. It is a Banach space on which G acts continuously by left translations. (For the sake of brevity one has chosen to work with spaces of continuous functions instead of measurable functions as in Chapter XII.)

13.9. Proposition. *Let ϕ be a positive spherical function on G and let $\mathcal{H}_\phi \subset C_\phi(G)$ be the Banach subspace of functions that satisfy the ϕ-mean-value property. Then there exists a positive isometry $f \to \hat{f}$ of $\mathcal{H}_\phi$ onto $C(G/P)$ that intertwines the left translation on $\mathcal{H}_\phi$ and the ℓ-twisted action on $C(G/P)$ for some $\ell \in P^*$. Furthermore, for all $f \in \mathcal{H}_\phi$, it follows that*

$$f(g) = \rho_\ell(g)m(\hat{f}) = \int \sigma_\ell(g,b)\hat{f}(g \cdot b)dm(b)$$
$$= \int \sigma_{\ell^{-1}\delta}(g^{-1},b)\hat{f}(b)dm(b),$$

where δ is the modular function of P.

To prove this result one needs the following analogue of Lemma 12.7.

13.10. Lemma. *There exists a positive contraction $\overline{\beta}$ of $C_\phi(G)$ into $C(G/P)$ such that $\overline{\beta}(\phi) = 1$ and $\overline{\beta}$ intertwines the left translation on $C_\phi(G)$ and the ℓ-twisted action on $C(G/P)$.*

Proof. The amenability of $P = MAN$ is replaced here by the fixed line property, established in Example 11.33. Because $(\delta_g * \phi)/\phi$ is bounded, G acts continuously on $C_\phi(G)$ by left translations. Hence, G acts also on the dual space and, in particular, on the cone of positive linear forms on $C_\phi(G)$. In the weak topology, the condition $\mu(\phi) = 1$ defines a compact base of this cone. Since P has the fixed line property, there exists a positive linear form β on $C_\phi(G)$, with $\beta(\phi) = 1$, and an exponential $\ell \in P^*$ such that, for all $f \in P$ and $f \in C_\phi(G)$,

$$\beta[\delta_t * f] = \ell(t^{-1})\beta[f].$$

Define $\overline{\beta}(f) \in C(K)$ by the formula $\overline{\beta}(f) = \beta[f_k] = \beta[\delta_{k^{-1}} * f]$. Then, because ℓ is trivial on $K \cap P$, $\overline{\beta}(f)$ is right $K \cap P$-invariant and belongs to $C(G/P)$. Moreover, $\|\overline{\beta}(f)\|_\infty \leq \|f\|_\phi \beta(\phi_k) = \|f\|_\phi$, $\overline{\beta}(\phi) = 1$, and $\overline{\beta}$ is positive. In addition, $\overline{\beta}(f_g) = \beta[f_{gk}]$. From the Iwasawa decomposition, it follows that $gk = (g \cdot k)\ t(g,k)$, with $g \cdot k \in K, t(g,k) \in T$. The P-invariance property of β implies that $\beta[f_{gk}] = \ell[t(g,k)]\beta[f_{g\cdot k}] = \sigma_\ell(g,b)\beta[f_{g\cdot k}]$. Hence, $\overline{\beta}$ intertwines left translation on $C_\phi(G)$ and the ℓ-twisted action on $C(G/P)$. □

13.11. Remark. The Iwasawa decomposition was used here only for simplicity. An analogous calculation is valid with the less precise decomposition $G = KP$ because ℓ is trivial on $K \cap P$.

Proof of Proposition 13.9.

(It is like the proof of Proposition 12.7.) Let $\psi \in C(G/P)$. Define f by the formula $f(g) = \rho_\ell(g)m(\psi)$. Then $f(gk) = f(g)$ because $\rho_\ell(gk) = \rho_\ell(g)\rho_\ell(k)$ and $\rho_\ell(k)m = \delta_k * m = m$. Moreover, if ν is K-bi-invariant then

$$\rho_\ell(\nu)m = \nu(\phi)m.$$

In particular, if $\nu = \tilde{m} * \delta_{g_1} * \tilde{m}$, it follows that $\int f(gk \cdot g_1)d\tilde{m}(k) = (f * \check{\nu})(g) = \rho_\ell(\delta_g * \nu)m(\psi) = \rho_\ell(g)\rho_\ell(\nu)m(\psi) = \rho_\ell(g)m(\psi)\phi(g_1)$. Hence, $f(g)\phi(g_1) = \int f(gkg_1)dk$.

Conversely, assume that f is such that

$$\phi(g_1)f(g) = \int f(gkg_1)dk = \int f_{gk}(g_1)dk.$$

Since the map $k \to f_{gk}$ from K to $C_\phi(G)$ is continuous, one may apply the map $\overline{\beta}$ to both sides of this equality. Hence, $f(g)\overline{\beta}(\phi) = \int \overline{\beta}[f_{gk}]dk$. The intertwining property of $\overline{\beta}$ implies $\int \overline{\beta}[f_{gk}]dk = \int \rho_\ell(g^{-1})\overline{\beta}(f_k)dk$. More explicitly, $f(g) = \int \sigma_\ell(g,b)\hat{f}(g \cdot b)dm(b)$ because $\delta_k * m = m$ and $\hat{f} = \overline{\beta}(f)$.

Making the change of variables $g \cdot b = b'$ and taking into account the fact, established in Lemma 11.41, that $\dfrac{dg \cdot m}{dm}(b) = \sigma_\delta(g^{-1}, b)$ (where δ is the modular function of P), it follows that

$$\begin{aligned} f(g) &= \int \sigma_\ell(g, g^{-1}g \cdot b)\hat{f}(g \cdot b)dm(b) = \int \sigma_\ell^{-1}(g^{-1}, g \cdot b)\hat{f}(g \cdot b)dm(b) \\ &= \int \sigma_\ell^{-1}(g^{-1}, b')\hat{f}(b')d(g \cdot m)(b') = \int (\sigma_\ell^{-1}\sigma_\delta)(g^{-1}, b')\hat{f}(b')dm(b') \\ &= \int \sigma_{\ell^{-1}\delta}(g^{-1}, b)\hat{f}(b)dm(b). \quad \square \end{aligned}$$

Proof of Theorem 13.1.

The first part follows directly from Proposition 13.3. To prove the second part one has to shift from ℓ to $\ell' = \ell^{-1}\delta$ because of the last formula in Proposition 13.9 that can be written as $f(g) = \int h_b^{\ell'}(g)\hat{f}(b)dm(b)$. Suppose that f is a minimal solution of the equation $Lf + \lambda f = 0$. Then, if $f(o) = 1$, $f(x) = \int h_b^\ell(x)d\nu(b)$, with ν a certain probability measure on $\mathcal{F}$.

To prove this, let ε_n be an approximate identity on G where ε_n has a continuous density with compact support. Then $\varepsilon_n \leq \alpha_n$ with $\alpha_n \in C_c(G, K)$. Let $f_n = \varepsilon_n * f$ and observe that by Lemmas 13.5 and 13.6 $0 \leq f_n \leq \alpha_n * f = r(\alpha_n)\phi$. Hence, $f_n \in \mathcal{H}_\phi$, because f satisfies the ϕ-mean-value property (Lemma 13.6).

If one replaces $\ell^{-1}\delta$ by ℓ in Proposition 13.9, it follows that $f_n(x) = \int h_b^\ell(x)\hat{f}_n(b)dm(b)$ with $\hat{f}_n \in C_\phi(G/P)$. In particular, the mass of the measure $\hat{f}_n(b)dm(b)$ is $f_n(o)$. Because the sequence $f_n(o)$ converges to $f(o) = 1$, the positive measures $\hat{f}_n(b)dm(b)$ have bounded mass and so one can extract a subsequence that converges weakly to a probability measure ν. Hence, $f(x) = \lim_n(\varepsilon_n * f)(x) = \int h_b^\ell(x)d\nu(b)$.

Furthermore, in view of Proposition 13.3, $-\lambda f = Lf = \int Lh_b^\ell d\nu(b) = r(\ell)\int h_b^\ell d\nu(b) = r(\ell)f$. Hence, $\lambda = -c(\ell)$ and $h_b^\ell \in \mathcal{H}^\lambda$. Since f is minimal and ν is a probability measure, it follows that $\nu = \delta_b$ for some $b \in \mathcal{F}$ and, so, $f = h_b^\ell$.

An extension of the main result

Let H denote a unimodular, locally compact, metrizable topological group, and let p be a probability measure on H. Let S denote the support of p and assume that $\cup_{n>0} S^n = H$. If p is well-behaved it follows, from Theorem 11.5, that the equation $f(e) = 1$ defines a Choquet simplex that is the compact base of the convex cone $\mathcal{H}^r$ of positive functions f on H satisfying $f * p = rf$. The problem is to describe the extremals of $\mathcal{H}^r$. It is clear from what has been already proved in this chapter that this can be done if the following properties are satisfied.

(1) p is a well-behaved K-bi-invariant probability measure.
(2) There exist a compact subgroup K and a closed subgroup P such that $H = KP$.
(3) P has the fixed line property.
(4) The convolution algebra $C_c(H, K)$ is commutative.

Clearly, in the case of a semisimple Lie group G and p well-behaved, K-bi-invariant, these conditions are satisfied, if K is a maximal compact subgroup and $P = MAN$ is a minimal parabolic subgroup. Other cases of interest are, for example, when H is reductive instead of semisimple, or when H is the local field analogue of a semisimple Lie group.

If H is of the form $H = KP$, as in Chapter XI, any exponential ℓ on P determines the cocycle $\sigma_\ell(g, b) = \ell[t(gk)]$, where $b \in H/P$ corresponds to $k \in K$ and $t(g)$ is the P-component of g in the decomposition $g = kt$, with $k \in K$ and $t \in P$. The function $h_b^\ell(g) = \sigma_\ell(g^{-1}, b)$ is well defined (see Remark 11.35(1)) and can be expected to play an essential role in the problem. Recall also from Chapter XI that the function Φ^ℓ was defined by $\Phi^\ell(g) = \int \sigma_\ell(g^{-1}, b)dm(b)$, where m is the unique normalized K-invariant measure on H/P. It was shown also that $h_b^\ell * p = \hat{p}(\ell)h_b^\ell$ with $\hat{p}(\ell) = \int \Phi^\ell(g^{-1})dp(g)$. Closely related calculations are found in [M1] where it is shown that Φ^ℓ is a spherical function.

The arguments used to prove Theorem 13.1 are now used to prove the following generalization of this result.

13.12. Theorem. *Let H be a unimodular, locally compact, metrizable group and p be a probability that together satisfy the conditions (1), (2), (3), and (4). Then,*

$$h_b^\ell * p = \hat{p}(\ell)h_b^\ell.$$

Every minimal solution of the equation $f * p = rf$ is of the form $f = h_b^\ell$ for some $b \in H/P, \ell \in P^*$ and $r = \hat{p}(\ell)$.

Proof. The first part has already been proved. With these hypotheses, it is clear that a uniform Harnack inequality is satisfied and that $\mathcal{H}^r$ has a compact base. Hence, Proposition 13.4 is true. Lemmas 13.5 and 13.6 hold in view of condition (4). Proposition 13.9 is of a group theoretical character and is based on the fixed line property of P. Hence, the proof of Theorem 13.1 applies in this context. $\square$

13.13. Examples. (1) If A is an abelian group and G is a semisimple Lie group, p is K-bi-invariant and well-behaved on $H = G \times A$, then H and p together satisfy conditions (1), (2), (3), and (4). Here P is replaced by $P \times A$ and the fixed line property is again valid for $P \times A$. Clearly $H/(P \times A) = G/P$ but $(P \times A)^* = P^* \times A^*$. Note that reductive groups are of this form.

(2) If G^a is a simply connected semisimple algebraic group defined over a local field $\mathbb{F}$ (see [M1], [M4]) and $G(\mathbb{F}) = H$ is the group of its $\mathbb{F}$-rational points, there exists a good maximal compact subgroup K (see [B16], [B17]) and a "minimal parabolic" subgroup $P = P(\mathbb{F})$ that has again the fixed line property (see Corollary 11.32) such that $H = KP$. Once again the algebra $C_c(H, K)$ is commutative. Hence, if p is well-behaved the conditions of Theorem 13.12 are satisfied. As a result, the minimal solutions are of the form h_b^ℓ with $\ell \in P^*$ and $b \in H/P$. In this case it was proved in [M1] that H/P is a boundary, Example 12.14(2). Hence, the situation is entirely analogous to that of a semisimple Lie group G.

(3) Products of groups of the type considered in (1) or (2) also satisfy the conditions of Theorem 13.12. If p is a K-bi-invariant well-behaved probability measure, then the extremals of the cone $\mathcal{H}^r$ are given by the formula $f = h_b^\ell$, where h_b^ℓ is now a product of corresponding functions for each of the factor groups.

13.14. Remark. Clearly, the equation $f * p = rf$ can be transferred to H/K via the H-invariant kernel on H/K defined by p.

The following definition is to be found in [M1].

13.15. Definition. Let H be a locally compact unimodular group, K a compact subgroup, ϕ a continuous complex function on H. Then ϕ is said to be **K-spherical** if, for all $g, g' \in H$,

$$\int \phi(gkg')dk = \phi(g)\phi(g').$$

Note that, from this definition, it follows that ϕ is K-bi-invariant. If $f \in C_c(H, K)$, then

$$\langle \phi, f \rangle = \int f(g)\phi(g)dg$$

defines a homomorphism of $C_c(H, K)$ into $\mathbb{C}^*$. It is also clear from Chapter XI that, if $H = KP$, where P is a closed subgroup and $\ell \in P^*$, the function $\Phi^\ell(g) = \int \ell[t(g^{-1}k)]dk$ is a positive spherical function.

Here one has, as a by-product, the following result.

13.16. Theorem. *Assume $H = KP$ is compactly generated and that conditions (2), (3), and (4) are satisfied. Then the positive K-spherical functions are given by the formula $\phi = \Phi^\ell$ with $\ell \in P^*$.*

Proof. Suppose ϕ is K-spherical and consider $q, q' \in C_c^+(H, K)$. By definition,

$$\langle \phi, q * q' \rangle = \langle \phi, q \rangle \langle \phi, q' \rangle.$$

In particular,

$$\langle \phi * \check{q}, q' \rangle = \langle \phi, q' \rangle \langle \phi, q \rangle.$$

Since ϕ is K-bi-invariant and q' is arbitrary in $C_c^+(H,K)$, it follows that

$$\phi * q' = \langle \phi, q' \rangle \phi.$$

Hence, ϕ is a positive eigenfunction for right convolution by $q' = \check{q}$ with eigenvalue $r = \langle \phi, \check{q} \rangle$. If the support of q' generates H, it follows from Theorem 13.12 that ϕ is a barycenter of the extremals h_b^ℓ, where $h_b^\ell(g) = \sigma_\ell(g^{-1}, b)$ with $b \in H/P$ and $\ell \in P^*$, i.e.,

$$\phi(g) = \int \sigma_\ell(g^{-1}, b) d\nu(\ell, b).$$

As ϕ is K-bi-invariant, $\phi(kg) = \phi(g) = \int \sigma_\ell(g^{-1}, k^{-1} \cdot b) d\nu(\ell, b)$. Integration with respect to k shows that

$$\phi(g) = \int \sigma_\ell(g^{-1}, b) dm(b) d\overline{\nu}(\ell) = \int \Phi^\ell(g) d\overline{\nu}(\ell).$$

Since each Φ^ℓ also defines a homomorphism of the algebra $C_c(H,K)$ into $\mathbb{C}$, it follows that $\phi = \Phi^\ell$ for some ℓ. □

13.17. Theorem. *Assume that H and p together satisfy the conditions (1), (2), (3) and (4). Denote by ℓ_0' the unique element of P^* such that $r(p) = \inf_{\ell \in P^*} \hat{p}(\ell) = \hat{p}(\ell_0')$ (see Theorem 11.40).*

Then, a positive solution f of the equation $f * p = r(p) f$ has the following unique integral representation as

$$f(g) = \int \sigma_{\ell_0'}(g^{-1}, b) d\nu(b) = \int h_b^{\ell_0'}(g) d\nu(b),$$

in terms of the functions $h_b^{\ell_0'}$, where ν is a positive measure on H/P.

Proof. Since, by Theorem 11.39, there is a unique ℓ such that $\hat{p}(\ell) = r(p)$, the result is a trivial corollary of Theorem 13.12.

Hence, the condition $f * p = r(p) f$ implies that $f(g) = \int \sigma_{\ell_0'}(g^{-1}, b) d\nu(b)$. The uniqueness comes from the fact that, by the transitivity of K on H/P, the functions $\sigma_{\ell_0'}(g^{-1}, b)$ are all extremal if one of them is extremal. □

13.18. Corollary. *There exists a unique solution of the equation $f * p = r(p) f$ that is left K-invariant. It is the function $f(g) = \Phi^{\ell_0'}(g) = \int \sigma_{\ell_0'}(g^{-1}, b) dm(b) = \int h_b^{\ell_0'}(g) dm(b)$.*

Proof. The condition $f(kg) = f(g)$ implies that

$$f(g) = \int \sigma_{\ell_0'}(g^{-1}, k \cdot b) d\nu(b),$$

for all $k \in K$. By the uniqueness of the representing measure, it follows that $k \cdot \nu = \nu$ for all $k \in K$. Hence, $\nu = m$. □

13.19. Proposition. *Assume that K is a maximal compact subgroup of a semisimple Lie group G and that p is a well-behaved K-bi-invariant probability measure. Then $\ell_0' = \delta^{1/2}$, where δ is the modular function of AN and $r_0 = r_0(p) = r(p)$. In addition, $\Phi^{\ell_0'}(g) = \int P^{1/2}(g,b)dm(b) = \Phi(g)$. The solutions of the equation $f * p = r_0 f$ are given by the formula*

$$f(g) = \int P^{1/2}(g,b)d\nu(b),$$

where ν is a uniquely determined positive measure on G/P.

The following lemma is used to prove this result.

13.20. Lemma. *If $P(g,b)$ is the Poisson kernel of G and*

$$\Phi(g) = \int P^{1/2}(g,b)dm(b),$$

then $\Phi \le \Phi^{\ell}$ for every positive spherical function Φ^{ℓ}, with equality if and only if $\ell = \delta^{1/2}$.

Proof. Consider the Iwasawa decomposition $G = KAN$ and the formula $\Phi^{\ell}(g) = \int \ell[t(g^{-1}k)]dk$. Let ρ denote one half the sum of the positive roots of the Lie algebra $\mathfrak{a}$ of A. Then, as has been mentioned before, the modular function δ of AN is $\delta = e^{-2\rho}$ and the Poisson kernel P is given by $P(g,b) = \delta[t(g^{-1}k)]$ where $b = k \cdot P$.

From [H3] it follows that for every $\ell \in P^* = A^*$, $\Phi^{e^{-\rho}\ell} = \Phi^{e^{-\rho}\ell \circ w}$, where $w \in W$ is any element of the Weyl group that acts on A^* by $\ell \to \ell \circ w$.

Now $e^{-\rho} = \left[\prod_{w\in W} e^{-\rho}\ell \circ w\right]^{\frac{1}{|W|}}$ because 0 is the isobarycenter of the W-orbit of any element of A^*. Hence, applying Hölder's inequality to the integral formula that expresses $\Phi = \Phi^{e^{-\rho}}$ in terms of minimal solutions, it follows that

$$\Phi(g) \le [\prod_{w\in W} \int (e^{-\rho}\ell \circ w)[t(g^{-1}k)]dk]^{\frac{1}{|W|}} \text{ and } \Phi(g) \le \Phi^{\ell}(g),$$

because $\Phi^{e^{-\rho}\ell} = \Phi^{e^{-\rho}\ell\circ w}$. The equality $\Phi = \Phi^{\ell}$ is well-known [H3] to be impossible if $\ell \ne e^{-\rho}$. This also follows directly by assuming equality in the Hölder inequality. □

Proof of Corollary 13.19.

From Lemma 13.20, it follows that

$$\hat{p}(e^{-\rho}) = \int \Phi(g^{-1})dp(g) \le \hat{p}(\Phi^{\ell}).$$

Hence, $\ell_0' = e^{-\rho} = \delta^{1/2}$. Since, by Proposition 11.43, $\hat{p}(\delta^{1/2}) = r_0$ it follows that $r_0 = \inf_{\ell} \hat{p}(\ell) = \hat{p}(\ell_0') = r(p)$.

The last formula and the uniqueness statement are special cases of Theorem 13.17.

Analytic determination of the minimal eigenfunctions of the Laplacian

Fixing the maximal compact subgroup K and a minimal parabolic subgroup P of G determines the Iwasawa decomposition $G = KAN$ corresponding to the Langlands decomposition of $P = MAN$, where M is the centralizer of A in K. As shown in Theorem 13.1, every exponential $\ell = e^{-(\rho+s)}$ on A yields the family of eigenfunctions h_b^ℓ of the Laplacian, where

$$h_b^\ell(g) = \sigma_\ell(g^{-1}, b),\ b \in \mathcal{F}.$$

These functions are indexed by the parameters $\ell \in A^*$ and $b \in \mathcal{F}$. By Theorem 13.1, every minimal eigenfunction is of this form. It remains to show which of these functions h_b^ℓ are minimal.

The first description of the minimal λ-eigenfunctions is an analytic one, stated earlier as Proposition 8.15, and due originally to Karpelevič [K3, p. 179, Theorem 12.2.1]. The proof given here is due to Guivarc'h [G14].

13.23. Theorem. *The set of normalised minimal λ-eigenfunctions of the Laplacian on $X = G/K$ is the set of functions $f_{v,b}$ defined by*

$$f_{v,b}(g \cdot o) = e^{-(\rho+v)}[a(g^{-1}k)] = e^{-(\rho+v)(H(g^{-1}k))},$$

where $v \in \overline{\mathfrak{a}_+^*}$ and $\|v\|^2 = \lambda_0 - \lambda$, and $k \in K$ is defined up to an element of M. Hence, this collection of functions, for $\lambda \leq \lambda_0$, is parametrized by $\overline{A_+^*} \times \mathcal{F}$. More specifically, if $v \in \overline{\mathfrak{a}_+^*}$, with $\|v\|^2 = \lambda_0 - \lambda$ and $b \in \mathcal{F}$, then (v, b) corresponds to $f_{v,b}$ and vice-versa.

The proof of this theorem makes use of intertwining operators, determined by elements of $\mathfrak{a}^*$, some of whose properties are now explained briefly (for further details see [G1, Chapter IV]).

If $v \in \mathfrak{a}^*$, let $E_v \subset C^\infty(G)$ be the subspace $E_v = \{f \in C^\infty(G) \mid f(gan) = f(g)e^{-(\rho+v)}(a)$ for all $g \in G, a \in A, n \in N\}$. The functions in E_v are clearly determined by their restrictions to K. The integral $c_w(v) = \int_{N/N_w} e^{-(\rho+v)}[a(nw)]d\dot{n}$ is finite if $v \in \overline{\mathfrak{a}_+^*}$ (see [G1, p. 180]), where $N_w = N \cap wNw^{-1}$, with $w \in W$, and the integration over the homogeneous space N/N_w is with respect to the N-invariant measure $d\dot{n}$. It follows that, if $f \in E_v$, the function $\bar{f}$, where

$$\bar{f}(g) = \int_{N/N_w} f(gnw)d\dot{n},$$

is well-defined. It is easy to see that $\bar{f} \in E_{w \cdot v}$ and that $\overline{\delta_g * f} = \delta_g * \bar{f}$. Hence, the map $f \to \bar{f}$ intertwines the natural representations of G on E_v and $E_{w \cdot v}$. This is of interest here because, if p is a fixed element of $C_c(G)$ and $r \in \mathbb{R}^+$, it implies that the map $f \to \bar{f}$ transforms solutions of the equation $p * f = rf$, with $f \in E_v$, into solutions of the same equation that belong to $E_{w\cdot v}$. Recall that $k(g)$ denotes the K-component of g in the Iwasawa decomposition $G = KAN$. To obtain an explicit formula, let μ_w^v denote the probability measure on N/N_w defined by

$$\mu_w^v(d\dot{n}) = \frac{1}{c_w(v)} e^{-(\rho+v)}[a(nw)]d\dot{n}.$$

Let ρ_w^v denote the image of μ_w^v by the map from N/N_w to K defined by $n \to k(nw)$. By definition, ρ_w^v is a probability measure. It determines the restriction of $\bar{f}$ to K, since

$$\bar{f}(k) = \int f(kk')d\rho_w^v(k').$$

Hence, one can consider the operator A_w^v on $C(K)$ defined by $A_w^v\phi = \phi * \check{\rho}_w^v$. It has the following intertwining property

$$(*) \qquad A_w^v[\rho_v(g)\phi] = \rho_{w\cdot v}(g)[A_w^v\phi],$$

where the representation $\rho_v(g)$ of G in $C(K) = C(G/AN)$ is defined by $[\rho_v(g)\phi](k) = e^{-(\rho+v)}[a(g^{-1}k)]\phi(g^{-1}\cdot k)$. Clearly, these representations and these operators are well defined on $C(\mathcal{F})$ and the intertwining property in eq. $(*)$ is valid. In particular, if $\phi = 1$, one obtains

$$\sigma_{\ell'}(g,k) = \int \sigma_\ell(g,kk')d\rho_w^v(k'),$$

where $\ell = e^{-(\rho+v)}, \ell' = e^{-(\rho+w\cdot v)}$, and $v \in \overline{\mathfrak{a}_+^*}$.

The proof of Theorem 13.23 makes use of the following two lemmas.

13.24. Lemma. *If $\ell = e^{-(\rho+v)}$ is an exponential on A with $v \notin \overline{\mathfrak{a}_+^*}$ and $\lambda = \lambda_0 - \|v\|^2$, the function $h_b^\ell = \sigma_\ell(g^{-1},b)$ is a non-minimal λ-eigenfunction of the Laplacian.*

Proof. It has been shown in Theorem 13.1 that h_b^ℓ is an eigenfunction. It follows from the above discussion that, if $v \notin \overline{\mathfrak{a}_+^*}$, there exists $w \in W$ with $w \neq e$, such that $w^{-1}\cdot v \in \overline{\mathfrak{a}_+^*}$. Hence,

$$\sigma_\ell(g^{-1},b) = \int \sigma_{\ell'}(g^{-1},kk')d\rho_w^v(k'),$$

with $\ell' = e^{-(\rho+w^{-1}\cdot v)}$ and $b = kM$.

The measure ρ_w^v is not concentrated on M because the support of μ_w^v is N/N_w. If not then, for all $n \in N$, $nw \in MAN$ and so $w \in MAN$, which implies $w = e$. As a result, $\sigma_\ell(g^{-1},b)$ is a non-trivial barycenter of the λ-eigenfunctions $\sigma_{\ell'}(g^{-1},b)$ and, so, h_b^ℓ is not minimal if $v \notin \overline{\mathfrak{a}_+^*}$.

13.25. Lemma. *If $\ell = e^{-(\rho+v)}, v \in \overline{\mathfrak{a}_+^*}$, the function h_b^ℓ is a minimal eigenfunction of L, for every $b \in \mathcal{F}$.*

Proof. One considers the spherical functions

$$\Phi_v(g) = \int e^{-(\rho+v)}[a(g^{-1}k)]dk$$

for $v \in \mathfrak{a}^*$, and one recalls that $\Phi_v = \Phi_{v'}$ if and only if $v' = w \cdot v$ for some $w \in W$. Furthermore, such a spherical function cannot be a non-trivial barycenter of other spherical functions because spherical functions correspond to homomorphisms of the convolution algebra $C_c(G, K)$ into $\mathbb{C}$ [H3].

Suppose that, for some $v \in \overline{\mathfrak{a}_+^*}$ and $b \in \mathcal{F}$, the function h_b^ℓ is not minimal, where $\ell = e^{-(\rho+v)}$. It follows from Theorem 13.1 that there is a probability measure ν on $A^* \times \mathcal{F}$ such that

$$\sigma_\ell(g,b) = \int \sigma_{\ell'}(g,b')d\nu(\ell',b').$$

Because the cone of positive λ-eigenfunctions is a Choquet simplex, this measure ν is concentrated on the set of minimal λ-eigenfunctions. Lemma 13.24 implies that ν is concentrated on $\overline{A_+^*} \times \mathcal{F}$. Since

$$\sigma_\ell(gk,b) = \sigma_\ell(g, k \cdot b), \ \ k \in K,$$

it follows that

$$\sigma_\ell(g, k \cdot b) = \int \sigma_{\ell'}(g, k \cdot b')d\nu(\ell',b').$$

Integrating this identity with respect to k implies that

$$\Phi_\ell(g) = \int \Phi_{\ell'}(g)d\bar{\nu}(\ell'),$$

where $\bar{\nu}$ is the projection of ν on $\overline{A_+^*}$. It follows that $\Phi_\ell = \Phi_{\ell'}$. Hence, $\bar{\nu} = \delta_\ell, \sigma_\ell(g,b) = \int \sigma_\ell(g,b')d\nu(b')$ and ν is concentrated on the minimal λ-eigenfunctions. Therefore, for some b', $\sigma_\ell(g',b')$ is minimal. Since $\sigma_\ell(gk,b) = \sigma_\ell(g,k \cdot b)$, the fact that at least one of these functions is minimal shows that every function $h_b^\ell, b \in \mathcal{F}$, is minimal. This contradicts the hypothesis that h_b^ℓ is non-minimal. □

Proof of Theorem 13.23.

These two lemmas imply that the minimal λ-eigenfunctions are the functions h_b^ℓ with $b \in \mathcal{F}, \ell \in \overline{A_+^*}, \ell = e^{-(\rho+v)}, v \in \overline{\mathfrak{a}_+^*}$, and $Lh_b^\ell = \lambda h_b^\ell$. It follows from Proposition 7.4 that $\lambda = \lambda_0 - \|v\|^2$. If ℓ is fixed, the stabilizer of h_b^ℓ in the projective action of G is equal to P_b, which equals kPk^{-1} if $b = kP$. Hence, these functions, are all different since, if $h_b^\ell = h_{b'}^{\ell'}$ for some $\ell, \ell' \in A^*$ and $b, b' \in \mathcal{F}$, then $\Phi_\ell = \Phi_{\ell'}$, which implies that $\ell = \ell'$. As a result, the family of minimal eigenfunctions, $\lambda \leq \lambda_0$, is parametrized by $\overline{A_+^*} \times \mathcal{F}$. □

THE BUSEMANN COCYCLE AND A GEOMETRICAL DETERMINATION OF THE MINIMAL EIGENFUNCTIONS OF THE LAPLACIAN

A more geometrical way of describing the minimal functions involves the Busemann function defined in § 3.1. Recall that, if $\overline{X}^c$ of X denotes the conical compactification of the symmetric space $X = G/K$, then G acts continuously on $\overline{X}^c$. Denote by d the Riemannian distance function on X associated with the Killing form of G and recall that the normalised Busemann function $d_\gamma(x) = d_z(x) = \tilde{d}(x,z)$, where $x \in X$ and $z = [\gamma] \in \partial\overline{X}^c$, is defined by the formula

$$\tilde{d}(x,z) = \lim_{t\to\infty} d(x,\gamma(t)) - t.$$

The function $\tilde{d}$ is normalised in order to have $\tilde{d}(o,z) = 0$, where the origin o in X is fixed and K is its stabilizer in G. This function $\tilde{d}$ has an important transformation property stated as eq. (*) in the next result (cf. eq. (†) in Remark 10.5(2))).

13.26. Proposition. *Let $d_z(x) = \tilde{d}(x,z)$, where d_z is the Busemann function corresponding to $z \in X(\infty)$. Then, if $g \in G$,*

$$(*) \qquad \tilde{d}(g^{-1}\cdot x, z) = \tilde{d}(x, g\cdot z) + \tilde{d}(g^{-1}\cdot o, z).$$

Hence, $\delta(g,z) \overset{\text{def}}{=} e^{-\tilde{d}(x,z)}$ is a cocycle, the so-called **Busemann cocycle,** *on $G \times X(\infty)$. This cocycle agrees with the Busemann cocycle δ_L corresponding to L, see Remark 10.5(2), if $z \in k\cdot L(\infty)$, where L is a unit vector in $\overline{\mathfrak{a}^+}$.*

Furthermore,

(1) $\tilde{d}(x, k\cdot z) = \tilde{d}(k^{-1}\cdot x, z)$ if $k \in K$.

In addition, if $z \in \overline{\mathfrak{a}(\infty)}$ then

(2) $\tilde{d}(n\cdot o, z) = o$ for all $n \in N$;
(3) $\tilde{d}(x, g\cdot z) = \tilde{d}(x, k(g)\cdot z)$; and
(4) $\tilde{d}(g\cdot o, k\cdot z) = \langle L, \log a(g^{-1}k)\rangle = B(L, H(g^{-1}k))$ if $z = L(\infty)$, where L is a unit vector in $\overline{\mathfrak{a}^+}$.

Proof. The eq. (*) follows from the identity $d(g^{-1}\cdot x,\gamma(t)) - d(o,\gamma(t) = \{d(x, g\cdot\gamma(t)) - d(o, g\cdot\gamma(t))\} + \{d(g^{-1}\cdot o,\gamma(t)) - d(o,\gamma(t))\}$.

As a result, $\tilde{d}$ is the logarithm of a cocycle δ on $G \times X(\infty)$. The fact that $\delta(g,z) = \delta_L(g,b)$, if $b = kM$ and $z = k\cdot L(\infty)$ with L a unit vector in $\overline{\mathfrak{a}^+}$, is a consequence of (1) and (3).

Statement (1) follows immediately from eq. (*) as K is the isotropy group of o.

It follows from eq. (*) that $h \to \tilde{d}(h\cdot o, z)$ is a homomorphism of P_z into $\mathbb{R}$. If $z \in \overline{\mathfrak{a}(\infty)}$, then P_z is a standard parabolic subgroup and so

$S = AN \subset P_z$. It follows from Remark 11.27(2) that any homomorphism of S into $\mathbb{R}$ is trivial on N. This proves (2).

Note that by eq. $(*)$, $\tilde{d}(x, kan\cdot z) = \tilde{d}(n^{-1}a^{-1}k^{-1}\cdot x, z) - \tilde{d}(n^{-1}a^{-1}\cdot o, z)$. By (2), $\tilde{d}(n^{-1}a^{-1}\cdot o, z) = \tilde{d}(a^{-1}\cdot o, z)$ and, in addition, $\tilde{d}(n^{-1}a^{-1}k^{-1}\cdot x, z) = \tilde{d}(a^{-1}k^{-1}\cdot x, z)$. Since $\tilde{d}(a^{-1}k^{-1}\cdot x, z) = \tilde{d}(k^{-1}\cdot x, a\cdot z) + \tilde{d}(a^{-1}\cdot o, z)$, the fact that $a \in P_z$ then implies that $\tilde{d}(x, kan\cdot z) = \tilde{d}(k^{-1}\cdot x, z) = \tilde{d}(x, k\cdot z)$. This proves (3).

To prove (4), let $x = g\cdot o$ and let $k_1 a^{-1} n^{-1}$ be the Iwasawa decomposition of $k^{-1}g$. Then, (2) and eq. $(*)$ imply that $\tilde{d}(g\cdot o, k\cdot L(\infty)) = \tilde{d}(na\cdot o, L(\infty)) = \tilde{d}(a, L(\infty)) = \tilde{d}(a(g^{-1}k), L(\infty))$. This reduces the computation to the usual Euclidean one (cf. the remark in § 3.1) relative to the inner product defined by the Killing form on $\mathfrak{a}$. This proves (4) since $a = a(g^{-1}k)$. □

Remark. With this proposition the discussion of the Busemann function, that began in the remark of § 3.1 and was continued following Proposition 8.15 and in Remark 10.5(2), is completed.

Since $v \in \overline{\mathfrak{a}^*_+}$ if and only if $v(H) = \langle V, H\rangle$ for a unique $V \in \overline{\mathfrak{a}^+}$, the properties of the Busemann cocycle imply the following result.

13.27. Lemma. *If $v \in \overline{\mathfrak{a}^*_+}$, then*

$$e^{-v(H(g^{-1}k))} = \{e^{-\tilde{d}(g\cdot o, k\cdot z)}\}^{\sqrt{\lambda_0-\lambda}},$$

for a unique $z = L(\infty) \in \overline{\mathfrak{a}^*(\infty)}$, where $\| v \| = \sqrt{\lambda_0 - \lambda}$.

It follows from this lemma that the minimal functions listed in Theorem 13.23 can also be described in terms of the geometrical data contained in the Busemann function and the basic cocycle $P^{1/2}(g,b) = e^{-\rho(H(g^{-1}k))}, b = kM$. Together with Theorem 13.23, this proves the main part of the next result that is the geometrical form of Karpelevič's result concerning the set of minimal solutions of $Lu + \lambda u = 0$, stated earlier as Proposition 8.15.

13.28. Theorem. *The set of normalised minimal λ eigenfunctions of the Laplacian on $X = G/K$ is the set of functions $h_{b,z}$ given by*

$$h_{b,z}(x) = P^{1/2}(x,b)\{e^{-\tilde{d}(x,z)}\}^{\sqrt{\lambda_0-\lambda}},$$

where $b \in \mathcal{F}$, $z \in \partial\overline{X}^c$, $P_b \subset P_z$, and $P(g\cdot o, b) = \dfrac{dg\cdot m}{dm}(b)$ is the Poisson kernel.

Hence, if $\lambda < \lambda_0$, the set of minimal λ-eigenfunctions is parametrized in a one-to-one way by the subset of points of $\partial(\overline{X}^{SF} \vee \overline{X}^c) = \partial((X \cup \partial X(\lambda_0)) \vee \overline{X}^c)$ whose stabilizer is a minimal parabolic subgroup.

Proof. It follows from Lemma 13.27 and Theorem 13.23 that, if $x = g\cdot o$ and $b = kM$, $h(x) = P^{1/2}(x,b)\{e^{-\tilde{d}(x,z)}\}^{\sqrt{\lambda_0-\lambda}}$ is minimal if $z = k\cdot L(\infty)$ with

$L \in \overline{\mathfrak{a}^+}$. As a result, if $\lambda \leq \lambda_0$ is fixed, $f_{v,b} = h_{b,z}$, where $v(H) = B(V, H)$ with $\|V\| = \sqrt{\lambda_0 - \lambda}$ and $z = L(\infty)$, with L the direction of V.

Since $L \in \overline{\mathfrak{a}^+}$ if and only if $P \subset P_{L(\infty)}$, it follows that $P_b \subset P_z$. Conversely, if $b = kM$ and $kPk^{-1} = P_b \subset P_z$ then $k^{-1}P_z k$ is a standard parabolic subgroup P^I. As a result, $z = k \cdot L(\infty)$ with $L \in C_I$. Hence, by Lemma 13.27 and Theorem 13.23, the function h is minimal and equals $h_{b,z}$.

The last statement can be proved by examining the stabilizer, in the twisted or projective action, of $f_{v,b}$. It suffices to consider the case of $b = \dot{e}$, since $S_\ell f_{v,b}(g \cdot o) = f_{v,b}(\ell^{-1}g \cdot o) = f_{v,\ell \cdot b}(g \cdot o)$ if $\ell \in K$. Since it is clear that $H(g'g^{-1}) = H(g^{-1})$ if $g' \in NM$, and that $H(g^{-1}a^{-1}) = H(g^{-1}) + H(a^{-1})$, it follows that P stabilizes $f_{v,\dot{e}}$ relative to the twisted action. The stabilizer is necessarily closed and so it is a standard parabolic group P^I. It follows, from the formula $S_\ell f_{v,b} = f_{v,\ell \cdot b}$ if $\ell \in K$, that $I = \emptyset$.

Conversely, suppose that a limit function h in $\partial X(\lambda)$ is not minimal. Then, by Theorem 8.2, $h = S_{ka_1} h_{I,v}$, where $a_1 \in \overline{A^{I,+}}$ and $v \in \overline{\mathfrak{a}^*_+}$ with $\|v\| = \sqrt{\lambda - \lambda_0}$. The type of argument just used shows that P^I is the stabilizer of $h_{I,v}$ relative to the twisted action. Since $I \neq \emptyset$ it follows that the stabilizer of h is not a minimal parabolic subgroup.

Since the compactifications $(X \cup \partial X(\lambda_0)) \vee \overline{X}^c$ and $\overline{X}^{SF} \vee \overline{X}^c$ are G-isomorphic, the last statement follows. $\square$

13.29. Remarks. (1) The factor $P^{1/2}(x, b)$, or $e^{-\rho}[a(g^{-1}k)]$, in the expression of $h_{b,z}(x)$ is related to the form of the Harish-Chandra isomorphism between the convolution algebras $C_c(G, K)$ and $C_c(A, W)$ [G1]. Here $C_c(A, W)$ is the subalgebra of W-invariant elements in $C_c(A)$.

(2) Combining the last parts of Theorems 13.23 and 13.28 gives an explicit correspondence ϕ between the product of the portion of the unit sphere in $\overline{\mathfrak{a}^+}$ with $\mathcal{F}$ and the subset of $\partial \overline{X}^c \times \mathcal{F}$ defined by the condition $P_b \subset P_z$. Clearly, ϕ is a G-equivariant homeomorphism.

§ **13.30.** Since, for $\lambda < \lambda_0$, the compactification $X \cup \partial X(\lambda)$ is G-isomorphic to $(X \cup \partial X(\lambda_0)) \vee \overline{X}^c$, it is also G-isomorphic to $\overline{X}^{SF} \vee \overline{X}^c$, where any compactification that is G-isomorphic to a maximal Satake–Furstenberg compactification, can play the role of $\overline{X}^{SF}$. In particular, it follows that $X \cup \partial X(\lambda)$ is G-isomorphic to $\overline{\mathcal{S}_0} \vee \overline{X}^c$.

Given a realization of the maximal Satake–Furstenberg compactification $\overline{X}^{SF}$, it is of interest to describe the subset of $\overline{X}^{SF} \times \overline{X}^c$ that corresponds to the ideal boundary $\partial X(\lambda)$. This will now be done for the case of $\overline{\mathcal{S}_0}$.

Let $j : X \to \overline{\mathcal{S}_0} \vee \overline{X}^c$ be the topological embedding defined by setting $j(x) = (K_x, x)$, where K_x is the isotropy group, or stabilizer, of $x \in X$. In view of Theorems 8.21 and 9.18, it follows that j has a unique continuous extension, also denoted by j, to $X \cup \partial X(\lambda)$ as a homeomorphism of this compactification onto the compactification that is the closure $\overline{j(X)}$ of $j(X)$ in $\overline{\mathcal{S}_0} \vee \overline{X}^c$.

13.31. Proposition. *Let $\partial j(X)$ denote the image of the ideal boundary $\partial X(\lambda)$ under j. Then*

(1) $(D^I, z) \in \partial j(X)$ *if and only if* $z \in \overline{C_I(\infty)}$.

(2) $\partial j(X) = \{(D, z) \subset (\overline{\mathcal{S}_0} \setminus \mathcal{S}_0) \times X(\infty) \mid D \subset P_z\}$.

(3) *if* $\xi = (D, z) \in \partial j(X)$, *the corresponding* λ*-eigenfunction* h_ξ *is* $h_\xi(x) = h^D(x)e^{-\sqrt{\lambda_0-\lambda}\,\vec{d}(x,z)}$, *where* h^D *is the unique* λ_0*-eigenfunction invariant under left translation by* D.

Proof. To verify (1), recall (see the proof of Theorem 8.21) that a point (D, z) is in $\partial j(X)$ if and only if there is an I-directional sequence $(y_n) \subset X$ (see Definition 8.6) such that D is its limit in $\overline{\mathcal{S}_0}$ and z is its limit in $\overline{X}^c$.

Note that, if $z \in \overline{C_I(\infty)}$, then there is an I-canonical sequence that converges to z in $\overline{X}^c$. It follows from this observation and Proposition 9.14 that $(D^I, z) \in \partial j(X)$ if $z \in \overline{C_I(\infty)}$.

Conversely, assume that $(D^I, z) \in \partial j(X)$. Let (y_n) be a J-directional sequence (y_n) that converges to D^I. If $y_n = k_n a_n \cdot o$, then $k_n \to k$ and $a_n^J \to a^J$. It follows that $D^I = (ka^J)D^J(ka^J)^{-1}$. Hence, by Corollary 9.17, $I = J, a^J = e$, and $k \in K^I M$. As a result, (a_n) is I-canonical and its limiting direction is z. This implies that $z \in \overline{C_I(\infty)}$.

Lemma 3.19 states that $P^I \subset P_z$ if and only if $z \in \overline{C_I(\infty)}$. Consequently, if $z \in \overline{C_I(\infty)}$, then $D^I \subset P_z$. It follows from the group action that $\partial j(X) \subset \{(D, z) \subset (\overline{\mathcal{S}_0} \setminus \mathcal{S}_0) \times X(\infty) \mid D \subset P_z\}$.

To show that $\partial j(X) \supset \{(D, z) \subset (\overline{\mathcal{S}_0} \setminus \mathcal{S}_0) \times X(\infty) \mid D \subset P_z\}$, it suffices to show that if $D \subset P_z$ then $P^I \subset P_z$, as this implies, by Lemma 3.19, that $z \in \overline{C_I(\infty)}$.

Since Corollary 14.30 shows that $P^I \subset P_z$ if $D^I \subset P_z$, this completes the proof of (2).

The final assertion follows from the group action once it is established for $(D^I, z) \in \partial j(X)$. Since $\xi = (D^I, z)$ is the limit of an I-canonical sequence with limiting direction z, Proposition 8.18 implies that the corresponding function in the Martin boundary is $h_\xi(g \cdot o) = h_I(\cdot o)e^{-v(H(g^{-1}))}$, where $v \in \overline{\mathfrak{a}_+^*}$ is represented by a vector in $\overline{\mathfrak{a}^+}$ with direction z. In view of Lemma 13.27, this means that $h_\xi(x) = h_I(x)\{e^{-\vec{d}(x,z)}\}^{\sqrt{\lambda_0-\lambda}}$. □

13.32. Corollary. *Let $\partial_e j(X)$ denote the subset of $\partial j(X)$ that corresponds to the minimal part of the Martin boundary $\partial X(\lambda)$. Then $\xi = (D, z) \in \partial_e j(X)$ if and only if, for some $g \in G$, $D = gMNg^{-1}$ and $P_z = gP^I g^{-1}$ for some proper subset I of Δ.*

Proof. Theorem 13.23 and Lemma 13.27, together with § 7.21 (m), imply that (D^I, z) corresponds to a minimal function if and only $I = \emptyset$ and the direction $z \in \overline{\mathfrak{a}^+}$. The result follows by group action. □

Remark. In the proof of Proposition 13.32, it was shown that the fiber in $\partial j(X)$ over the group $D^I \in \overline{\mathcal{S}_0}$ with respect to projection onto the first coordinate of $\overline{\mathcal{S}_0} \times \overline{X}^c$ onto $\overline{\mathcal{S}_0}$ is the closed simplex $\overline{C_I(\infty)}$.

Minimal eigenfunctions for random walks

It has been shown in Theorem 13.12 that the minimal r-eigenfunctions are of the form h_b^ℓ for $\ell \in A^*$ and $b \in \mathcal{F}$. Let $\ell = e^{-(\rho+v)}, v \in \mathfrak{a}^*$. Then Lemmas 13.24 and 13.25 hold for solutions of convolution equations, except that in Lemma 13.24 there is no connection between the eigenvalue and the norm $\|v\|$ of v. Since Proposition 13.26 is purely geometric, these lemmas imply the following result.

13.33. Theorem. *Let p be a well-behaved probability on G. Let $\ell \in A^*$ denote $e^{-(\rho+v)}$ for $v \in \mathfrak{a}^*$. Then the set of minimal solutions for a convolution equation $f * p = rf$, where $r \geq r_0(p)$, is the set of functions given by the following formula:*

$$f(g \cdot o) = e^{-(\rho+v)}[a(g^{-1}k)],$$

where $\hat{p}(\ell) = r, v \in \overline{\mathfrak{a}_+^*}$, and $k \in K$ is defined up to an element of M.

13.34. Remarks. (1) Note that there is no connection between the eigenvalue and the norm of the linear functional v as in the case of the Laplacian.

(2) It follows from Theorem 13.30 that, if p is fixed, the set of minimal r-eigenfunctions for all $r \geq r_0(p)$ is independent of p. It is equal to the set of all the minimal eigenfunctions of the Laplacian and this fact can be proved a priori. Here, this fact has been proved using a more general approach that extends easily to the situation where K-bi-invariance of p is not satisfied (see [F4]).

(3) When K-bi-invariance is not assumed, the lemmas used to prove (2) have the same form as here (see [F4]). This means that the minimal r-eigenfunctions can be written as $f(g) = e^{-(\rho+v)}[a(g^{-1}k)]\frac{\psi_v(g^{-1}\cdot k)}{\psi_v(k)}$, where ψ_v is a positive and continuous function on $\mathcal{F}$ that is the unique solution of an integral equation that involves p and v. A similar use of the intertwining operator also shows that $v \in \overline{\mathfrak{a}_+^*}$ and $k \in K$ are defined up to an element of M. This gives a more precise form of the description of the minimal eigenfunctions than that given in [F4] and this result can be used to give a proof of a result analogous to Theorem 13.12.

CHAPTER XIV

RANDOM WALKS AND GROUND STATE PROPERTIES

The main questions previously examined can also be considered in the general framework of random walks. If one takes into account the results in Chapters IX and X, this leads to new proofs and new formulations of many of the results discussed earlier.

The aim of this chapter is to prove a version of Theorem 7.22 using the methods of convolution equations, i.e., random walks (see Theorem 14.4). As a corollary, one obtains another proof of Theorem 7.33, that, taken together with Theorem 9.18, identifies the Martin compactification of X at the bottom of the positive spectrum with the maximal Satake–Furstenberg compactification (see Corollary 14.22).

Using this approach, one shows that the Martin compactification at the bottom of the positive spectrum for the convolution operator associated with a well-behaved, K-bi-invariant measure p on a semisimple Lie group G is G-isomorphic to the compactification obtained by attaching the Furstenberg boundary to G in a canonical way (Corollary 14.29).

The essential reason that this result holds is because the set of $r_0(p)$-minimal eigenfunctions coincides with the set of λ_0-minimal eigenfunctions of the Laplacian on X. Alternatively, it is due to the ground state property (Proposition 14.20) and Theorem 14.4.

BASIC DEFINITIONS AND PROPERTIES

Given the probability measure m_I on $\mathcal{F}$, i.e., the unique $K^I M$-invariant probability measure on the orbit $P^I \cdot \dot{e} = K^I \cdot \dot{e}$, and the (minimal) functions $h_b, b \in \mathcal{F}$, where $h_b(g \cdot o) = P^{1/2}(g, b)$, one may define the function h_I directly as $\int P^{1/2}(g,b) dm_I(b)$. (Recall that $h_I(g \cdot o) = \int P^{1/2}(g,b) dm_I(b)$ by Theorem 7.22.) In particular, if $I = \Delta$, then h_Δ is the spherical function Φ, where

$$\Phi(g \cdot o) = \int P^{1/2}(g,b) dm(b).$$

This is the unique normalized λ_0-eigenfunction of the Laplacian L that is K-left invariant. In addition, the formula for h_I given in Theorem 7.22 follows from this definition: namely, if $g \cdot o = n_I a_I g_1 \cdot o$ with $g_1 \in G^I$, then

$$h_I(g \cdot o) = h_I(n_I a_I g_1 \cdot o) = e^{\rho(\log a_I)} \Phi^I(g_1 \cdot o),$$

where Φ^I is the spherical function associated with the symmetric space X^I (§ 2.13). The formula reflects the factorization of $P^{1/2}(n_I a_I g_1, b_1)$, where $b_1 = k_1 \cdot \dot{e} \in K^I \cdot \dot{e}$ as

$$P^{1/2}(n_I a_I g_1, b_1) = e^{\rho(\log a_I)} \Big[\frac{d(g_1 \cdot m_I)}{dm_I}(b_1)\Big]^{1/2} = e^{\rho(\log a_I)} e^{-\rho^I(H(g_1^{-1} k_1))}.$$

(see eq. (7.23) and the fact that $H((n_I a_I g_1)^{-1}k_1) = H(g_1^{-1}k_1) - \log a_I$).

The uniqueness of the function Φ with these properties follows from Corollary 13.19.

If $g = kan$, with $k \in K^I M, a \in A_I$, and $n \in N_I$, belongs to $R^I = K^I M A_I N_I$, and $b_1 = k_1 \cdot \dot{e} \in P^I \cdot \dot{e}$, then $\sigma(g^{-1}, b_1) = P^{1/2}(g, b_1) = e^{-\rho(H(g^{-1}k_1))} = e^{-\rho(H(a^{-1}))}$ is independent of b_1 since K^I normalizes N_I and centralizes A_I. It follows that $S_g m_I = g \cdot m_I = m_I$ because $A_I N_I$ acts trivially on $P^I \cdot \dot{e}$. Since the normalizer of $D^I = K^I M N_I$ is R^I, one can define the function h^D, for $D \in \overline{S}_0$ with $D = gD^I g^{-1}$, to be the λ_0-eigenfunction represented by the measure $S_g m_I$ if $g \in G$.

14.1. Remarks. (1) When $D = gKg^{-1}$, the function $h^D = S_g \Phi$.

(2) The stabilizer in G of h_I under the twisted action is R^I since the proof of Proposition 7.29 makes no use of the Laplacian and applies without change.

The aim of this chapter is to prove the following result (that is in effect another version of Theorem 7.22) by the methods of convolution kernels. The connection with eigenfunctions of the Laplacian will be made in Lemma 14.21 with another proof of Theorem 7.22 as an immediate corollary. As a result, one gets in fact a more general result (Theorem 14.4) that is valid for random walks.

14.2. Theorem. *For D in $\overline{S}_0$, the function h^D is the unique positive normalized λ_0-eigenfunction of the Laplacian that is invariant under left translation by D.*

It is clear, by definition of h^D, that it suffices to prove the theorem when $D = D^I$. The proof will follow from some propositions of a more general character that involve positive measures on a locally compact group H. These results will be applied to the case of $H = G$ or $H = M_I A_I = G^I M A_I = Z(I)$ (see Propositions 2.15 and 2.16).

If p is a well-behaved, K-bi-invariant measure on G, then it follows from Propositions 11.43, 11.44, and 13.20 that $r_0 = \int \Phi(g^{-1}) dp(g) = r_0(p)$ is the spectral radius in $L^2(G)$ of the convolution operator defined by p.

14.3. Remarks. (1) If Φ' is any spherical function and D is any G-invariant differential operator on G/K, the fact that $D(\Phi' * p) = D\Phi' * p = \chi(D)(\Phi' * p)$ implies that $\Phi' * p = r\,\Phi'$ when p is well-behaved and K-bi-invariant on G (see Helgason [H3, Corollary 2.3, p. 402]). Since $\Phi'(o) = 1$, it follows that $r = \int \Phi'(g^{-1}) dp(g) = \int \Phi'(g) dp(g)$.

(2) If p is well-behaved and K-bi-invariant on the semisimple group G, then $*p - Id$ is the analogue of the Laplacian L, as pointed out in Remark 11.4(2), and so a direct analogue of $1 - \lambda_0$ is given by $r_0 = \int \Phi(g^{-1}) dp(g) = \int \Phi(g) dp(g)$, where Φ is the spherical function defined by eq. (s) of § 7.21. This is so since the spherical nature of Φ and the K-bi-invariance of p imply, by (1), that $\Phi * p = r_0 \Phi$, i.e., $\Phi(*p - Id + \lambda_0 Id) = 0$. Commutativity of convolution for K-bi-invariant functions implies that $\Phi * p = p * \Phi = r_0 \Phi$.

Formulated in terms of convolution, Theorem 14.2 will be a consequence of the following result.

14.4. Theorem. *For D in S_0, the function h^D is the unique positive normalized r_0-eigenfunction of the operator defined by right convolution with p that is also invariant under left translation by D.*

The plan of the proof is to use the decompositions $G = P^I K$, $P^I = N_I A_I M_I$, and the properties of the group $H = A_I M_I = G^I M A_I$ (see Proposition 2.15) to reduce the equations to equations on H, M_I, and G^I, where uniqueness properties of what will be referred to as the ground state are available. This technique is called the descent method in [G1]. First, the properties of the group $A_I M_I$ and of the decomposition $G = P^I K$ will be examined in a more general context.

CONVOLUTION

§ **14.5.** Let H denote a locally compact metrizable group with a compact subgroup K. If f is a non-negative Borel function on H and μ is a K-right invariant bounded measure, then $f * \mu$ is a K-right invariant function. With the natural identification of functions on H/K with K-right invariant functions on H, one may view $f * \mu$ as a function on H/K whenever f is a non-negative Borel function on H, i.e., if π: $H \to H/K$ is the canonical projection, $f = F \circ \pi$, where F is a Borel function on H/K.

Let $\bar{\mu}$ on H/K denote the image of μ under π and define a convolution on H/K by setting $F * \mu$ to be the function for which $(F * \mu) \circ \pi = (F \circ \pi) * \mu = f * \mu$, i.e., $(F * \bar{\mu})(g \cdot o) = \int F(g \cdot h^{-1} \cdot o) d\mu(h)$. Usually, in what follows, no distinction will be made between f and F, p and $\bar{p}$.

There is a natural bijection between the K-right invariant measures μ on H and measures ν on H/K: to μ associate its image $\bar{\mu}$ under π and, if ν is a measure on H/K, define $\tilde{\nu}(A) = \int [\int 1_A(gk) d\tilde{m}(k)] d\nu(gK)$, for $A \in \mathfrak{B}(H)$, where $\tilde{m}$ is the Haar measure on K. If μ, η are measures on H and η is K-right invariant, then $\mu * \eta$ is also K-right invariant on H. In this way, the convolution $\nu * \gamma$ of two measures ν and γ on H/K may be defined as follows: one first lifts them to H as $\tilde{\nu}$ and $\tilde{\gamma}$, and then defines $\nu * \gamma$ as the image $(\tilde{\nu} * \tilde{\gamma})^-$ of $\tilde{\nu} * \tilde{\gamma}$ under π. Note that, if μ is a measure on H and ν is a measure on H/K, the measure $\mu * \nu$ defined at the beginning of Chapter XI coincides with $\overline{\mu * \tilde{\nu}}$, i.e., the two definitions of $\mu * \nu$ agree.

These operations on measures can also be carried out by defining an **averaging kernel** Π from H/K to H by $\Pi(gK, A) = \tilde{m}(g^{-1}A)$ and a **projection kernel** π from H to H/K by $\pi(g, A) = \delta_g(\pi^{-1}(A))$: then $\bar{\mu} = \mu\pi$ and $\tilde{\nu} = \nu\Pi$. Hence, $\nu * \gamma = (\nu\Pi * \gamma\Pi)\pi$. Recall (see [M7, p. 173]) that a **kernel** $\Gamma(x, A)$ is measurable in x and a measure in A.

A K-bi-invariant measure p on H defines a natural kernel P: for any positive measure μ, let $\mu P = \mu * p$; and for any positive function f on H, set $Pf = f * \check{p}$, i.e. $Pf(g) = \int f(gh) dp(h)$; then $P(g, A) = p(g^{-1}A)$.

Since it is K-right invariant, the measure p is completely determined by its image $\bar{p}$ on H/K. There is a kernel $\bar{P}$ associated with $\bar{p}$: for a positive function F on H/K, set $\bar{P}F(g \cdot o) = \int F(g \cdot x)d\bar{p}(x)$ and observe that the K-left invariance of $\bar{p}$ implies the integral defines a function on H/K. (Note that $\bar{P}F = F * \check{\bar{p}}$.) Furthermore, $\bar{P}(g \cdot o, B) = \bar{p}(g^{-1} \cdot B) = \int 1_B(g \cdot x)d\bar{p}(x)$ and, so, if ν is any positive measure on H/K, $\nu\bar{P}(B) = \int d\nu(g)\bar{p}(g^{-1} \cdot B)$. In addition, these kernels are interlaced by the natural projection π. In other words, $P(F \circ \pi) = (\bar{P}F) \circ \pi$ and $\overline{\mu P} = \bar{\mu}\bar{P}$.

Specializing to the case of a connected semisimple Lie group G , the Iwasawa decomposition $G = KAN$ implies that $G/K = NA \cdot o$. Hence, measures on $G/K = NA{\cdot}o$ can be identified with measures on the subgroup $S = NA$ of G. Any measure ν on S can be viewed as a measure on G such that $\nu(A) = \nu(A \cap S)$, for any Borel subset A of G. Then, $\tilde{\nu} = \nu * \tilde{m}$. This identification of G/K with S gives a formula for $F * \nu$ involving integration on S when ν, viewed on G/K, is K-left invariant, i.e., when $\tilde{\nu}$ is K-bi-invariant (here F is Borel and non-negative). If q is the K-bi-invariant measure on G such that $\check{q} = \tilde{\nu}$, then

$$(*) \qquad (F * \nu)(s) = \int F(st)d\bar{q}(t).$$

To verify this note that, if $f = F \circ \pi$ then, by definition of $F * \nu$ and q, $(F * \nu)(s) = (f * \check{q})(s) = \int f(sg)dq(g)$. Let $\bar{q}$ be the image of q under π on S. Then $q = \bar{q} * \tilde{m}$. Hence, $\int f(sg)dq(g) = \int\int f(stk)d\bar{q}(t)d\tilde{m}(k)$. Since f is K-right invariant, this double integral equals $\int f(st)d\bar{q}(t) = \int F(st)d\bar{q}(t)$.

This observation concerning the computation of $F * \nu$ has an important corollary that is stated as the next lemma.

14.6. Lemma. *Let G denote a connected semisimple Lie group. Let χ be a locally bounded K-right invariant function on G whose restriction to $S = NA$ is a character, i.e., a homomorphism of S into $\mathbb{C}$. Then, χ is an eigenfunction of any right convolution operator defined by a K-bi-invariant measure p of compact support, with eigenvalue $\int \chi(g^{-1})dp(g)$.*

Proof. Let $q = \check{p}$. Then, if $\bar{\chi}$ is the restriction of χ to S and $\nu = \bar{p} = \check{\bar{q}}$, it follows from eq. $(*)$ that $(\chi * p)(s) = (\chi * \check{q})(s) = (\bar{\chi} * \nu)(s) = \int \bar{\chi}(st)d\bar{q}(t)$.

Since $\bar{\chi}$ is a character on S, it follows that

$$\begin{aligned}\int \bar{\chi}(st)d\bar{q}(t) &= \bar{\chi}(s)\int \bar{\chi}(t)d\bar{q}(t) \\ &= \chi(s)\int \chi(g)dq(g) = \chi(s)\int \chi(g^{-1})dp(g).\end{aligned}$$

Hence, $(\chi * p)(sk) = (\chi * p)(s) = \{\int \chi(g^{-1})dp(g)\}\chi(sk)$. □

SPHERICAL FUNCTIONS AND MINIMAL EIGENFUNCTIONS

§ **14.7.** For later use in this chapter some results from Chapters XI and XIII are repeated here. As usual, let $a(g)$ denote the A-component of g, i.e., $a(g) = \exp H(g)$, in the Iwasawa decomposition $G = KAN$.

If $v \in \mathfrak{a}^*$ define the positive character or exponential $\ell \in A^*$ by setting $\ell(\exp H) = e^{-(v+\rho)(H)}$ on $A = \exp \mathfrak{a}$ and let ℓ denote $e^{-(v+\rho)}$. To each exponential ℓ there is associated the cocycle $\sigma_\ell(g,b) = \ell[a(g^{-1}k)] = e^{-(v+\rho)(H(g^{-1}k))}$ and corresponding function $h_b^\ell(g) = \sigma_\ell(g,b)$ on G. The integral over K of these functions is the spherical function $\Phi^\ell = \Phi_v$, where $\Phi^\ell(g) = \int h_b^\ell(g)dm(b) = \int \ell[a(g^{-1}k)]dk = \int e^{-(v+\rho)(H(g^{-1}k))}dk$, and dk denotes the Haar measure $\tilde{m}$ on K. Note that the spherical function Φ defined in § 7.21 equals $\Phi_0 = \Phi^{e^{-\rho}}$.

Let p denote a well-behaved K-bi-invariant positive bounded measure on G. The functions h_b^ℓ are all eigenfunctions of the convolution operator $*p$. More specifically, it follows from Proposition 11.36, Definition 11.37, and Lemma 11.38 that $h_b^\ell * p = \hat{p}(\ell)h_b^\ell$, where $\hat{p}(\ell) = \int \Phi^\ell(h^{-1})dp(h)$. This function $\hat{p}$ on A^* is called the Laplace transform of p (see Definition 11.37). Note that $\hat{p}(e^{-\rho}) = r_0$. Further, r_0 is the infimum over all exponentials of the Laplace transform of p (see Theorem 11.40) and in view of Theorem 13.12, the interval $[r_0(p), \infty) = \{\hat{p}(\ell) \mid \ell \in A^*\}$.

The positive functions f for which $f * p = \hat{p}(\ell)f$ form a convex cone with compact base that is a Choquet simplex. Every minimal element is of the form h_b^ℓ, for some $\ell \in A^*$. In particular, if $h_b = h_b^{e^{-\rho}}$, then $h_b * p = r_0 h_b$, for all $b \in \mathcal{F}$, and not only is every minimal solution of the equation $f * p = r_0 f$ of the form h_b, but the minimal solutions are exactly the functions $h_b, b \in \mathcal{F}$, as shown in Proposition 13.19.

GROUND STATE PROPERTIES

The ground state Φ defined in § 7.21 plays a crucial role in the identification, in Theorem 9.18, of the Martin boundary $\partial X(\lambda_0)$ with the Furstenberg boundary. Viewed as a measure, it satisfies the convolution equation $\Phi * p = r_0\Phi$, as observed in Remark 14.3.(2). In the following, the formal aspect of the ground state will be discussed in the context of locally compact groups and convolution operators defined by well-behaved measures. It is important and useful to have results analogous to those for the Laplacian L in generalized horocyclic coordinates that indicate the form of L acting on functions on certain submanifolds, i.e., that depend only upon some of the generalized horocyclic coordinates. (Recall that this is how the function h_I was determined in Theorem 7.22.)

14.8. Definition. Assume that H is a locally compact group, p is a well-behaved, positive, bounded measure on H and Λ is a closed subgroup. Let $r(\Lambda, p)$ denote the infimum of the positive numbers r such that there exists some Λ-left invariant measure μ with $\mu * p = r\mu$. Let $r(p)$ denote $r(\{e\}, p)$.

14.9. Remarks. (1) There are measures μ for which $\mu * p = r\mu$, since the right invariant Haar measure on H has this property with r equal to the total mass of p. The convex cone of positive measures ν such that $\nu * p \leq r\nu$ has a compact base that is a Choquet simplex (see Theorem 11.5). In particular, since p is well-behaved, if $\mu * p = r\mu$, then $r = 0$ if and only if $\mu = 0$. As a result, it is easy to show from compactness arguments that the infimum $r(\Lambda, p)$ is attained (see Theorem 11.10).

(2) If Λ is compact, then $r(\Lambda, p) = r(p)$ as any solution of $\mu * p = r\mu$ can be averaged over Λ to produce a left Λ-invariant solution. Also, $r(p) \leq r(\Lambda, p)$ for any closed subgroup Λ.

(3) When p is a probability measure, the number $r(p)$ gives the rate of decay of the probability of returning at time n to a given neighbourhood of the identity of G.

It was shown in Proposition 11.19 that $r(p) = \overline{\lim}_n p^n(e)^{1/n} \leq r_0(p)$, where $r_0(p)$ is the spectral radius in $L^2(H)$ of the convolution operator defined by p. In particular, if H is semisimple, $r(p) = r_0(p) = r_0$, as shown in Proposition 14.12. (This was also proved in Corollary 13.19.)

14.10. Definition. The pair (Λ, p) is said to satisfy the **ground state property** if there is a unique normalized, Λ-left invariant measure μ such that

$$\mu * p = r(\Lambda, p)\mu.$$

14.11. Remarks. (1) Because p is well-behaved, the measure μ has a strictly positive continuous density f with respect to the right invariant Haar measure that satisfies $f * p = r(\Lambda, p) f$ (see Chapter XI). The normalization is given by $f(e) = 1$. The corresponding measure is denoted by $\phi_{\Lambda,p}$ and will be identified with its density in what follows.

(2) Since $\tilde{m} * p = p * \tilde{m}$, where $\tilde{m}$ is the Haar measure on K, if p is K-bi-invariant, it follows from the uniqueness that $\phi_{K,p}$ is also K-bi-invariant: $\mu * p = r\mu$ implies that $r(\mu * \tilde{m}) = (\mu * p) * \tilde{m} = (\mu * \tilde{m}) * p$ and so $\phi_{K,p}$ is K-right invariant.

(3) It is easy to show that if $H = \mathbb{R}^d$ and p is well-behaved, then every pair (Λ, p) has the ground state property. If $\hat{p}$ denotes the Laplace transform of p, then $\phi_{\Lambda,p}$ is the unique exponential ℓ such that $\hat{p}(\ell) = r(\Lambda, p)$. The uniqueness follows from the strict convexity of $\log \hat{p}(\ell)$, as in the proof of Theorem 11.40.

In the case of a semisimple group G it follows, from Chapter XIII, that for any well-behaved, K-bi-invariant measure p and maximal compact subgroup K, the pair (K, p) has the ground state property as stated in the next result.

14.12. Proposition. *Let K be a maximal compact subgroup of a connected semisimple Lie group G. If p is K-bi-invariant, then the pair (K, p) has the ground state property. Moreover,*

$$r(K, p) = r(p) = r_0, \quad \text{and } \phi_{K,p} = \Phi.$$

Proof. Corollary 13.19 implies that $r(p) = r_0(p) = r_0 = \inf_{\ell \in A^*} \hat{p}(\ell)$. By definition, $r(p) \le r(K, p)$. Conversely, if μ is a solution of the equation $\mu * p = r(p)\mu$, one has $(\tilde{m} * \mu) * p = r(p)\tilde{m} * \mu$. Hence, $r(K,p) \le r(p)$. The ground state property follows from Corollary 13.18, as Φ is the unique positive K-bi-invariant solution of $f * p = r_0 f$ on G. □

In order to prove Proposition 14.20, it is useful to consider what happens to the ground state property if one restricts to a subgroup or forms a factor group. Note that this formal discussion corresponds to the study of the behavior of L on functions that, in generalized horocyclic coordinates, do not depend upon all the variables.

The first case to consider is that of a semidirect product (cf. Proposition 7.10, in particular when $I = \emptyset$).

14.13. Proposition. *Assume that $H = Z \ltimes V$ is the semidirect product of the normal subgroup V by the subgroup Z. Assume V is unimodular and that $\Lambda \supset V$ is a subgroup such that $\bar{\Lambda} \stackrel{def}{=} \Lambda \cap Z$ is compact. Let $\bar{p}$ denote the projection of p onto Z, and η Haar measure on V. Then $r(\Lambda, p) = r(\bar{\Lambda}, \bar{p})$. Furthermore, if $(\bar{\Lambda}, \bar{p})$ has the ground state property, then so does (Λ, p) and $\phi_{\Lambda,p} = \eta * \phi_{\bar{\Lambda},\bar{p}}$. In terms of densities, $\phi_{\Lambda,p} = \phi_{\bar{\Lambda},\bar{p}} \circ \tau$, where τ is the projection of H on Z.*

Proof. Let μ be a Radon measure on H that is V-left invariant. Then it can be written in a unique way as $\eta * \bar{\mu}$, with $\bar{\mu}$ a Radon measure on Z. Let $A \in \mathfrak{B}(V)$ and $B \in \mathfrak{B}(Z)$. Since $A \to \mu(A \cdot B)$ is a V-left invariant measure, $\mu(A \cdot B) = \eta(A)\gamma(B)$, where $B \to \gamma(B)$ is the measure $\bar{\mu}$. Since the sets $A \cdot B$ generate $\mathfrak{B}(H)$, as one sees by identifying H with $V \times Z$, and $\mu(A \cdot B) = \eta(A)\bar{\mu}(B)$, it follows that $\mu = \eta * \bar{\mu}$.

Furthermore, the Λ-invariance of μ is equivalent to the $\bar{\Lambda}$-invariance of $\bar{\mu}$ because $\Lambda = V \cdot \bar{\Lambda}$: if $g = vz, v \in V, z \in Z$, then $\mu(gA \cdot B) = \mu(zA \cdot B) = \eta(zAz^{-1})\bar{\mu}(zB)$; since $\eta(zAz^{-1}) = \gamma(z)\eta(A)$, where γ is an exponential on Z, it follows from the compactness of $\bar{\Lambda}$ that, for $g = vz \in \Lambda$, one has $\mu(gA \cdot B) = \mu(A \cdot B)$ if and only if $\bar{\mu}(z \cdot B) = \bar{\mu}(B)$.

The following calculation implies that $\eta * \bar{\mu} * p = \eta * \bar{\mu} * \bar{p}$: if $\omega = uz, u \in V, z \in Z$, then for $x \in V, y \in Z, 1_{A \cdot B}(xy\omega) = 1_A(xyuy^{-1})1_B(yz)$. As a result, the unimodularity of V implies that $\int 1_{A \cdot B}(xy\omega)d\eta(x) = \eta(A)1_B(yz)$. It follows from this that

$$\begin{aligned}\iiint 1_{A \cdot B}(xy\omega)d\eta(x)d\bar{\mu}(y)dp(\omega) &= \eta(A)\iint 1_B(yz)d\bar{\mu}(y)d\bar{p}(z)\\ &= \eta(A)(\bar{\mu} * \bar{p})(B).\end{aligned}$$

Hence, if μ is Λ-left invariant, one has $\mu * p = r\mu$ if and only if $\bar{\mu} * \bar{p} = r\bar{\mu}$: note that $(\mu * p)(A \cdot B) = r\mu(A \cdot B) = r\eta(A)\bar{\mu}(B)$, and that $(\mu * p)(a.B) = (\eta * \bar{\mu} * \bar{p})(a.B) = \eta(A)(\bar{\mu} * \bar{p})(B)$. Hence, $r(\Lambda, p) = r(\bar{\Lambda}, \bar{p})$.

Also if $r = r(\bar{\Lambda}, \bar{p})$, the ground state property (Definition 14.10) implies that the $\bar{\Lambda}$-left invariant solution $\bar{\mu}$ to $\bar{\mu} * \bar{p} = r\bar{\mu}$ is unique. As a result,

$\mu = \eta * \bar{\mu}$ is the unique Λ-left invariant solution of $\mu * p = r(\Lambda, p)\mu$. The formula for $\phi_{\Lambda,p}$ follows. In terms of densities, this implies that $\phi_{\Lambda,p}(vz) = \phi_{\bar{\Lambda},\bar{p}}(z), v \in V, z \in Z$, i.e., for each $z \in Z$, $\phi_{\Lambda,p}$ is constant on the fibers Vz with value $\phi_{\bar{\Lambda},\bar{p}}(z)$. □

The next two situations that will be considered involve the following formal lemma. Whenever one has $H = LN$ with L and N closed subgroups such that $L \cap N$ is compact, there is a general procedure for transporting measures on H to measures on L. It is stated as the following lemma.

14.14. Lemma. *For a positive measure μ on $H = LN$, let $\bar{\mu}$ be its image on the homogeneous space $Y = L/L \cap N$ under the map $\tau : H \to L/L \cap N$ given by $\tau(h) = \ell(L \cap N)$ if $h = \ell n$. Denote by $\Pi(\mu)$ the unique $L \cap N$-right invariant measure on L whose image on Y under the canonical map $\pi : L \to L/L \cap N$ equals $\bar{\mu}$. Then Π is a kernel from H to L. Furthermore, if $f \geq 0$ is a measurable function on Y, then for any positive measure μ on H, one has*

$$\int (f \circ \tau) d\mu = \int (f \circ \pi) d\Pi(\mu).$$

In addition, if Λ is a subgroup of L and μ is Λ-left invariant, then $\Pi(\mu)$ is also Λ-left invariant.

Proof. Let $T(g, A) = m'(g^{-1}A)$, where m' denotes the normalized Haar measure on $L \cap N$ and $A \in \mathfrak{B}(H)$. This kernel is the composition $\Pi_1 P_2$ of the projection kernel P_2 from H to Y and the averaging kernel Π_1 that lifts measures from Y to P (see § 14.5). It follows that $\Pi(\mu) = \mu T$. □

The first situation to examine is when $H = PK$ with K a compact subgroup of H and P a closed subgroup, i.e., $N = K$ and $L = P$.

14.15. Corollary. *Assume that $H = PK$, where K is a compact subgroup of H and P is a closed subgroup of H. Then H/K can be identified with $Y = P/K'$, where $K' = K \cap P$, by the map $gK \to bK' = (gK) \cap P$ if $g = bk, b \in P, k \in K$.*

Let $\tilde{m}$ denote the Haar measure of K and $\tilde{m}'$ the Haar measure of $K' = K \cap P$. Then,

(1) $\Pi(g, X) = \tilde{m}'(b^{-1}X)$ *if* $g = bk, b \in P, k \in K$;
(2) *if Λ is a subgroup of P and μ is Λ-left invariant, then $\Pi(\mu)$ is also Λ-left invariant;*
(3) *if μ, ν are measures on H, with $\tilde{m} * \nu = \nu$, then $\Pi(\mu * \nu) = \Pi(\mu) * \Pi(\nu)$;*
(4) *if η denotes a right Haar measure on H, then $\Pi(\eta)$ is a right Haar measure on P;*
(5) *if a well-behaved measure p on H is K-bi-invariant, then $\Pi(p)$ is well-behaved on P and K'-bi-invariant with a continuous density p' constant on the cosets bK' equal to the value of p on $gK, g = bk$;*
(6) *if p is a well-behaved measure on H that is K-bi-invariant and $\mu * p = r\mu$, then $\Pi(\mu) * \Pi(p) = r\Pi(\mu)$.*

Proof. (1) is an immediate consequence of the proof of Lemma 14.14. The equivariance of the projection and lifting kernels implies (2). In (3), the hypothesis implies that $\mu * \nu$ is K-right invariant and is completely determined by its image $\overline{\mu * p}$ under $\tau : H \to Y$. This image is by definition $\bar{\mu} * \bar{\nu}$. The averaging kernel lifts this convolution on Y to the convolution of $\Pi(\mu)$ and $\Pi(\nu)$ on P.

To verify (4), note that any right Haar measure on P is K'-right invariant and so is the lift of its image on Y.

To determine the density of p, disintegrate it (i.e., look at a regular conditional expectation of p given τ) over its image on Y. This implies (5). Given (3), (6) is immediate. □

Lemma 14.14 and Corollary 14.15 imply the following result.

14.16. Proposition. *Assume that $H = PK$, where K is a compact subgroup and P is a closed subgroup. Let p denote a positive, bounded K-bi-invariant measure and let D denote a closed subgroup of P. Then $r(D,p) = r(D,\Pi(p))$. Furthermore, (D,p) has the ground state property if and only if $(D,\Pi(p))$ has the same property.*

Moreover, when the ground state property holds, then there is a function $\bar{\phi}_{D,p}$ on $Y = H/K = P/K'$ such that $\phi_{D,p} = \bar{\phi}_{D,p} \circ \tau$ and $\phi_{D,\Pi(p)} = \bar{\phi}_{D,p} \circ \pi$, where $\tau : H \to Y = H/K$ and $\pi : P \to Y = P/K'$ are the canonical projections. Equivalently, in terms of measures, $\Pi(\phi_{D,p}) = \phi_{D,\Pi(p)}$ and $\phi_{D,p} = \phi_{D,\Pi(p)} * \tilde{m}$.

Proof. It follows from Corollary 14.15(2) and (3) that if μ is a D-left invariant measure on H such that $\mu * p = r\mu$, then $\Pi(\mu) * \Pi(p) = r\Pi(\mu)$.

Conversely, if ν is a D-left invariant measure on P that satisfies $\nu * \Pi(p) = r\nu$, then $(\nu * \tilde{m}) * p = r(\nu * \tilde{m})$, where $\tilde{m}$ is the Haar measure on K. This holds since ν is K'-right invariant (see § 14.5) and so can be viewed as a measure on $Y = P/K' = H/K$: the corresponding K-right invariant measure on H is $\nu * \tilde{m}$; finally, $(\nu * \tilde{m}) * p$ can be identified with the same convolution on Y as $\nu * \Pi(p)$. This shows that (i) $r(D,p) = r(D,\Pi(p))$ and (ii) (D,p) has the ground state property if and only if $(rD,\Pi(p))$ has this property.

Finally, by Corollary 14.15(3) and the uniqueness of the ground state, one has $\Pi(\phi_{D,p}) = \phi_{D,\Pi(p)}$. The uniqueness of the ground state means that $\phi_{D,p} * \tilde{m} = \phi_{D,p}$ and $\phi_{D,\Pi(p)} * \tilde{m}' = \phi_{D,\Pi(p)}$, since $p * \tilde{m} = \tilde{m} * p$ and $\Pi(p) * \tilde{m}' = \tilde{m}' * \Pi(p)$, where $\tilde{m}$ is the Haar measure on K and $\tilde{m}'$ is the Haar measure on K'. In terms of densities, this means that $\phi_{D,p}$ is K-right invariant and $\phi_{D,\Pi(p)}$ is K'-right invariant. Hence, both densities determine the same function $\bar{\phi}_{D,p}$ on Y. □

14.17. Definition. Let K be a closed subgroup of the locally compact group S. Then K is said to satisfy the **strong ground state property** if (i) the ground state property holds for every pair (K,p), where p is a well-behaved K-bi-invariant probability measure and, furthermore, (ii) the measure $\phi_{K,p}$ is independent of p. In this case, this measure, which will be denoted by ϕ_K, is called the **K-ground state of S.**

Remark. If G is a connected semisimple Lie group, it follows from Proposition 14.12 that any maximal compact subgroup has the strong ground state property and $\phi_K = \Phi$. The same is true for groups of Euclidean motions with $\phi_K = 1$.

14.18. Proposition. *Assume that L, S and M are closed subgroups of H with (i) $H = SM$, (ii) $S \cap M \subset L \subset S$, and (iii) LM a compact subgroup of H. If L has the strong ground state property as a subgroup of S, then LM has the strong ground state property (as a subgroup of H).*

Proof. Let ϕ_L be the L-ground state of S and let ρ denote the Haar measure of LM. Let p be a LM-bi-invariant measure on $H = SM$. In order to apply Proposition 14.16, let $P = S, K = LM$ and $D = L$. Let Π denote the corresponding kernel (see Lemma 14.14). The measure $\Pi(p)$ is $LM \cap S = L(S \cap M)$-bi-invariant. Hence, it is L-bi-invariant. The ground state property is valid for $(L, \Pi(p))$ and, therefore, by Proposition 14.16, for (L, p). Furthermore, by Proposition 14.16, since $\phi_{L,\Pi(p)} = \phi_L$, the (L, p) ground state $\phi_{L,p} = \phi_L * \rho$, which is independent of p. □

The second situation to consider involves another application of Lemma 14.14. Here $H = SA$, where S and A are again closed subgroups of H with $S \cap A$ compact, i.e., $L = S$ and $N = A$. In addition A is assumed to be central and to be compactly generated in H.

14.19. Proposition. *Let S, A and K be closed subgroups of the locally compact group H such that (i) $H = SA$, (ii) A is central, compactly generated in H and (iii) $S \cap A \subset K \subset S$. Assume that K is a compact subgroup of S with the strong ground state property, relative to S. Then, for any well-behaved K-bi-invariant probability measure p on H, the pair (K, p) has the ground state property. Furthermore, there exists an exponential ℓ on H, that is trivial on S, such that $\phi_{K,p} = \phi_K \cdot \ell^{-1}$, where ϕ_K denotes the K-ground state ϕ_K on S extended to H by setting $\phi_K(sa) = \phi_K(s)$.*

Proof. All measures in what follows will be taken to be K-left invariant. The proof is divided into three parts:

Part (a): Since the convex cone of positive measures μ such that $\mu * p = r\mu$ is a Choquet simplex, it follows that there are minimal solutions of the equation $\mu * p = r\mu$, i.e., extremal elements in this simplex. Let μ be a minimal solution of the equation. Since A is central in H, it follows that $\delta_a * \mu = \mu * \delta_a$, for all $a \in A$. Hence, $r(\delta_a * \mu) = \mu * \delta_a * p$. In addition, because p is well-behaved, there exist constants $C(a) > 0$ and an integer $k(a) \geq 2$ such that $\delta_a * p \leq C(a)p^{k(a)}$. Hence, $\mu * \delta_a * p \leq C(a)\mu * p^{k(a)} = C(a)r^{k(a)}\mu$ and, so, $r\delta_a * \mu \leq C(a)r^{k(a)}\mu$.

The minimality of μ and the fact that $r(\delta_a * \mu) = (\delta_a * \mu) * p$ implies that, for some positive constant $\ell(a)$, one has $\delta_a * \mu = \ell(a)\mu$. Since ℓ is clearly a character, it is an exponential.

The compactness of the subgroup $S \cap A$ implies that ℓ is trivial on it, i.e., $\ell(S \cap A) = \{1\}$. As a result, one may extend ℓ as an exponential to H by setting $\ell(h) = \ell(a)$ if $h = sa$, with $s \in S$ and $a \in A$.

Let f be the density of μ relative to the right Haar measure η. Since A is central, $\delta_a * \mu = \ell(a)\mu$ implies that, for all measurable sets E,

$$\int 1_E(ya)f(y)d\eta(y) = \int 1_E(ay)f(y)d\eta(y) = \ell(a) \int 1_E(x)f(x)d\eta(x).$$

Hence, it follows that, for all $a \in A$, one has $f(xa^{-1}) = \ell(a)f(x)$. If $x = sa$, this implies $f(s) = \ell(a)f(sa)$. Let $\psi(g) = f(s)$, if $g = sa$, $s \in S$ and $a \in A$: the function ψ is well-defined as $g = s_1a_1 = s_2a_2$ implies that $a_2a_1^{-1} \in S \cap A$ and so $f(s_1) = \ell(a_1)f(g) = \ell(a_2)f(g) = f(s_2)$. Hence, one has $f = \psi\ell^{-1}$, i.e., $f(sa) = \ell^{-1}(a)\psi(s)$.

Since A is central and ℓ is trivial on any compact subgroup of H, it follows that, for any $g \in H$, the function $h \to \psi(gh^{-1})$ is A-right invariant. Hence, it factors through the maps τ and Π of Lemma 14.14. As a result, for any measure ν on H, $\int \psi(gh^{-1})d\nu(h) = \int \psi(st^{-1})d\Pi(\nu)(t)$ if $g = sa, s \in S$ and $a \in A$. From this it follows that one may reduce the convolution equation $\mu * p = r\mu$ on H to the convolution equation $\psi * p^\ell = r\psi$, where $p^\ell = \Pi(\ell p)$ and ψ is a function on S.

More explicitly, if F is the density of μ and $g = sa$, it follows that

$$\begin{aligned} r\ell^{-1}(g)\psi(s) &= \int f(gh^{-1})dp(h) \\ &= \int \psi(gh^{-1})\ell^{-1}(gh^{-1})dp(h) \\ &= \ell^{-1}(g) \int \psi(gh^{-1})\ell(h)dp(h) \\ &= \ell^{-1}(g) \int \psi(st^{-1})dp^\ell(t). \end{aligned}$$

Part (b): This last equation, with $r = r(K,p)$, implies that $r(K,p^\ell) \leq r(K,p)$. Conversely, if ϕ is K-left invariant on S and $\phi * p^\ell = r(K,p^\ell)\phi$, then $(\phi\ell^{-1}) * p = r(K,p^\ell)(\phi\ell^{-1})$, where ϕ also denotes its extension to H defined by $\phi(g) = \phi(s)$ if $g = sa$ (here one makes use of the fact that $K \supset S \cap A$) and, so, since $f = \phi\ell^{-1}$ is K-left invariant, it follows that $r(K,p) = r(K,p^\ell)$ for every exponential ℓ on H that is trivial on S.

Part (c): To conclude the proof, it will suffice to show that the equation $f * p = r(K,p)f$ has a unique minimal solution. Let f, f' be two minimal solutions of the equation $f * p = r(K,p)f$ with $f(e) = f'(e) = 1$. Then $f = \psi\ell^{-1}$ and $f' = \psi'\ell'^{-1}$. The strong ground state property of K in S implies that $\psi = \psi' = \phi_K$. It remains therefore to show that the exponential is unique.

Let $0 < \alpha < 1$ and $\alpha' = 1 - \alpha$ and let γ denote the exponential $\ell^\alpha \ell'^{\alpha'}$. Consider the function $g = f^\alpha f'^{\alpha'} = \phi_K\gamma^{-1}$. Since p is well-behaved,

Hölder's inequality implies that $g * p \leq r(K,p)g$, with equality if and only if $f = f'$.

The strong ground state property implies that $\phi_K * p = r(K,p)\phi_K$ and so $g*p = r(K,p)g$. In Part (b) it was shown that $r(K,p) = r(K,p)$. Hence, $f = f'$. □

To prove Theorems 14.2 and 14.4, that give the relation between λ_0-eigenfunctions and distal subgroups, one makes use of these general facts to verify the following result, the terminology being that used at the beginning of this chapter.

14.20. Proposition. *Let K be a maximal compact subgroup of G that contains M. Then the subgroup $K^I M = K \cap M_I$ of $M_I = G^I M$ has the strong ground state property. If p is a K-bi-invariant, well-behaved probability measure on G, the pair (D^I, p) has the ground state property.*

Proof. As observed in § 2.13, $K^I = G^I \cap K$ is a maximal compact subgroup of the semisimple group G^I. By Corollary 2.16, $M_I = G^I M$ and so $K^I M = K \cap M_I$ (the centralizer $M(I)$ in K of $\mathfrak{a}_I$ by Proposition 2.15). The strong ground state property holds for K^I as a subgroup of G^I by Proposition 14.12. Hence, by Proposition 14.18, it holds for $K^I M$ as a subgroup of M_I.

Since $P^I \supset P$ and $G = PK$, it follows that $G = P^I K$. Let Π denote the kernel from G to P^I defined in Lemma 14.14, with $H = G, L = P^I$, and $N = K$. Since, by the Langlands decomposition (Corollary 2.16), $P^I = M_I A_I N_I$, one has $D^I = K^I M N_I \subset P^I$. By Proposition 14.16, (D^I, p) has the ground state property if $(D^I, \Pi(p))$ has the ground state property. Since P^I is the normalizer of $\mathfrak{n}_I$ in G, it follows that $P^I = (M_I A_I) \ltimes N_I$ is the semidirect product of $M_I A_I$ and N_I with N_I normal in P^I. One applies Proposition 14.13 with $Z = M_I A_I$, $V = N_I$, and $D^I = \Lambda \supset V = N_I$. Note that $\bar{\Lambda} = D^I \cap M_I A_I = K^I M$ is compact. It follows that $(D^I, \Pi(p))$ has the ground state property if $(K^I M, \overline{\Pi(p)})$ has the ground state property relative to $M_I A_I$, where $\overline{\Pi(p)}$ is the projection of $\Pi(p)$ on $Z = M_I A_I$, and $\bar{\Lambda} = \Lambda \cap Z = D^I \cap (M_I A_I) = K^I M$. Since A_I is central in $M_I A_I$, it follows from Proposition 14.18 and the strong ground state property of $K^I M \subset M_I$, that $(K^I M, \overline{\Pi(p)})$ has the ground state property relative to $M_I A_I$. □

Proof of Theorem 14.4.

From Proposition 14.20, it follows that it is enough to show that $r_0 = r(D^I, p)$. To begin with, $r_0 \leq r(D^I, p)$ since by Remark 14.9(2), $r(p) \leq r(D^I, p)$, and $r(p) = r_0$ by Proposition 14.12. Since h_I is D^I-invariant and $h_I * p = r_0 h_I$, the opposite inequality $r_0 \geq r(D^I, p)$ is a consequence of Definition 14.8. □

RANDOM WALKS, EIGENFUNCTIONS OF THE LAPLACIAN AND $X \cup \partial X(\lambda_0)$

As shown in Corollary 14.22, given Theorem 14.4 it is relatively easy to establish the identification, as G-spaces, of the Martin compactification $X \cup \partial X(\lambda_0)$, where $X = G/K$, and the maximal Satake–Furstenberg compactification $\overline{X}^{SF}$ (previously established by Theorems 4.33, 7.33, and 9.18). The first step is to show how the eigenfunctions of the Laplacian are also eigenfunctions of convolution operators defined by a well-behaved K-bi-invariant positive bounded measure p.

14.21. Lemma. *Assume $\lambda \le \lambda_0$. Then there exists a continuous function p^λ with compact support such that $p^\lambda(o) > 0, p^\lambda(k \cdot x) = p^\lambda(x)$ for all $k \in K$ and $x \in X$, with the property that $Lf + \lambda f = 0$ implies*

$$f(g \cdot o) = \int f(g \cdot x) p^\lambda(x) dx = \int f(g \cdot x) dp^\lambda(x), i.e., f * \check{p}^\lambda = f,$$

where dx denotes the volume measure on X.

Proof. Let Φ' be a (positive) spherical function such that $L\Phi' + \lambda\Phi' = 0$. Define the differential operator $D\phi = \frac{1}{\Phi'}\{L(\phi\Phi')\} + \lambda\phi$. Then D is K-left invariant and $D1 = 0$. The heat kernel associated with D is given by $e^{-\lambda t}\dfrac{1}{\Phi'(x)}p^t(x,y)\Phi'(y)$, where $p^t(x,y)$ is the heat kernel associated with L. It defines a Markovian semigroup and, hence, a corresponding diffusion or "Brownian motion" with infinitesimal generator D. Let the origin o be the starting point of the diffusion.

Denote by q_r the exit law of this diffusion on the geodesic sphere of radius r and center at the origin o, i.e., q_r is the harmonic measure for the geodesic ball of radius r associated with o. If $Lf + \lambda f = 0$ and $u = f/\Phi', Du = 0$ and $u(o) = \int u(x) dq_r(x)$. In other words, $f(o) = \int f(x)\Phi'(x)^{-1} dq_r(x)$.

The measure $dq^\lambda(x)$ defined by integrating the measures $\Phi'(x)^{-1}dq_r(x)$ in r over the interval [1, 2] has a bounded density $q^\lambda(x)$ relative to the volume measure dx that is supported by the shell defined by $1 \le d(o,x) \le 2$. It is K-left invariant by definition. Furthermore, $f(o) = \int f(x) q^\lambda(x) dx$.

Since, by group invariance, the function $f_g(x) = f(g \cdot x)$ is also a λ-eigenfunction

$$f(g \cdot o) = \int f(g \cdot x) q^\lambda(x) dx.$$

Lifting the functions f and q^λ to the group G as K-right invariant functions, it follows that $f * \check{q}^\lambda = f$. Also since q_r is K-bi-invariant as a function on G, the same is true of q^λ.

Define $p^\lambda = q^\lambda * q^\lambda$. Then $f * p^\lambda = f$ and p^λ is continuous with compact support and $p^\lambda(o) > 0$. Hence, the corresponding measure $dp^\lambda(g) = p^\lambda(g) d_r g$ on G is well-behaved. □

Remarks. (1) If $\overline{N}$ is the convolution kernel on G/K defined by $\check{p}^{\lambda}$ (see § 14.5), then $Lf + \lambda f = 0$ implies $\overline{N}f = f$.

(2) If $\lambda = \lambda_0$ this lemma is an immediate consequence of the fact that, for any well-behaved probability p on G, the minimal $r_0(p)$-eigenfunctions coincide with the minimal λ_0-eigenfunctions of L. This fact, proved in Proposition 13.19, was pointed out in §14.7.

Proof of Theorem 14.2.

Observe that the function $\Phi(g) = \int P^{1/2}(g \cdot o, b) dm(b)$ satisfies $Lu + \lambda_0 u = 0$. The function h^D is a D-left invariant solution of the same equation. Lemma 14.21 implies that there is a K-bi-invariant, well-behaved, bounded measure $p = \check{p}^{\lambda_0}$ on G such that $Lu + \lambda_0 u = 0$ implies $u * p = u$.

Since $\Phi * p = \Phi$, it follows that $r_0(p) = 1$ (see § 14.7) and so by Theorem 14.4, there is a unique positive normalized solution of the convolution equation $u * p = u$ that is D-left invariant. Therefore, h^D is the unique normalized positive solution of the equation $Lu + \lambda_0 u = 0$. □

Given Theorem 14.2, one has the following simple corollary of Lemma 14.21 (already established, using other methods, by Theorem 9.18).

14.22. Corollary. *If $y_n = g_n \cdot o \in X \to D$ in $\partial\overline{S_0}$, then $\dfrac{G^{\lambda_0}(x, y_n)}{G^{\lambda_0}(o, y_n)} \to h^D$. Hence, the Martin compactification $X \cup \partial X(\lambda_0)$ is isomorphic to the maximal Satake–Furstenberg compactification $\overline{X}^{SF}$.*

Proof. It suffices to consider the case of $D = D^I$. Since each normalized Green function is $g_n K g_n^{-1}$-invariant, it follows from Lemma 9.7 that any limit function along a subsequence is necessarily D^I-invariant. As it is normalized, by Theorem 14.2, it equals h_I.

Hence, by Definition 3.27, the Martin compactification $X \cup \partial X(\lambda_0)$ dominates the compactification $\overline{S_0}$ and, hence, the compactification $\overline{X}^{SF}$ (an observation that was first made by Olshanetsky in [O1]) as, by Theorem 9.18, $\overline{S_0}$ and $\overline{X}^{SF}$ are isomorphic G-compactifications.

To show that the Martin compactification is isomorphic to $\overline{S_0}$, it suffices to show that $h^D = h^{D'}$ implies $D = D'$. Since the distal part of the stabilizer of h^D is D, this follows immediately from the identification in Remark 14.1(2) of the stabilizer of h^D as $gR^I g^{-1}$ if $D = gD^I g^{-1}$. □

Remark. It is important to emphasize that the proof of Corollary 14.22 is purely probabilistic in the sense that it uses only the structure theory of the semisimple Lie group and convolution operators. The proof follows directly from the approach used in Chapters XI and XIII to compute the minimal eigenfunctions.

THE MARTIN COMPACTIFICATION OF X DETERMINED BY A RANDOM WALK

Let H denote a locally compact, metrizable group and K a compact subgroup. Let p be a well-behaved K-bi-invariant measure on H. If $\bar{p}$ is its image on $X = H/K$ by $\pi : H \to H/K$, let $\bar{C}^r$ denote the convex cone of Radon measures μ on X such that $\mu * \bar{p} \leq r\mu$, where the convolution is defined in § 14.5. This is the same as the convex cone of images of measures ν in C^r, i.e., those measures ν on H such that $\nu * p \leq r\nu$. The image of the compact base C_1^r is a compact base $\bar{C}_1^r$ of $\bar{C}^r$: it consists of those μ such that $\tilde{\mu}(\check{p}) = 1$, i.e., for which $\tilde{\mu} \in C_1^r$ (where $\tilde{\mu}$ is defined in § 14.5).

If $V^r = \sum_{n=0}^{\infty} r^{-n}p^n = \delta_e + \sum_{n=1}^{\infty} r^{-n}p^n$ is the potential kernel of the random walk on H, its image $\bar{V}^r$ is the potential kernel of the random walk on X.

As in the case of the random walk on H, one defines the Martin kernel $\bar{K}_{\bar{g}}^r$ for the random walk on X, where $\bar{g} = g \cdot o$, by the formula

$$\bar{K}_{\bar{g}}^r = \frac{\delta_g * \bar{V}^r}{(\delta_g * V^r)(\check{p})}.$$

Since $\overline{\delta_g * \nu} = \delta_g * \bar{\nu}$ for any measure ν on H and $g \in H$, it follows that $\bar{K}_{\bar{g}}^r$ is the image of K_g^r by π as

$$K_g^r = \frac{\delta_g * V^r}{(\delta_g * V^r)(\check{p})}.$$

In exactly the same way as for the random walk on H, the map $g \cdot o \to \bar{K}_{\bar{g}}^r$ embeds X into $\bar{C}_1^r$.

14.23. Definition. The closure of the $\{\bar{K}_{\bar{g}}^r \mid g \in H\}$ in $\bar{C}_1^r$ relative to the weak topology is called the (p,r)-**Martin compactification of** X and is denoted by $\tilde{X}(p,r)$. Identifying X with $\{\bar{K}_{\bar{g}}^r \mid g \in H\}$, $\tilde{X}(p,r)$ will also be denoted as $X \cup \partial X(p,r)$. The boundary $\partial X(p,r)$ will be called the (p,r)-**Martin boundary of** X. (If (p,r) is determined by the context the reference to (p,r) will usually be omitted.)

As V^r has a density $V^r(x)$ outside e, the Martin kernel K_y^r has a density $K^r(x,y)$ on $H \backslash \{y\}$ (see Definition 11.11) given by

$$K^r(x,y) = \frac{\delta_y * V^r}{\delta_y * V^r(\check{p})}(x).$$

Since the density $V^r(x) = \sum_{n=1}^{\infty} r^{-n}p^n(x)$, the density of $\delta_g * V^r$ on $H\backslash\{y\}$ is $\sum_{n=1}^{\infty} r^{-n}p^n(g^{-1}x)$. It follows from this that $K^r(xk,y) = K^r(x,y)$ and $K^r(x,yk) = K^r(x,y)$ if $k \in K$. Consequently, $K^r(x,y)$ is K-right invariant as a function of x and can be viewed as a function on X. With this identification, it follows that

$$K_g^r = \frac{\delta_g}{\delta_g * V^r(\check{p})} + K^r(\cdot, g),$$

$$\bar{K}_{\bar{g}}^r = \frac{\delta_{\bar{g}}}{\delta_g * V^r(\check{p})} + K^r(\cdot, g).$$

Note that, since the image of $F(\pi(g))d\eta(g)$ is $F(x)d\bar{\eta}(x)$, it follows that $K^r(\cdot, g)$, when viewed as a function on X, is the density of $\bar{K}_{\bar{g}}^r$ on $X \backslash \{g \cdot o\}$ relative to the image $\bar{\eta}$ of right Haar measure η on H.

These observations imply the following result.

14.24. Theorem. *A sequence $(g \cdot o)_{n\geq 1}$ converges to ν in $\tilde{X}(p,r)$ if and only if the densities $K^r(\cdot, g_n)$, viewed as functions on X converge uniformly on compact sets to the density of the limit measure ν.*

A sequence $(g_n)_{n\geq 1} \subset H$ converges in $\tilde{H}(p,r)$ to μ, a K-right invariant measure that has a density h, with respect to η, if and only if the sequence $(g \cdot o)_{n\geq 1}$ converges in $\tilde{X}(p,r)$ to ν, where ν has a density $\bar{h}$ with respect to $\bar{\eta}$.

In addition, $h = \bar{h} \circ \pi$ and $\nu = \bar{\mu}$. As a result, $\partial H(p,r)$ can be identified with $\partial X(\bar{p}, r)$. In particular, the minimal points can be identified and $\mathcal{H}_1^r$ can be identified with $\bar{\mathcal{H}}_1^r$, the set of functions F on X such that $F * \bar{p} = F$ and $F(o) = 1$. More specifically, this correspondence is given by $F \to F \circ \pi = f$. As a result, this correspondence interlaces the continuous action of G on $\partial H(p,r)$ and on $\partial X(p,r)$.

In order to explain the relation between $\tilde{G}(p,r)$ and $\tilde{X}(p,r)$, it is useful to explain how to use a proper map to attach two locally compact metrizable spaces.

14.25. Definition. *Let X and Y be locally compact metrizable spaces and denote by ϕ a continuous and proper map of Y onto X. If $\overline{X} = X \cup \partial\overline{X}$ is a compactification of X let $\bar{\phi}$ denote the map of $Y \cup \partial\overline{X}$ (disjoint union) onto $\overline{X}$ defined by*

$$\bar{\phi}(y) = \phi(y) \quad \textit{if} \quad y \in Y$$
$$\bar{\phi}(y) = y \quad \textit{if} \quad y \in \partial\overline{X}.$$

Define the (compact) space $Y \cup_\phi \partial\overline{X}$ to be the set $Y \cup \partial\overline{X}$ endowed with the topology $\mathcal{T}$ for which the open sets are of the form $C = A \cup \bar{\phi}^{-1}(B)$ where $A \subset Y$ is open in Y and $B \subset \overline{X}$ is open in $\overline{X}$.

This definition is justified by the following lemma that is proved in Appendix A (see Lemma A.20).

14.26. Lemma. *The collection $\mathcal{T}$ of sets C is a compact topology on $Y \cup_\phi \partial\overline{X}$ with Y a dense open subspace. Hence, it is a compactification of Y.*

A sequence $(y_n) \subset Y \cup_\phi \partial\overline{X}$ converges to $z \in \partial\overline{X}$ if and only if $\bar{\phi}(y_n)$ converges to $\bar{\phi}(z) = z$.

Moreover, the topology on $Y \cup_\phi \partial\overline{X}$ is uniquely defined by these two conditions. Finally, if X, Y are H-spaces and ϕ is H-equivariant, then $Y \cup_\phi \partial\overline{X}$ is a H-space and $\bar{\phi}$ is H-equivariant.

This lemma and Theorem 14.24 imply the following result.

14.27. Corollary. *If ϕ is the projection $g \to g \cdot o$ from H to X, then one has*

$$\tilde{H}(p,r) = H \cup_\phi \partial X(p,r).$$

Specializing to the case of G and any well-behaved K-bi-invariant probability p on G, one gets the following result.

14.28. Theorem. *If $y_n \in X$ converges to $D \in \partial\overline{S}_0$, then $\bar{K}^{r_0}(x, y_n)$ converges to $\frac{1}{r_0}h^D(x)$ uniformly on compact subsets of X. Hence, $\tilde{X}(p,r)$ is G-isomorphic to $\overline{X}^{SF}$.*

It follows from this that, for any p, the Martin compactification $\tilde{G}(p, r_0)$ is the compactification obtained by attaching the boundary $\partial\overline{X}^{SF}$.

14.29. Corollary. *Denote by ϕ the canonical map from G to $X = G/K$. Then one has $\tilde{G}(p, r_0) = G \cup_\phi \partial\overline{X}^{SF}$.*

Remark. In the semisimple case, as pointed out in § 2.7, the Killing form B defines a positive definite quadratic form B_θ on $\mathfrak{g}$. This determines a left invariant Riemannian metric on G. The associated Laplace-Beltrami operator L_θ determines a family of Martin compactifications. It is known that, since the metric is K-bi-invariant, every positive eigenfunction on G is necessarily K-right invariant. The compactification $\tilde{G}(p, r_0)$ coincides with the Martin compactification $G \cup \partial G(\lambda_0)$ for L_θ at the bottom of the positive spectrum.

An application to parabolic subgroups

These methods not only may be used to prove Theorem 14.4 and its Brownian motion counterpart, Theorem 14.2, but they also can be used to do something new, namely to characterize the parabolic subgroups that contain the group N^I or the group D^I. One has the following geometrical result, the first part of which also follows from Borel-Tits [B9, Proposition 4.4b)], that is relevant to the study of the fibration of $X \cup \partial X(\lambda)$ over $\overline{X}^{SF}$.

14.30. Corollary. *Let Q denote a parabolic subgroup of G. Then $Q \supset N_I$ if and only if Q contains a minimal parabolic subgroup that is contained in P^I. Furthermore, if $D^I \subset Q$, then $P^I \subset Q$.*

Proof. Let gPg^{-1} be a minimal parabolic subgroup contained in $P^I \cap Q$. Then $N_I \subset P \subset g^{-1}P^Ig$. It follows, from Proposition 2.18, that $g^{-1}P^Ig = P^I$ and that $g \in P^I$. Hence $Q \supset gPg^{-1} \supset gN_Ig^{-1} = N_I$, where the last equality holds since, by Theorem 2.8, P^I is the normalizer in G of $\mathfrak{n}_I$.

Now assume that Q contains N_I. First, note that, in view of the Iwasawa decomposition, every minimal parabolic subgroup is of the form kPk^{-1}. Furthermore, the map $kPk^{-1} \to kM$ from the set of minimal parabolic subgroups to the Furstenberg boundary $\mathcal{F} = K/M$ is a bijection (obvious if one views $\mathcal{F}$ as G/P): the inverse map embeds $\mathcal{F}$ homeomorphically into $\overline{S_0}$.

Consider the set $E = \{b = kM \in \mathcal{F} \mid kPk^{-1} \subset Q\}$. This a closed set in $\mathcal{F}$ that is stable under Q and, hence, under N_I. By considering the twisted action, see eq. (10.2), of N_I on measures, the fixed point property of Tychonoff–Schauder (see Chapter XI) implies the existence of a measure ν on E such that, for some exponential c on N_I, one has $S_u\nu = c(u)\nu$ if $u \in N_I$.

Let h be the solution of $Lh + \lambda_0 h = 0$ with representing measure ν, i.e., $h(x) = \int P^{1/2}(x,b)d\nu(b)$. Then, $S_u h = c(u)h$, for all $u \in N_I$.

Assume, for the moment, that $c(u) = 1$ for all $u \in N_I$. Then the eigenfunction given by $h'(x) = \int_{K^I M} h(k \cdot x)d\eta(k)$, where η is normalized Haar measure on $K^I M$, is D^I-left invariant. Note that it is $K^I M$-left invariant. Since N_I is normal in D^I and $c(u) = 1$ if $u \in N_I$, it follows that $h(kuk^{-1} \cdot x) = h(x)$ if $u \in N_I$ and $k \in K^I M$. Consequently, if $k \in K^I M$ and $u \in N_I$, then $h(k \cdot x) = h(ku \cdot x)$. Hence, $h'(u \cdot x) = \int_{K^I M} h(ku \cdot x)d\eta(k) = h(x)$ if $u \in N_I$. As a result, the function h' is D^I-left invariant. It follows from Theorem 14.2 that $h' = h^I$, since $h'(e) = h(e) = 1$.

The representing measure of h' on $\mathcal{F}$ is $\int(\delta_k * \nu)dm_I(k)$. Hence, $m_I = \int(\delta_k * \nu)dm_I(k)$, which implies that the support of ν is contained in P^I/P, i.e., the orbit $K^I \cdot \dot{e} \subset K/M = \mathcal{F}$.

On the other hand, this support is contained in E. As a result, $E \cap P^I/P \neq \emptyset$. It follows that $Q \supset gPg^{-1}$ for some $g \in P^I$.

It remains, therefore, to show that the exponential c is trivial on N_I. Consider the cone C of positive λ_0-eigenfunctions and the set $\mathcal{E}$ of exponentials c' on N_I for which there exists an eigenfunction $h' \in C$ with $h'(u^{-1} \cdot x) = c'(u)h'(x)$ for all $u \in N_I$ and $x \in X$.

The group P^I acts by conjugacy on the normal subgroup N_I and also on the set $\mathcal{E}$, as it acts on C by left translation. The compactness of the base of the cone C implies the compactness of $\mathcal{E}$ and the relative compactness of the P^I-orbits of elements of $\mathcal{E}$ (see Lemma 11.30). The orbit of 1 is the unique relatively compact P^I-orbit in the vector space of exponentials on N_I. This shows that $\mathcal{E} = \{1\}$ and, hence, in particular that $c(u) = 1$, for all $u \in N_I$.

To prove the last statement, note that if $Q \supset D^I$, the set E is also $K^I M$-left invariant. Since $E \cap P^I/P \neq \emptyset$, this implies that $E \supset P^I/P$, and so $Q \supset P^I$. $\square$

CHAPTER XV

EXTENSION TO SEMISIMPLE ALGEBRAIC GROUPS DEFINED OVER A LOCAL FIELD

In this chapter it is indicated how to extend the main results about random walks from Chapters IX, XII, XIII and XIV to the case of a semisimple algebraic group defined over a non-archimedean local field $\mathbb{F}$. More explicitly, the bounded harmonic functions are described, the minimal eigenfunctions are determined, and the Martin compactification at the bottom of the positive spectrum is described. As the general outline of the proofs of the relevant results have already been given (see Chapters IX, XII, XIII, and XIV), the additional indications will be very brief. The aim here is to explain and state the results. Also, answers are given to some of the questions raised by Cartier in [C1]. This chapter should be considered as a first step toward the formulation and understanding of the general theory of Martin compactifications for random walks on semisimple groups defined over a local field. The results are very similar to those in the case of the real field. However, to carry out the proofs used in the real case, it is necessary to have structural information about the group of rational points over the field $\mathbb{F}$. Most of this information can be found in [M4, pp. 8–56].

Some notations and fundamental properties

Let $\mathbb{F}$ denote a locally compact and non-discrete field that is equipped with an **ultrametric absolute value** $x \to |x|$. If dx denotes a Haar measure on the additive group of $\mathbb{F}$, the absolute value of $y \in \mathbb{F}$ is the constant such that $d(yx) = |y|dx$. Let π denote a fixed uniformizer of $\mathbb{F}$. Then, every non-zero element of $\mathbb{F}$ can be written uniquely in the form $x = \pi^n x_0$ with $n \in \mathbb{Z}, |x_0| = 1$. Finally, let Ω denote a minimal algebraically closed field that contains $\mathbb{F}$.

Consider a semisimple algebraic group $G^a \subset G\ell(d, \Omega)$ that is simply connected, defined over $\mathbb{F}$, and $\mathbb{F}$-isotropic. Let $G(\mathbb{F})$ denote the subgroup of its $\mathbb{F}$-rational points. It is known that $G(\mathbb{F})$ is a unimodular, non-amenable, locally compact group that is compactly generated [M4, p. 53]. If $H^a \subset G^a \subset G\ell(d, \Omega)$ is an algebraic subgroup of G^a, let $H(\mathbb{F})$ denote the subgroup of its $\mathbb{F}$-rational points. Since it is of interest to consider subgroups H of $G(\mathbb{F})$, such as the subgroup $H(\mathbb{F})$ of H^a, the same terminology will be used for them as for algebraic subgroups of G^a.

Denote by $T^a \subset G^a(\mathbb{F})$ a maximal $\mathbb{F}$-split torus and by $P^a \subset G^a$ a minimal parabolic subgroup containing T^a. N^a will denote the unipotent radical of P^a and $\mathfrak{n}$ the Lie algebra of N^a. Let $N = N(\mathbb{F})$, $P = P(\mathbb{F})$, and $T = T(\mathbb{F})$. The set of positive roots of T^a in $\mathfrak{n}$ will be denoted by Σ^+ and $\Delta \subset \Sigma^+, \Delta = \{\alpha_1, \alpha_2, \cdots, \alpha_r\}$ will denote the subset of simple roots.

The normalizer of T^a acts on T^a as a finite automorphism group W, the Weyl group. W also acts on the group $C(T)^a$ of rational characters of T^a and $C(T)^a$ is isomorphic to $\mathbb{Z}^r$. Clearly, $\Delta \subset C(T)^a$. The set of characters that are linear combinations with positive coefficients of elements of Δ is denoted by $C^+(T)^a$.

It is known that the compact homogeneous space $\mathcal{F} = G(\mathbb{F})/P$ is the set of $\mathbb{F}$-rational points of a projective manifold [B8]. Denote by M the maximal compact subgroup of the centralizer Z of T in $G(\mathbb{F})$, i.e., $Z = MT$ [M4]. Let K denote a maximal compact subgroup that contains M and is special [B16]. It is known that the Weyl group W can be represented as a subgroup of K. This implies that the convolution algebra of K-bi-invariant functions is commutative.

In order to give the analogues of the Iwasawa and Cartan decompositions, let $\mathbb{F}_0 = \{\pi^n, n \in \mathbb{Z}\} \subset \mathbb{F}, \hat{\mathbb{F}} = \{\pi^n, n \in \mathbb{N}\} \subset \mathbb{F}$ and denote by T_0 the set $\{s \in T \mid \chi(s) \in \mathbb{F}_0 \text{ for all } \chi \in C(T)^a\}$, and by $\overline{T}^+$ the set $\{s \in T(\mathbb{F}) \mid \chi(s) \in \hat{\mathbb{F}} \text{ for all } \chi \in C^+(T)^a\}$. Then the following result holds [M4, Theorem 2.21, p. 51].

15.1. Theorem. *The group $G(\mathbb{F})$ has the following decompositions:*

$$G(\mathbb{F}) = KT_0N,$$
$$G(\mathbb{F}) = K\overline{T}^+K.$$

The components of $g \in G(\mathbb{F})$ in T_0 or $\overline{T}^+$ are uniquely defined.

As in the real field situation, $P = MTN$. It follows from Corollary 11.32 that P has the fixed line property. Hence, P is amenable.

Standard parabolic subgroups can be defined as in the real case: for any $I \subset \Delta$, denote by T_I the set $\{s \in T(\mathbb{F}) \mid \alpha(s) = 1, \text{ for all } \alpha \in I\}$ and by $Z(I)$ the centralizer of T_I in $G(\mathbb{F})$. The standard parabolic subgroup P^I is defined as $P^I = Z(I)N$. Then P^I is the semi-direct product of $Z(I)$ and its unipotent radical N_I. If $K^I = K \cap Z(I)$, K^I is a maximal compact subgroup of $Z(I)$, and from [B11, Lemma 4.7] it follows that $Z(I) = K^I\overline{T}^+K^I$. Every parabolic subgroup is conjugate to some P^I and, as shown in [B16], there exist a finite number of conjugacy classes of maximal compact subgroups. Furthermore, from [B9] one knows that $P \cap Z(I)$ is a minimal parabolic subgroup of the reductive group $Z(I)$ in the sense that $Z(I)/P \cap Z(I)$ is the set of $\mathbb{F}$-rational points of a projective manifold. It is easy to show that K^I is transitive on $\mathcal{F}^I = Z(I)/P \cap Z(I) \subset \mathcal{F}$ and that $P \cap Z(I)$ has the fixed line property.

Denote by m (respectively, m_I) the unique K-invariant (respectively, K^I-invariant) probability measure on $\mathcal{F}$ (respectively, $\mathcal{F}^I$). Define the Poisson kernel $P(g, b)$ to be $\frac{dg \cdot m}{dm}(b)$ for $g \in G(\mathbb{F})$ and $b \in \mathcal{F}$. If $x = g \cdot o \in X = G(\mathbb{F})/K$, define $P(x, b) = P(g \cdot o, b)$ to be $P(g, b)$. It is a basic fact, to be found in [M1], that the convolution algebra $C_c(G(\mathbb{F}), K)$ is commutative if K is a special maximal compact subgroup of a simply connected semisimple algebraic group defined over a local field $\mathbb{F}$. As in Chapter IX one considers the space of closed subgroups of $G(\mathbb{F})$ and the closure of the $G(\mathbb{F})$-orbit of K: $\overline{S}_0 = \overline{\{gKg^{-1}; g \in G(\mathbb{F})\}}$. Denote by D^I the closed subgroup $D^I = K^I N_I$ (in particular $D^\phi = MN$). Given a well-behaved and K-bi-invariant probability measure p on $G(\mathbb{F})$, consider the number $r(p) = \overline{\lim}_n p^n(e)^{1/n}$ defined in Definition 11.18. As indicated in Chapter XIV the convolution operator $f \to f * p$ acts as a transition operator on $X = G(\mathbb{F})/K$ and commutes with the $G(\mathbb{F})$-action. Hence, the equation $f * p = rf$ can be considered on $G(\mathbb{F})$ as well as on $G(\mathbb{F})/K$. Moreover, the potential kernel of this transition operator can be written as $y \to \delta_g * \bar{V}^r$ when $y = g \cdot o$ and $\bar{V}^r$ is the projection of V^r on $G(\mathbb{F})/K$. Hence, one may also consider the Martin kernel $\bar{K}_y^r = \dfrac{\delta_g * \bar{V}^r}{\delta_g * \bar{V}^r(\check{p})}$ and the Martin compactification of X relative to p.

Extension of the main results of Chapters XII, XIII, XIV

15.2. Theorem. (see Corollary 9.15). *The closure $\overline{S}_0$ of the orbit of K under $G(\mathbb{F})$ is the union of a finite number of orbits. Each of these orbits is the orbit of one D^I for some I. In particular, the orbit of $D^\phi = MN$ is the unique compact orbit and is isomorphic to $\mathcal{F} = G(\mathbb{F})/P$.*

The proof is based on the use of sequences $(a_n) \subset \overline{T}^+$, the study of the limiting behavior of the group $a_n K a_n^{-1}$ and of the measures $a_n \cdot m$. For related statements see [G2] and [G10].

15.3. Theorem. (See Theorem 12.13) *The stabilizer of the group D^I is $R^I = D^I T_I$. The stabilizer of the measure m_I on $\mathcal{F}$ is equal to R^I.*

This statement corresponds to Lemma 9.13.

15.4. Definition. The map from $X = G(\mathbb{F})/K$ into $\overline{S}_0$ defined by $g \to gKg^{-1}$ is an embedding, denoted by i. The closure $\overline{i(X)}$ of the image is defined to be the compactification $\overline{X}^{SF}$ of X.

15.5. Theorem. (See Theorem 12.10). *The bounded solutions of the equations $f * p = f$ are given by the Poisson formula*

$$f(g) = g \cdot m(\hat{f}) = \int P(g,b)\hat{f}(b)dm(b)$$

with a unique $\hat{f} \in L^\infty(\mathcal{F})$.

If $g_n \in G(\mathbb{F})$ is a sequence such that the measures $g_n \cdot m$ converge weakly to the Dirac measure δ_b ($b \in \mathcal{F}$), the sequence of functions on K given by $\phi_n(k) = f(kg_n)$ converges weakly in $L^2(K)$ to $\hat{f}(k \cdot b)$.

Proof. Since p is well-behaved and K-bi-invariant, and the conditions (1), (2) and (3) of Theorem 12.13 are satisfied, the result follows because of the choice of K and the fact that P has the fixed line property.

15.6. Theorem. (See Propositions 13.19 and 14.12). *The spectral radius in $L^2(G(\mathbb{F}))$ of the convolution operator $*p$ is equal to*

$$r_0(p) = r(p) = \int \Phi(g^{-1})dp(g) = r_0,$$

where $\Phi(g) = \int P^{1/2}(g,b)dm(b)$. The minimal $r_0(p)$-eigenfunctions are the functions $P^{1/2}(g,b)$ $(b \in \mathcal{F}$. In particular, there exists a unique K-invariant $r_0(p)$-eigenfunction. It is the function $\Phi(g) = \int P^{1/2}(g,b)dm(b)$.

Proof. The conditions (1), (2), (3), and (4) of Theorem 13.12) are valid here. Hence, the positive spherical functions are of the form $\phi^\ell(g) = \int \ell[a(g^{-1}k)]dk$ where $a(g)$ is the T_0 component in the Iwasawa decomposition of g and ℓ is an exponential on T_0 (see Theorem 13.16). The proof of Lemma 13.20 applies in this context and implies that $\Phi \leq \phi^\ell$. Hence, the Corollaries 13.18 and 13.19 are valid. □

In order to describe the minimal r-eigenfunctions for $r > r_0(p)$, one needs to introduce new notations, similar to those of Chapter XIII. Observe that every character on P^a or T^a is uniquely defined by its restriction to T_0. On the other hand, $T_0^* = T^*$ can be identified with $C(T)^a \otimes \mathbb{R}$: to every element $\chi \in C(T)^a$ one associates the exponential $s \to |\chi(s)|$ and $s \to \log|\chi(s)|$ is the corresponding element of the vector space $C(T)^a \otimes \mathbb{R}$. Furthermore, one can fix a scalar product $\langle u, v\rangle$ on $C(T)^a \otimes \mathbb{R}$ that is W-invariant and set, for every $u \in C(T)^a \otimes \mathbb{R}$,

$$\breve{u}(v) = \frac{2\langle u, v\rangle}{\langle u, u\rangle}.$$

Then $\breve{u}$ is a linear form on $C(T)^a \otimes \mathbb{R}$ and is identified with an element of the vector space $T_0 \otimes \mathbb{R}$. In particular, for every root α, $\breve{\alpha} \in T_0 \otimes \mathbb{R}$ is well defined. It is known that the subset $T_+^* = \{\ell \mid \log(\breve{\alpha}_i) \geq 0, 1 \leq i \leq r\}$ of T^* is a fundamental domain for the action of W on T_0^*. This fundamental domain does not depend on the choice of the scalar product on $C(T)^a \otimes \mathbb{R}$. Denote by δ the modular function of P (see Lemma 11.41) and recall that $P(g,b) = \delta[a(g^{-1}k)]$, where $k \in K$ corresponds to $b \in \mathcal{F} = K/M$.

Then Theorem 13.12 and the method of proof of Theorem 13.23 imply the following result.

15.7. Theorem. *The minimal solutions of the equation $f * p = rf$ are given by*

$$f(g) = P^{1/2}(g,b)\sigma_\ell(g^{-1},b)$$

with a unique $\ell \in T_+^*$ and $\hat{p}(\ell\delta^{1/2}) = r$.

Recall that G^a is said to be split over $\mathbb{F}$ if G^a has a maximal torus that is defined and split over $\mathbb{F}$, i.e., isomorphic over $\mathbb{F}$ to $(\Omega^\times)^r$.

Consider now the Martin compactification of $G(\mathbb{F})$ and X relative to p. One has the following theorem.

15.8. Theorem. *Suppose G^a is split over $\mathbb{F}$. Consider a sequence $(g_n) \subset G(\mathbb{F})$ such that $g_n K g_n^{-1}$ converges to $D^I \in \partial \overline{S_0}$. Then, the Martin kernels $K_{g_n}^{r_0}$ and $\overline{K_{g_n}^{r_0}}$ converges to $\frac{1}{r_0} h_I(x) = \frac{1}{r_0} \int P^{1/2}(x,b) dm_I(b)$. The Martin compactification $\tilde{X}(p, r_0)$ is equal to $\overline{X}^{SF}$. The Martin compactification $\tilde{G}(\mathbb{F})(p, r_0)$ is equal to $G(\mathbb{F}) \cup \partial \overline{X}^{SF}$ endowed with the natural topology defined by the surjection of $G(\mathbb{F})$ onto $G(\mathbb{F})/K$ (see Corollary 14.27).*

The proof is based on the following proposition that is an analogue of Proposition 14.20. It is proved in the same way using the descent method.

15.9. Proposition. *Assume G^a is split over $\mathbb{F}$. If p is a well-behaved K-bi-invariant probability measure on $G(\mathbb{F})$, the pair (D^I, p) has the ground state property. The corresponding function is $h_I(g) = \int P^{1/2}(g,b) dm_I(b)$.*

15.10. Lemma. *Assume that G^a is split over $\mathbb{F}$. Denote by $\pi_I(p)$ the probability measure on the group P^I associated with p, where $G(\mathbb{F}) = KP^I$, and by p_I its projection on $Z(I)$ modulo N_I. Then, the pairs $(D^I, \pi_I(p))$ and (K^I, p_I) have the ground state property.*

Proof. Consider the root subgroups $U_\alpha \subset G(\mathbb{F})$ associated with the roots $\alpha \in \Sigma$ [M4]. There exists an isomorphism θ_α of the additive group of $\mathbb{F}$ onto U_α that is defined over $\mathbb{F}$ and such that for every $s \in T$ and $x \in \mathbb{F}$, one has $\theta_\alpha[\alpha(s)x] = s\theta_\alpha(x)s^{-1}$. It is known that the subgroups U_α generate $G(\mathbb{F})$ and that there exists an anti-automorphism σ of $G(\mathbb{F})$ such that $\sigma(U_\alpha) = U_{-\alpha}, \sigma(K) = K, \sigma(s) = s$ for every $s \in T$ [M1]. The commutativity of $C_c(G(\mathbb{F}), K)$ follows immediately from this fact because $\sigma(KsK) = KsK$ and every element of $C_c(G(\mathbb{F}), K)$ is a linear combination of the indicator functions of the sets KsK. By definition, the centralizer of T_I is σ-invariant as is T_I. It follows that $\sigma(K^I) = K^I$. Since $Z(I) = K^I T K^I$, every K^I-invariant function on $Z(I)$ is also σ-invariant. Hence, the commutativity of $C_c(Z(I), K^I)$ follows. Now the conditions (2), (3), and (4) of Theorem 13.12 are valid for $Z(I)$ because $Z(I) = K^I[P \cap Z(I)]$, $P \cap Z(I)$ has the fixed line property, and the convolution algebra $C_c(Z(I), K^I)$ is commutative. Furthermore, p_I is clearly K^I invariant as is p and $\pi_I(p)$. From Corollary 13.18, applied to the group $Z(I)$, one gets the result for the pair (K^I, p_I). Because $D^I \cap Z(I) = K^I$, Proposition 14.13 implies the validity of the same property for the pair $(D^I, \pi_I(p))$.

Proof of Proposition 15.9.

It is a direct consequence of Proposition 14.16. Clearly, the function

$$h_I(g) = \int P^{1/2}(g,b) dm_I(b)$$

satisfies $h_I * p = r_0 h_I$. Furthermore, the D^I-invariance of m_I implies that h_I is D^I-invariant under left translations. On the other hand, Lemma 15.10 and Proposition 14.16 imply that (D^I, p) has the ground state property. The last assertion of the theorem follows. □

Proof of Theorem 15.8.

It is the same as that of Corollary 14.22 and Theorem 14.28, since, from Theorem 15.3, the stabilizer of the measure m_I in the twisted action of $G(\mathbb{F})$ (see Chapters IX and X) is equal to $R^I = D^I T_I$. In other words, the property $S_g h_I = h_I$ implies $g \in R^I$ and $g D^I g^{-1} = D^I$ (see Chapters IX and X). Hence, the cluster values of the Martin kernel $\bar{K}_{y_n}^{r_0}$, for $y_n \in G(\mathbb{F})/K$, are in one-to-one correspondence with the conjugates of the various subgroups D^I and these are the boundary points of the space $\overline{S}_0 = \overline{X}^{SF}$. Finally, $\tilde{X}(p, r_0) = \overline{X}^{SF}$. From Corollary 14.29 it follows that $\tilde{G}(\mathbb{F})(p, r_0) = G(\mathbb{F}) \cup \partial \overline{X}^{SF}$, with the natural topology induced by the canonical map from $G(\mathbb{F})$ to $G(\mathbb{F})/K$. □

15.11. Remarks. (1) In the case where G^a is split over $\mathbb{F}$, one can give a more explicit expression for the Poisson kernel $P(g, b) = \delta[a(g^{-1}k)]$, where $k \in K$ corresponds to $b \in K/M$ and δ is the modular function of $P = NTM$. Clearly, the exponential δ is determined by its values on T_0. Consequently, the definition of the Haar measure of N by the isomorphisms θ_α implies that

$$\delta^{-1}(s) = \prod_{\alpha \in \Sigma^+} |\alpha(s)|,$$

where $|\alpha(s)|$ is the absolute value of $\alpha(s)$ defined at the beginning of this chapter. Hence, the expression for $P(x, b)$ is completely similar to that used in the case of the real field in terms of the sum of the positive roots.

A similar formula is valid in the non-split case and is obtained in the same way. One has to take into account the multiplicity of the exponential $|\alpha(s)|$ when α is a root.

(2) As in the real case [F3], the scope of the Poisson formula can be extended to cover the case where p is not K-bi-invariant but has a density with support S such that $G(\mathbb{F}) = \cup_{n>0} S^n$.

APPENDIX A

COMPACTIFICATIONS OF FLATS

In this section it will be shown that the polyhedral and the Karpelevič compactifications of a flat exist. First one shows that there exists a compactification of V with the characteristic properties stated in Theorem 3.29. This implies that the polyhedral compactification of a flat exists.

A.1. Proposition. *For each open subset $O \subset V$, let $O^* = O \cup \{y^C(\infty) \mid$ for some $\varepsilon > 0$ and $c \in C, c + C + B^\perp(y^C;\varepsilon) \subset O\}$, where $B^\perp(y^C;\varepsilon)$ is the open ball in $V(C)^\perp$ centered at y^C with radius ε. Then the sets O^*, O open in V, form a base for a topology $\mathcal{T}$ on $V \cup \Delta^*(V)$. This topology has the following properties:*

1. *the subspace V is open and dense,*
2. *every fundamental sequence converges relative to $\mathcal{T}$,*
3. *two fundamental sequences have the same limit if and only if their formal limits agree, and*
4. *it is Hausdorff and metrizable and, hence, compact.*

Proof. If $y^C(\infty) \in O_1^* \cap O_2^*$, there exist $c_1, c_2 \in C$ and $\varepsilon > 0$ such that $c_i + C + B^\perp(y^C;\varepsilon) \subset O_i$. Hence, $c_1 + c_2 + C + B^\perp(y^C;\varepsilon) \subset O_1 \cap O_2$, and so $O_1^* \cap O_2^* \subset (O_1 \cap O_2)^*$. Since the opposite inclusion is obvious, $O_1^* \cap O_2^* = (O_1 \cap O_2)^*$, which implies that the sets O^* form a base for a topology. It is evident that V is an open dense subset since if O is open and relatively compact in V, then $O^* = O$.

The definition of the sets O^* implies that every fundamental sequence converges in this topology to its formal limit.

Since there are a finite number of cones and each cone has a countable dense set, it follows that a countable base for the topology of V and a countable number of the sets $\{c + C + B^\perp(y^C;\varepsilon)\}^*$ form a countable base for the topology $\mathcal{T}$. If $\mathcal{T}$ is Hausdorff then property (3) holds. To prove that it is compact and metrizable, it suffices, in view of Remark 3.26 and Urysohn's metrization theorem to show that $\mathcal{T}$ is Hausdorff and regular. The argument uses the following lemma.

A.2. Lemma. *Let C be a cone of the polyhedral cone decomposition. Then, if $c \in C$, $d(nc + C, \overline{C} \backslash C) \to \infty$ as $n \to \infty$.*

Proof. It is an immediate consequence of the fact that the cone $C \overset{\text{def}}{=} \{v \in V \mid \ell_i(v) > 0, 1 \le i \le m,\ \ell_i(v) = 0, m+1 \le i \le p\}$, where each ℓ_i is a linear functional on V. $\square$

Continuation of the proof of Proposition A.1.

First, to show that the topology is Hausdorff, it suffices to show that distinct points in $\Delta^*(V)$ have disjoint neighborhoods. Suppose that $y_1^{C_1}(\infty) \neq y_2^{C_2}(\infty)$. If $C_1 = C_2$, it is clear that they have disjoint neighborhoods. Assume that $C_1 \neq C_2$. If $\overline{C_1} \cap \overline{C_2} = \emptyset$ then, again, it is clear that the points have disjoint neighborhoods since the distance between the tail ends of the cones C_1 and C_2 that lie outside a ball of radius N centered at the origin goes to infinity. If $\overline{C}_1 \cap \overline{C}_2 \neq \emptyset$ then one of three possibilities hold: (i) $\overline{C}_1 \supset C_2$; (ii) $C_1 \subset \overline{C_2}$; or (iii) $\overline{C}_1 \cap C_2 = C_1 \cap \overline{C}_2 = \emptyset$. In the first case C_2 is a face of C_1. It follows from Lemma A.2 that for $c_i \in C_i$ and large n, $\{nc_1 + C_1 + B^\perp(y_1^{C_1}; \varepsilon)\} \cap \{c_2 + C_2 + B^\perp(y_2^{C_2}; \varepsilon)\} = \emptyset$.

It remains to consider case (iii), as the argument for (ii) is the same as for (i). In this case, the distance between $nc_1 + C_1$ and $nc_2 + C_2$ goes to infinity.

This follows from Lemma A.2: if $V(C_1) = V(C_2)$ it is clear since this distance is at least as large as the distance of $nc_1 + C_1$ to $\overline{C}\backslash C$; if $V(C_1) \neq V(C_2)$, project $V(C_1)$ onto $V(C_2)^\perp$ and apply Lemma A.2 to the image of the cone C_1; this shows that the distance of $nc_1 + C_1$ to $V(C_2)$ tends to infinity as n tends to infinity.

Since the distance between $nc_1 + C_1$ and $nc_2 + C_2$ goes to infinity, for any $\varepsilon > 0$, one has $nc_1 + C_1 + B^\perp(y_1^{C_1}; \varepsilon) \cap nc_2 + C_2 + B^\perp(y_2^{C_2}; \varepsilon) = \emptyset$ for large n.

Let $O(c, \varepsilon) = c + C + B^\perp(y^C; \varepsilon)$ with $c \in C$. Then, to prove regularity, it is enough to show that $\overline{O(2c, \varepsilon)^*} \subset O(c, 2\varepsilon)^*$. Since $2c + \overline{C}^V + \overline{B^\perp(y^C, \varepsilon)}^V = \overline{2c + C + B^\perp(y^C, \varepsilon)}^V$, where the superscript ^{V} indicates closure in V, it follows that $\overline{O(2c, \varepsilon)^*} \cap V \subset O(c, 2\varepsilon)^*$.

If $y_1^{C_1}(\infty) \in \overline{O(2c, \varepsilon)^*} \cap \Delta^*(V)$, and if $c_1 \in C_1$, then for all n, $\{nc_1 + C_1 + B^\perp(y_1^{C_1}; \frac{1}{n})\} \cap O(2c, \varepsilon) \neq \emptyset$ since $\{nc_1 + C_1 + B^\perp(y_1^C; \frac{1}{n})\}^* \cap O(2c, \varepsilon)^* = \left(\{nc_1 + C_1 + B^\perp(y_1^{C_1}; \frac{1}{n})\} \cap O(2c, \varepsilon)\right)^*$. This implies that $\overline{C_1}^V \cap \overline{C}^V \neq \emptyset$: if $z_n \in \{nc_1 + C_1 + B^\perp(y_1^{C_1}; \frac{1}{n})\} \cap O(2c, \varepsilon)$, then $\| z_n \| \to \infty$ and for some subsequence (z_{n_k}), the limiting direction is in $\overline{C_1}^V \cap \overline{C}^V$.

There are three cases to consider as above: (i) C is a face of C_1, i.e., $\overline{C}_1 \supset C$; (ii) C_1 is a face of C, i.e., $C_1 \subset \overline{C}$; or (iii) $\overline{C}_1 \cap C = C_1 \cap \overline{C} = \emptyset$.

The first case (i) cannot occur as, by Lemma A.2, the distance of $nc_1 + C_1$ from C tends to infinity. In case (ii) let $z_n = nc_1 + c_1(n) + u(n) = 2c + c(n) + v(n) \in \{nc_1 + C_1 + B^\perp(y_1^{C_1}; \frac{1}{n})\} \cap O(2c, \varepsilon)$ with $c_1(n) \in C_1, c(n) \in C$, $u(n) \in B^\perp(y_1^{C_1}; \frac{1}{n})$ and $v(n) \in B^\perp(y^C; \varepsilon)$. Now $u(n) = s(n) + v(n)$ with $s(n) \in V(C)$ — since $V(C_1)^\perp \supset V(C)^\perp$ — and so $z_{n,C} = nc_1 + c_1(n) + s(n) = 2c + c(n)$ and $\| v(n) - y \| < \varepsilon$, $\| s(n) - \overline{y}_1 \| < \frac{1}{n}$, where $\overline{y}_1$ is the projection of y_1 onto $V(C)$.

The cone $C \stackrel{\text{def}}{=} \{v \in V \mid \ell_i(v) > 0, 1 \leq i \leq m,\ \ell_i(v) = 0, m+1 \leq i \leq m+n\}$ where each ℓ_i is a linear functional on V and C_1 is determined by setting a certain number of the linear functionals $\ell_i, 1 \leq i \leq m$ equal to zero. Assume $\ell_{i_0}, 0 \leq i_0 \leq m$, vanishes on C_1. Then $\ell_{i_0}(z_{n,C}) = \ell_{i_0}(s(n) > \ell_{i_0}(2c)$.

Let D_n^1 denote the projection of $B^{\perp}(y_1, \frac{1}{n}) \subset V(C_1)^{\perp}$ onto $V(C)$ and let D_n^2 denote its projection onto $V(C)^{\perp}$. It will suffice to show that for large n, $nc_1 + C_1 + D_n^1 + D_n^2 \subset O(c, 2\varepsilon)$.

To see this observe that, since the minimum value δ of $\ell_i(c)$ for the linear functionals $\ell_i, 1 \leq i \leq m$, is positive and $s(n) \in D_n^1$, it follows that for large n, $s \in D_n^1$ implies $\ell_i(s) > \ell_i(c)$ if ℓ_i vanishes on C_1. Hence, $nc_1 + C_1 + D_n^1 \subset c + C$ if n is large. Since $\| v - y \| < 2\varepsilon$ if $v \in D_n^2$, it follows that $nc_1 + C_1 + D_n^1 + D_n^2 \subset O(c, 2\varepsilon)$.

Finally, case (iii) cannot occur since the distance of $nc_1 + C_1$ from $\overline{C}$ tends to infinity. □

The next result is used in the proof that every unbounded sequence in V has a fundamental sequence (see Proposition 3.25 (2)) and, as such, is part of the proof of Proposition A.1 (see Remark 3.26).

A.3. Lemma. *Let (y_n) be a sequence in V such that $\| y_n \| \to \infty$. Let $C \in \Pi$. Assume that $(\frac{1}{\|y_n\|})y_n \to L \in C$ and $y_n^C \to y^C$. Then*

(1) $y_{n,C} \in C$ eventually; and

(2) $y_{n,C} \overset{C}{\to} \infty$.

Proof. Since $y_n = y_{n,C} + y_n^C = \| y_{n,C} \| (\frac{1}{\|y_{n,C})\|})y_{n,C} + \| y_n^C \| (\frac{1}{\|y_n^C\|})y_n^C$, it follows that (i) $\| y_{n,C} \| / \| y_n \| \to 1$ and, hence, (ii) $(\frac{1}{\|y_{n,C}\|})y_{n,C} \to L$.

The cone $C = \{x \mid \ell_i(x) > 0, 1 \leq i \leq m, \text{ and } \ell_i(x) = 0, \ m+1 \leq i \leq p\}$, where $\ell_i, 1 \leq i \leq p$, is a set of linear functionals. It follows from (ii) that eventually $\ell_i(y_{n,C}) > 0, 1 \leq i \leq m$, which proves (1) since $V(C) = \{x \mid \ell_i(x) = 0, \ m+1 \leq i \leq m+n\}$.

If $c \in C$, then $c + C = \{x \mid \ell_i(x) > \ell_i(c), 1 \leq i \leq m, \text{ and } \ell_i(x) = 0, \ m+1 \leq i \leq m+n\}$. It follows from (i), that eventually, $y_{n,C} \in c + C$, i.e., (2) holds. □

It remains to show that the Karpelevič compactification exists (see Theorem 5.20). In order to prove that the topological space $\mathfrak{a} \cup \mathcal{K}(\mathfrak{a})$ (see Theorem 5.20) is compact, it suffices to prove two things: first, the topology has a countable base and, second, the topology is regular. Given this, the topology is metrizable and so, by Remark 3.26, it is compact in view of Lemma 5.13 (see also Proposition 3.25).

Let π be an ordered partition of Δ and denote by $\mathcal{K}(\pi)$ the set of all the formal limit points with associated partition π. It has a countable dense set since (i) the intersection of each of the cones $C_{I_1}, C_{I_{i+1}}^{I_i}$ with the unit sphere has a countable dense set and (ii) $\mathfrak{a}^I$ also has a countable dense set. The union of these countable sets as π varies, together with the union of their images under the Weyl group, is a countable dense subset E of $\mathcal{K}(\mathfrak{a})$. The union over the points $Ad(k_i)(D_0(\infty), D_1(\infty), \ldots, D_\ell(\infty), H^I) \in E$ of the countable neighborhood base $Ad(k_i)O(D_0, D_1, \ldots, D_\ell, H^I; 1-\frac{1}{n}, n, \frac{1}{n})^*$ together with a countable base for $\mathfrak{a}$ of open balls (say) gives a countable base for the topology on $\mathfrak{a} \cup \mathcal{K}(\mathfrak{a})$. This is clear if one recalls from the triangle inequality on the unit sphere that the angle subtended by D' and D is less than the sum of the angles subtended by D' and D'' and by D'' and D. Hence, $\mathrm{Cone}(D'; \sqrt{\frac{1}{2}(1+\eta)}) \subset \mathrm{Cone}(D;\eta)$ if $D' \in \mathrm{Cone}(D; \sqrt{\frac{1}{2}(1+\eta)})$.

Since $\mathcal{K}(\mathfrak{a}) = \cup_{k \in M'} \cup_\pi k \cdot \mathcal{K}(\pi)$ and the Weyl group is finite, it follows that $\mathfrak{a} \cup \mathcal{K}(\mathfrak{a})$ has a countable base.

The regularity follows from the next lemma.

A.4. Lemma. *Let* $(D_0, D_1, \ldots, D_\ell, H^I)(\infty) \in \mathcal{K}(\mathfrak{a})$ *and* $0 < \eta < 1, d > 0, \epsilon > 0$. *Then,* $\overline{O(D_0, D_1, \ldots, D_\ell, H^I; \eta, d, \epsilon)^*} \subset O(D_0, D_1, \ldots, D_\ell, H^I; \frac{1}{2}(1+\eta), \frac{1}{2}d, 2\epsilon)^*$.

Proof. The closures in $\mathfrak{a}$ of the sets $O(D_j; \eta, d)$ for $0 \leq j \leq \ell$ and of the set $N(H^I; \epsilon)$ are subsets of $O(D_j; \frac{1}{2}(1+\eta), \frac{1}{2}d)$ and of $N(H^I; 2\epsilon)$ respectively. Hence, the closure of $O(D_0, D_1, \ldots, D_\ell, H^I; \eta, d, \epsilon)$ in $\mathfrak{a}$ is a subset of $O(D_0, D_1, \ldots, D_\ell, H^I; \frac{1}{2}(1+\eta), \frac{1}{2}d, 2\epsilon)$. As a result, it suffices to verify the inclusion on $\mathcal{K}(\mathfrak{a})$.

Let $z_o = (D_0^o(\infty), D_1^o(\infty), \ldots, D_{\ell^o}^o(\infty), {H_o}^{I^o})$ and assume that $z_o \in \overline{O^*}$, where $O = O(D_0, D_1, \ldots, D_\ell, H^I; \eta, d, \epsilon)$. Then, by Lemma 5.13, there is a K-fundamental sequence $(Ad(k_i)H_n^o)_{n \geq 1}$ of points in O which converges in the topology of $\mathfrak{a} \cup \mathcal{K}(\mathfrak{a})$ to z_o. It follows from Lemma 5.19 that $(H_n^o)_{n \geq 1}$ converges to z_o and, since this sequence is in O, every root in I is bounded on (H_n^o). Hence, $I \subset I^o$. It is therefore enough to show that, for each i, $0 \leq i \leq \ell$, there is a basic neighborhood of z_o contained in $O(D_i; \frac{1}{2}(1+\eta), \frac{1}{2}d) = \mathfrak{a}_{I_i} \oplus \mathrm{Cone}(D_i; \frac{1}{2}(1+\eta), \frac{1}{2}d) \cap \mathfrak{a}^{I_i}$ and also that there is a basic neighborhood contained in $N(H^I; 2\epsilon) = \mathfrak{a}_I \oplus B^\perp(H^I; 2\epsilon)$, where the ordered partition π corresponding to the point $(D_0, D_1, \ldots, D_\ell, H^I)(\infty)$ is $J_0 \cup J_1 \cup \cdots \cup J_\ell \cup I$.

To determine a basic neighborhood of z_o contained in $O(D_i; \frac{1}{2}(1+\eta), \frac{1}{2}d)$, let H_n^i denote the projection of H_n^o onto $\mathfrak{a}^{I_i}$. Then $(H_n^i)_{n \geq 1}$ is a K-fundamental sequence on $\overline{\mathfrak{a}^{I_i,+}}$ with formal limit

$$z_o' = (D_0'(\infty), D_1'(\infty), \ldots D_m'(\infty), H^{I'}).$$

The corresponding ordered partition π' of I_i is determined by the ordered partition π^o corresponding to z_o. Since $I \subset I^o$ it follows that $I \subset I'$ and that the first set in π', i.e., J_0', is the intersection $J_k^o \cap I_i$ where k is the smallest integer j with $J_j^o \cap I_i \neq \emptyset$ and $\pi^o = J_0^o \cup J_1^o \cup \cdots \cup J_{\ell^o}^o \cup I^o$. Hence, the primary direction D_0' of the sequence $(H_n^i)_{n \geq 1}$ is the projection onto $\mathfrak{a}^{I_i}$ of D_k^o.

Since D_0' belongs to the closure of $\mathrm{Cone}(D_i; \eta, d) \cap \mathfrak{a}^{I_i}$, it follows that $\mathrm{Cone}(D_0'; \delta', d') \cap \mathfrak{a}^{I_i} \subset \mathrm{Cone}(D_i; \frac{1}{2}(1+\eta), \frac{1}{2}d) \cap \mathfrak{a}^{I_i}$, for some $d' > 0$ and δ', $0 < \delta' < 1$. It will now be shown that $\mathfrak{a}_{I_k^o} \oplus \mathrm{Cone}(D_k^o; \delta, d) \cap \mathfrak{a}^{I_k^o} = O(D_k^o; \delta, d)$ is a subset of $\mathfrak{a}_{I_i} \oplus \mathrm{Cone}(D_0'; \delta', d') \cap \mathfrak{a}^{I_i}$, for some $d > 0$ and

$\delta,\ 0<\delta<1$. This implies that $O(D_k^o;\delta,d)\subset O(D_i,\frac{1}{2}(1+\eta),\frac{1}{2}d)$ and, so, a basic neighborhood of z_o is a subset of $O(D_i,\frac{1}{2}(1+\eta),\frac{1}{2}d)$.

To complete the argument, first note that $I_k^o\supset I_i$ since $I_i\cap J_\ell=\emptyset$ if $0\le\ell<k$. If $J_k^o\subset I_i$ then $J_0'=J_k^o$ and $D_k^o=D_0'$. In this case, for any $d'>0$ and $\delta',\ 0<\delta'<1$, one has

$$\mathrm{Cone}(D_k^o;\delta',d')\cap\mathfrak{a}^{I_k^o}\subset\mathfrak{a}_{I_i}^{I_k^o}\oplus\mathrm{Cone}(D_0';\delta',d')\cap\mathfrak{a}^{I_i}.$$

If $J_k^o\subsetneq I_i$ then $J_k^o\supsetneq J_0'$ and $D_k^o=\lambda D_0'+L_k^o$, where $L_k^o\in\mathfrak{a}_{I_i}^{I_k^o}$ and $\lambda\in\mathbb{R}^+$.

Sublemma. *If $0<\delta'<1$ and $d'>0$, then there exist $d>0$ and $\delta,\ 0<\delta<1$, such that*

$$\mathrm{Cone}(D_k^o;\delta,d)\cap\mathfrak{a}^{I_k^o}\subset\mathfrak{a}_{I_i}^{I_k^o}\oplus\mathrm{Cone}(D_0';\delta',d')\cap\mathfrak{a}^{I_i}.$$

Proof of the sublemma. The intersection of $\mathfrak{a}_{I_i}^{I_k^o}\oplus\mathrm{Cone}(D_0';\delta',d')\cap\mathfrak{a}^{I_i}$ with the unit sphere in $\mathfrak{a}^{I_k^o}$ is open and contains D_k^o. As a result, for some $\delta,\ 0<\delta<1$, one has $\mathrm{Cone}(D_k^o;\delta)\cap\mathfrak{a}^{I_k^o}\subset\mathfrak{a}_{I_i}^{I_k^o}\oplus\mathrm{Cone}(D_0';\delta')\cap\mathfrak{a}^{I_i}$. From this it follows that, given $d'>0$, for sufficiently large d one has the result.

It follows then that $O(D_k^o;\delta,d)=\mathfrak{a}_{I_k^o}\oplus\mathrm{Cone}(D_k^o;\delta,d)\cap\mathfrak{a}^{I_k^o}$ is a subset of $\mathfrak{a}_{I_i}\oplus\mathrm{Cone}(D_0';\delta',d'))\cap\mathfrak{a}^{I_i}$ for sufficiently large d and δ close to 1, $0<\delta<1$.

For the final step of the proof of Lemma A.4, it suffices to show that $N(H^I;2\epsilon)=\mathfrak{a}_I\oplus B^\perp(H^I;2\epsilon)$ contains a basic neighborhood of $z_o=(D_0^o(\infty),D_1^o(\infty),\ldots,D_\ell^o(\infty),H_o{}^{I^o})$— recall that $B^\perp(H^I;2\epsilon)=\{H\mid\parallel H-H^I\parallel<2\epsilon\}\cap\mathfrak{a}^I$.

Since $I^o\supset I$, the projection of the points H_n^o onto $\mathfrak{a}^I$ gives a sequence in $B^\perp(H^I;\epsilon)$ with limit point the projection of $H_o{}^{I^o}$ onto $\mathfrak{a}^I$. It follows that $B^\perp(H_o{}^{I^o};\epsilon)\subset\mathfrak{a}_I^{I^o}\oplus B^\perp(H^I;2\epsilon)$, which implies that $N(H_o{}^{I^o};\epsilon)=\mathfrak{a}_{I^o}\oplus B^\perp(H_o{}^{I^o};\epsilon)\subset\mathfrak{a}_I\oplus B^\perp(H^I;2\epsilon)=N(H^I;2\epsilon)$. □

This completes the proof that $\mathfrak{a}\cup\mathcal{K}(\mathfrak{a})$ is a compactification of $\mathfrak{a}$ (Theorem 5.20). □

Compactifications of X

The two compactifications of a flat, polyhedral and Karpelevič, and the action of K determine two compactifications of X: the dual cell compactification and Karpelevič's compactification.

To determine a compactification of X, i.e., of $\mathfrak{p}$, that satisfies the conditions of Theorem 3.38 or equivalently, Theorem 3.39, one proceeds by investigating how the polyhedral compactifications of the various flats through the base point o fit together. While this discussion could be carried out on the space X itself, it is somewhat simpler to do it in $\mathfrak{p}$.

A.5. Theorem. *Define the equivalence relation* $\sim$ *on* $K \times \{\mathfrak{a} \cup \Delta^*(\mathfrak{a})\}$ *by setting* $(k_1, x_1) \sim (k_2, x_2)$ *if* $Ad(k_1)x_1 = Ad(k_2)x_2$ *(see Remark 3.33). Let* $[k, x]$ *denote the equivalence class of* (k, x)*. Then the quotient space* $Q \stackrel{\text{def}}{=} [K \times \{\mathfrak{a} \cup \Delta^*(\mathfrak{a})\}]/\sim$ *is a compact Hausdorff space.*

This compact space Q is in fact a compactification of $\mathfrak{p}$ if one identifies $X \in \mathfrak{p}$ with $[k, H]$ where $Ad(k)H = X$ (see (2.2)). One then shows that this compactification has the desired properties.

Assume Theorem A.5 and let $\pi : K \times \{\mathfrak{a} \cup \Delta^*(\mathfrak{a})\} \to Q$ denote the canonical map: $\pi(k, x) = [k, x]$. Then, for any $k \in K$, the image $\pi(\{k\} \times \{\mathfrak{a} \cup \Delta^*(\mathfrak{a})\})$ in Q of $\{k\} \times \{\mathfrak{a} \cup \Delta^*(\mathfrak{a})\}$, namely, the set $\{[k, x] \mid x \in \mathfrak{a} \cup \Delta^*(\mathfrak{a})\} = \{[k, H] \mid H \in \mathfrak{a}\} \cup \{[k, (C(\infty), y^C)] \mid (C(\infty), y^C) \in \Delta^*(\mathfrak{a})\}$ is homeomorphic to $Ad(k)\mathfrak{a} \cup \Delta^*(Ad(k)\mathfrak{a}) = Ad(k)\{\mathfrak{a} \cup \Delta^*(\mathfrak{a})\}$ as π restricted to $\{k\} \times \{\mathfrak{a} \cup \Delta^*(\mathfrak{a})\}$ is bijective, continuous and the image is compact. It will be convenient to let $Ad(k)\mathfrak{a}$ also denote $\{[k, H] \mid H \in \mathfrak{a}\}$ and similarly, to let $Ad(k)\mathfrak{a} \cup \Delta^* Ad(k)\mathfrak{a} = Ad(k)\{\mathfrak{a} \cup \Delta^*(\mathfrak{a})\}$ denote $\{[k, x] \mid x \in \mathfrak{a} \cup \Delta^*(\mathfrak{a})\}$.

It follows that, in the quotient space Q, the sets $\mathfrak{a}$ and $Ad(k)\mathfrak{a}$ are dense in $\mathfrak{a} \cup \Delta^*(\mathfrak{a})$ and $Ad(k)\{\mathfrak{a} \cup \Delta^*(\mathfrak{a})\}$ respectively. Still assuming Theorem A.5, one has the following result which shows that Q is a compactification of $\mathfrak{p}$ that satisfies the conditions of Theorem 3.38.

A.6. Proposition.

(1) *The map* $i : \mathfrak{p} \to Q$ *defined by* $i(X) = [k, H]$ *if* $X = Ad(k)H$ *embeds* $\mathfrak{p}$ *as a dense open subspace of this quotient space ;*
(2) (Q, i) *is a* K*-compactification of* $\mathfrak{p}$*;*
(3) *the closure in* Q *of the image of* $\mathfrak{a}$ *under* i *is isomorphic to the polyhedral compactification of* $\mathfrak{a}$*; and*
(4) *the closure of* $\mathfrak{a} \cap Ad(k)\mathfrak{a}$ *in* Q *is the intersection of the closures of* $\mathfrak{a}$ *and of* $Ad(k)\mathfrak{a}$*.*

Proof. From Proposition 2.2 it follows that every $X \in \mathfrak{p}$ is of the form $Ad(k)H, H \in \mathfrak{a}$. Obviously, $(k_1, H_1) \sim (k_2, H_2)$ if $X = Ad(k_1)H_1 = Ad(k_2)H_2$. Hence, the map $i : \mathfrak{p} \to Q$ is an injection. It is also a topological embedding as compactness implies that the unit ball in $\mathfrak{p}$ is homeomorphic to the image under π of $K \times$ {the unit ball in $\mathfrak{a}$}. The image of $\mathfrak{p}$ is open as it is the complement in Q of the image $\pi(K \times \Delta^*(\mathfrak{a}))$ of the compact set $K \times \Delta^*(\mathfrak{a})$. The image is dense because it is the image of $K \times \mathfrak{a}$ under the quotient map π and $K \times \mathfrak{a}$ is dense in $K \times \{\mathfrak{a} \cup \Delta^*(\mathfrak{a})\}$. Hence, (Q, i) is a compactification of $\mathfrak{p}$. It is obviously a K-compactification since K acts on $K \times \{\mathfrak{a} \cup \Delta^*(\mathfrak{a})\}$ by left translation of the first component k of (k, x).

Property (3) follows from the observations made above.

To verify (4), assume x is a boundary point in $\bar{\mathfrak{a}} \cap \overline{Ad(k)\mathfrak{a}}$. Then $x = Ad(k_i)(C_I(\infty), H^I) = Ad(kk_j)(C_{I'}(\infty), H'^{I'})$ with $k_i, k_j \in M'$ and $H^I \in \overline{\mathfrak{a}^{I,+}}, H'^{I'} \in \overline{\mathfrak{a}^{I',+}}$. It follows from Lemma 3.36 (3) that $I = I'$, $H^I = H'^I =$

H and $k_i^{-1}kk_j \in (K^I \cap aK^Ia^{-1})M$, where $a = \exp H, H = H^I = H'^I$, equivalently — $k_i^{-1}kk_j \in K^JM$ where $H \in C_J^I$.

It follows that if $c \in C_I$, the sequence $(Ad(k_j)(nc+H)) = (Ad(kk_\ell)(nc+H)) \subset \mathfrak{a} \cap Ad(k)\mathfrak{a}$ is C_I-fundamental with formal limit x and so $x \in \overline{\mathfrak{a}} \cap \overline{Ad(k)\mathfrak{a}}$. □

To complete the proof of Theorem 3.38, it remains to verify Theorem A.5. As is well known (see, for example, Bourbaki [B13, Ch. I, p. 97]), the point to check is that $\sim$ is a closed equivalence relation, i.e., if F is a closed set, then the set of points equivalent to some point of F is also closed. To prove this, extensive use is made of the various types of singularities of the points in X (see the comments following Definition 3.11) and in $\Delta^*(X)$.

Let F denote a closed subset of $K \times \Delta^*(\mathfrak{a})$ and denote by $F^{\sim}$ the set of all points that are equivalent to some point of F. This set will be called the **saturation of** F (with respect to $\sim$).

The proof that the saturation of a closed set is closed makes use of the various kinds of singularities to "stratify" the set and to identify the equivalence classes of a point with a subgroup of K, which depends upon the type of singularity. The relation between an equivalence class and a subgroup of K is stated as the following lemma.

A.7. Lemma. *Let $C = Ad(k_i)C_I$ and $C' = Ad(k_j)C_{I'}$ be two cones of the polyhedral decomposition of $\mathfrak{a}$.*

(1) *Let $x = Ad(k_i)H \in C$ and $x' = Ad(k_j)H' \in C'$. If $(k,x) \sim (k',x')$, then*
(a) $I = I'$,
(b) $H = H'$, and
(c) $k'k_j \in kk_iK^IM$.

In other words, if $H \in C_I$, then

$$[k, Ad(k_i)H] = \{(k', Ad(k_j)H) \mid k' = kk_i\ell k_j^{-1},\ \ell \in K^IM,\ k_j \in W\}.$$

(2) *Let $x = (C(\infty), y)$, and $x' = (C'(\infty), y')$. If $(k,x) \sim (k',x')$, then,*
(a) $I = I'$ and
(c) $k'k_j \in kk_iK^IM$.
In addition, if $y \in Ad(k_i)\overline{\mathfrak{a}^{I,+}}$ and $y' \in Ad(k_j)\overline{\mathfrak{a}^{I,+}}$ then
(d) there is a unique $H \in \overline{\mathfrak{a}^{I,+}}$ with $y = Ad(k_i)H, y' = Ad(k_j)H$ and $k'k_j \in kk_i(K^I \cap aK^Ia^{-1})M$, where $a = \exp H \in \overline{A^{I,+}}$.

Equivalently, $H \in C_J^I \subset \overline{\mathfrak{a}^{I,+}}$ implies that

$$[k,(Ad(k_i)C_I(\infty), Ad(k_i)H)] =$$
$$\{(k', (Ad(k_j)C_I(\infty), Ad(k_j)H) \mid k' = kk_i\ell k_j^{-1},\ \ell \in K^JM,\ k_j \in W\}.$$

Proof. In case (1), $Ad(kk_i)H = Ad(k'k_j)H'$. It follows from Propositions 2.18, 3.9 and 3.16 that $I = I'$ and $\ell = (kk_i)^{-1}(k'k_j) \in K^I M$. Since $kk_i \exp H \cdot o = k'k_j \exp H' \cdot o$, it follows from § 2.3 that $H = H'$.

In case (2), the result is essentially a restatement of Lemma 3.36 (3). □

Notice that the equivalence classes of a point (k, x) have essentially the same description regardless of whether $x \in \mathfrak{a}$ or $x \in \Delta^*(\mathfrak{a})$: in both cases a subgroup $K^I M$ is involved.

The key to proving that the saturation of a closed set is closed is the next result. First observe that $\mathfrak{a} \cup \Delta^*(\mathfrak{a})$ is the orbit under the Weyl group of $\overline{\mathfrak{a}^+} \cup \Delta^*(\overline{\mathfrak{a}^+})$, where $\Delta^*(\overline{\mathfrak{a}^+})$ consists of all the points $(C_I(\infty), H^I)$ with $I \subset \Delta$ and $H^I \in \overline{\mathfrak{a}^{I,+}}$. Recall from § 3.2 that the Weyl group W has a complete set of representatives $\{k_1, k_2, \ldots, k_{|W|}\}$ where $|W|$ is the cardinality of W

A.8. Proposition. *Let $F \subset K \times \{\mathfrak{a} \cup \Delta^*(\mathfrak{a})\}$ be a closed set and let $I \subset \Delta$. Denote by F^I the subset of F consisting of the points $(k, Ad(k_i)y)$ where $y \in \overline{\mathfrak{a}^+} \cup \Delta^*(\overline{\mathfrak{a}^+})$ is such that $Ad(\ell)y = y$ for all $\ell \in K^I M$.*

Then F^I is closed and its saturation $F^{I\sim}$ contains the image of F^I under the map $(k_j, \ell, (k, Ad(k_i)y)) \to (kk_i\ell k_j^{-1}, Ad(k_j)y)$ of $W \times K^I M \times F^I \to K \times \{\mathfrak{a} \cup \Delta^*(\mathfrak{a})\}$.

Proof. Let $(k_n, x_n) = (k_n, Ad(k_{i_n}) \in F^I, n \geq 1$ and assume that $(k_n, x_n) \to (k, x)$, i.e., $k_n \to k$ and $x_n \to x$. Since there are only a finite number of elements in the Weyl group, one can assume, by passing to a subsequence, that $i_n = i$, for all $n \geq 1$. If $x_n = Ad(k_i)y_n$, it follows immediately that $(k, x) \in F^I$ as $y_n \to y$ and $x = Ad(k_i)y$.

If $(k, x) = (k, Ad(k_i)y) \in F^I$ and $k' = kk_i\ell k_j^{-1}$ with $\ell \in K^I M$, then $(k, x) \sim (k', x')$ where $x' = Ad(k_j)y$. As a result,

$$[k, Ad(k_i)y] \supset \{(kk_i\ell k_j^{-1}, Ad(k_j)y) \mid \ell \in K^I M, k_j \in W\}. \quad \square$$

To saturate a closed set F, one proceeds by degrees of singularity. First, let F_o be the compact image of $W \times M \times F$ under the map

$$(k_j, \ell, (k, Ad(k_i)y)) \to (kk_i\ell k_j^{-1}, Ad(k_j)y).$$

By Proposition A.8, the set $F_o \subset F^\sim$. Let $\alpha \in \Delta$ and let $I = \{\alpha\}$. Let $F' = F \cup F_o$. Then F' contains the equivalence classes $[k, x]$ for all the elements $(k, x) = (k, Ad(k_i)y) \in F$ such that $[k, x] = \{(kk_i\ell k_j^{-1} \mid \ell \in M, k_j \in W\}$: in some sense they are the subset of "regular " elements of F. While the equivalence classes of the "non-regular" points do not necessarily lie in F' at least part of each equivalence class is to be found in F' as $F' \subset F^\sim$

The "non-regular" elements lie in one of a finite number of closed subsets of F': to each simple root α corresponds the set $F^{\{\alpha\}} = \{(k, Ad(k_i)y) \in F' \mid Ad(\ell)y = y \text{ for all } \ell \in K^{\{\alpha\}}M\}$. The saturation of F is the union of the saturations of each of the closed sets $F^{\{\alpha\}}$ together with the subset of "regular" points in F', i.e., the set $F' \backslash (\cup_{\alpha\in\Delta} F^{\{\alpha\}})$.

It suffices therefore to show that each of the sets $F^{\{\alpha\}}$ has a closed saturation. Now by Proposition A.8, the saturation of $F^{\{\alpha\}}$ contains the image $F_o^{\{\alpha\}}$ of $W \times K^{\{\alpha\}}M \times F^{\{\alpha\}}$ under the map $(k_j, \ell, (k, Ad(k_i)y)) \to (kk_i\ell k_j^{-1}, Ad(k_j)y)$. Here the set of " regular" points in the union of $F^{\{\alpha\}'} = F^{\{\alpha\}} \cup F_o^{\{\alpha\}}$ is the union of the equivalence classes of the points $(k, Ad(k_i)y) \in F^{\{\alpha\}}$ such that $[k, Ad(k_i)y] = \{(kk_i\ell k_j^{-1}, Ad(k_j)y) \mid \ell \in K^{\{\alpha\}}M, k_j \in W\}$. The remaining points have a "singularity" determined by one of the subgroups $K^{\{\alpha,\beta\}}M$, where $\beta \neq \alpha$ is a second simple root. Again by Proposition A.8, the set $F^{\{\alpha,\beta\}} = \{(k, Ad(k_i)y) \in F^{\{\alpha\}'} \mid Ad(\ell)y = y \text{ for all } \ell \in K^{\{\alpha,\beta\}}M\}$ is a closed set and the saturation of $F^{\{\alpha\}}$ is the union of the set of "regular " points in $F^{\{\alpha\}}$ and the saturations of the finite collection of sets $F^{\{\alpha,\beta\}}, \beta \in \Delta\backslash\{\alpha\}$. It is clear that in this way, in a finite number of steps, one may show that the saturation of any closed set is closed.

This completes the proof of Theorem A.5.

THE PROOF OF THEOREM 5.31

To determine a compactification of X that satisfies the conditions of Theorem 5.31, in other words, to determine the Karpelevič compactification of X, one may proceed, as in the case of the dual cell compactification, by investigating how the Karpelevič compactifications of the various flats through the base point o fit together. This is the idea behind the following theorem.

While this discussion could be carried out on the space X itself, it is somewhat simpler (as in the other case) to do it in $\mathfrak{p}$. Note that as before, the Karpelevič compactification of $Ad(k)\mathfrak{a}$ can be viewed as the image of the Karpelevič compactification $\mathfrak{a} \cup \mathcal{K}(\mathfrak{a})$ under $Ad(k)$. In other words, if $z_0 = Ad(k_i)(D_0(\infty), D_1(\infty), \ldots, D_\ell(\infty), H^I) \in \mathcal{K}(\mathfrak{a})$ define $Ad(k)z_0$ to be $(Ad(kk_i)D_0(\infty), Ad(kk_i)D_1(\infty), \ldots, Ad(kk_i)(D_\ell(\infty), Ad(kk_i)H^I)$, a point of $\mathcal{K}(Ad(k)\mathfrak{a})$. In other words, $Ad(k)\{\mathfrak{a} \cup \mathcal{K}(\mathfrak{a})\} = Ad(k)\mathfrak{a} \cup \mathcal{K}(Ad(k)\mathfrak{a})$.

A.9. Theorem. *Define the equivalence relation* $\sim_{\mathcal{K}}$ *on* $K \times \{\mathfrak{a} \cup \mathcal{K}(\mathfrak{a})\}$ *by setting* $(k_1, x_1) \sim_{\mathcal{K}} (k_2, x_2)$ *if* $Ad(k_1)x_1 = Ad(k_2)x_2$ *where the action of* K *on* $\mathcal{K}(\mathfrak{a})$ *is defined above. Let* $[k, x]$ *denote the equivalence class of* (k, x). *Then the quotient space* $Q_{\mathcal{K}} \stackrel{\text{def}}{=} [K \times \{\mathfrak{a} \cup \Delta^*(\mathfrak{a})\}]/\sim_{\mathcal{K}}$ *is a compact Hausdorff space.*

As before, one can see that this compact space $Q_{\mathcal{K}}$ is in fact a compactification of $\mathfrak{p}$ if one identifies $X \in \mathfrak{p}$ with $[k, H]$ where $Ad(k)H = X$. One then shows that this compactification has the properties of the Karpelevič compactification. In particular, one has the following result that shows that $Q_{\mathcal{K}}$ is a compactification of $\mathfrak{p}$ in which the closure of a flat through o is the Karpelevič compactification of the flat.

A.10. Proposition.

(1) *The map* $i : \mathfrak{p} \to Q_{\mathcal{K}}$ *defined by* $i(X) = [k, H]$ *if* $X = Ad(k)H$ *embeds* $\mathfrak{p}$ *as a dense open subspace of this quotient space;*
(2) $(Q_{\mathcal{K}}, i)$ *is a* K*-compactification of* $\mathfrak{p}$*;*
(3) *the closure in* Q *of the image of* $\mathfrak{a}$ *under* i *is isomorphic to the Karpelevič compactification of* $\mathfrak{a}$*; and*
(4) *the closure of* $\mathfrak{a} \cap Ad(k)\mathfrak{a}$ *in* $Q_{\mathcal{K}}$ *is the intersection of the closures of* $\mathfrak{a}$ *and of* $Ad(k)\mathfrak{a}$*.*

Proof. The proofs of (1), (2) and (3) are the same as in the case of Proposition A.6 for the corresponding statements.

Assume that $z \in \mathcal{K}(\mathfrak{a}) \cap \mathcal{K}(Ad(k)\mathfrak{a})$. Then

$$z = Ad(k_i)(D_0(\infty), D_1(\infty), \ldots, D_\ell(\infty), H^I)$$

and

$$z = Ad(kk_j)(D_0(\infty), D_1(\infty), \ldots, D_\ell(\infty), H^I)$$

with $k_i, k_j \in W$. As a result, $Ad(k_i)C_I = Ad(kk_j)C_I \subset \mathfrak{a} \cap Ad(k)\mathfrak{a}$ and $Ad(k_i)H^I = Ad(kk_j)H^I \in \mathfrak{a} \cap Ad(k)\mathfrak{a}$.

The sequence $H_n = \sum_{i=0}^{\ell} H_n^i + H^I$ defined in the proof of Proposition 5.15 has the property that $Ad(k_i)H_n$ is K-fundamental in $\mathfrak{a}$ and in $Ad(k)\mathfrak{a}$ and converges to z. This proves (4). □

To prove that $Q_{\mathcal{K}}$ is compact, it is necessary, as before, to show that the saturation of a closed set is closed. This fact is based on the following analogue of Lemma A.7 (2) that describes the equivalence classes of points $(k, z), z \in \mathcal{K}(\mathfrak{a})$ in terms of the subgroups $K^I M$.

A.11. Lemma. *Let* $z, z' \in \mathcal{K}(\mathfrak{a})$*. If*

$$z = Ad(k_i)(D_0(\infty), D_1(\infty), \ldots, D_\ell(\infty), H^I),$$

$$z' = Ad(k_j)(D_0'(\infty), D_1'(\infty), \ldots, D_\ell'(\infty), H'^{I'}),$$

then $(k, z) \sim_{\mathcal{K}} (k'z')$ *implies that*

(1) $\ell = \ell'$*;*
(2) $I = I'$*;*
(3) $D_i = D_i', 0 \le i \le \ell$*; and* $H^I = H'^I$*.*

Furthermore, if $H^I \in C_J^I$*, then* $(kk_i)^{-1}k'k_j \in K^J M$*. Hence,*

$$[k, z] = \{(k', z') \mid k' = kk_i \ell k_j^{-1}, \ell \in K^J M\}.$$

Proof. (1) and (3) follow by definition. As in the proof of Proposition 5.15, for a suitable decreasing sequence of integers n_i, $L = n_0 D_0 + n_1 D_1 +$

$\dots n_\ell D_\ell \in C_I$ and $L' = n_0 D_0' + n_1 D_1' + \dots n_\ell D_\ell' \in C_{I'}$. Hence, it follows that $Ad(k)C_I = C_{I'}$ and so $I = I'$. Finally, § 2.3 implies that $H^I = H'^I$.

If $H^I \in C_J^I$ then, for large n, $nL + H^I \in C_J$ and so $Ad(kk_i)(nL + H^I) = Ad(k'k_j)(nL + H^I)$. This implies that $(kk_i)^{-1}k'k_j \in K^J M$. The last statement is now obvious. □

Combining this result with Lemma A.7(1), one sees that the equivalence classes $[k, x]$ relative to $\sim_{\mathcal{K}}$ with $x \in \mathfrak{a} \cup \mathcal{K}(\mathfrak{a})$ are determined by the subgroups $K^I M$ in exactly the same way as for the equivalence relation $\sim$ considered in Theorem A.5.

As a result, the analogue of Proposition A.8 holds, where $\mathcal{K}(\overline{\mathfrak{a}^+})$ is the set of points $z \in \mathcal{K}(\mathfrak{a})$ of the form $z = (D_0(\infty), D_1(\infty), \dots, D_\ell(\infty), H^I)$.

A.12. Proposition. *Let $F \subset K \times \{\mathfrak{a} \cup \mathcal{K}(\mathfrak{a})\}$ be a closed set and let $I \subset \Delta$. Denote by F^I the subset of F consisting of the points $(k, Ad(k_i)y)$ where $y \in \overline{\mathfrak{a}^+} \cup \mathcal{K}(\overline{\mathfrak{a}^+})$ is such that $Ad(\ell)y = y$ for all $\ell \in K^I M$.*

Then F^I is closed and its saturation $F^{I\,\sim_{\mathcal{K}}}$ relative to $\sim_{\mathcal{K}}$ contains the image of F^I under the map $(k_j, \ell, (k, Ad(k_i)y)) \to (kk_i\ell k_j^{-1}, Ad(k_j)y)$ of $W \times K^I M \times F^I \to K \times \{\mathfrak{a} \cup \mathcal{K}(\mathfrak{a})\}$.

As a result, by the argument used in the proof of Theorem A.5, one has the following result.

A.13. Corollary. *The equivalence relation $\sim_{\mathcal{K}}$ on $K \times \{\mathfrak{a} \cup \mathcal{K}(\mathfrak{a})\}$ is closed.*

This completes the proof of Theorem 5.31.

The topology of $\Delta^*(V)$

In Proposition A.1 a compact metrizable topology $\mathcal{T}$ is defined on $V \cup \Delta^*(V)$ with basic open sets the sets $O^* = O \cup \{y^C(\infty) \mid$ for some $\varepsilon > 0$ and $c \in C, c + C + B^\perp(y^C; \varepsilon) \subset O\}$, where O is an open subset of V. This induces a compact metrizable topology on the closed subset $\Delta^*(V)$.

It was pointed out following Definition 3.23 that the dual cell complex $\Delta^*(V)$ can be assembled inductively by adding cells to the skeletons. It is not hard to see (and this will be indicated below) that another compact metrizable topology $\mathcal{T}'$ on $\Delta^*(V)$ is defined by this procedure. To verify that it coincides with the topology given by Proposition A.1, it is enough to show that each set $O^* \cap \Delta^*(V)$ belongs to $\mathcal{T}'$. In Taylor [T4], where use was made of filters to define the ideal boundary points, this followed automatically, as the proof of Proposition A.1 in that paper (see Theorem 3.2 of [T4]) used the inductive procedure directly.

First, some comments about the definition of $\mathcal{T}'$. Let E_1 and E_2 be two compact metric spaces and let $\phi : A \to E_2$ be a continuous injection of a closed set $A \subset E_1$ into E_2. Let $\phi(A)$ denote the image of A under ϕ. Denote by $E_1 \cup_\phi E_2$ the space obtained by attaching E_2 to E_1 along ϕ: it is defined to be the quotient of the disjoint sum $E_1 + E_2$ under the equivalence relation for which the only non-trivial equivalence classes are the sets $\{a, \phi(a)\}, a \in A$; this is a closed equivalence relation since if $C \subset E_1 + E_2$ is closed and $C_i = C \cap E_i$, the saturation of C is $C_1 \cup \phi^{-1}(\phi(A) \cap C_2) + \phi(A \cap C_1) \cup C_2$. Consequently, $E_1 \cup_\phi E_2$ is a compact space. It is also metrizable as it is not hard to see that it has a countable base (alternatively, one may construct a metric explicitly from a metric on each of the E_i).

This shows that the inductive construction of $\Delta^*(V)$, outlined following Definition 3.23, defines a compact metrizable topology $\mathcal{T}'$ on $\Delta^*(V)$.

It is convenient to view A as a subset of $E_1 \cup_\phi E_2$, in other words to identify a with $\{a, \phi(a)\}$ for all $a \in A$. Let U_i be open subsets of E_i such that $\phi(U_1 \cap A) = U_2 \cap \phi(A)$. Then, $U_1 + U_2$ is saturated and its image U in $E_1 \cup_\phi E_2$ is an open set containing a. Furthermore, every open set in $E_1 \cup_\phi E_2$ that contains $a \in A$ contains a set of this type.

A.14. Proposition. *If O is open in V, then $O^* \cap \Delta^*(V)$ belongs to $\mathcal{T}'$.*

Proof. The argument is by induction. One shows that if the intersection of O^* with the $(k-1)$-skeleton plus several k-dimensional cells is open in that subspace, then it remains open after adding one more k-dimensional cell.

For each k list the cones of co-dimension k that belong to the polyhedral cone decomposition. Let $\Delta^*_{(k-1),i}(V)$ denote the $(k-1)$ skeleton together with the first i of the k-dimensional cells attached, and let $\Delta^*_{(k-1),0}(V)$ denote the $(k-1)$-skeleton. Let $O^*_{(k-1),i} = O^* \cap \Delta^*_{(k-1),i}(V)$. Then, $O^*_{0,0}$ is a subset of the finite 0-skeleton $\Delta^*_{0,0}(V)$ and, hence, is an open subset of this skeleton. Assume that $O^*_{(k-1),i}$ is open in $\Delta^*_{(k-1),i}(V)$ and let C_0 be the next cone of co-dimension k in the list of such cones. The map ϕ that attaches $V(C_0)^\perp \cup \Delta^*(V(C_0)^\perp)$ is defined on the subset A of the $(k-1)$-skeleton $\Delta^*_{(k-1),0}(V)$ that corresponds to the polyhedral boundary $\Delta^*(V(C_0)^\perp)$ of $V(C_0)^\perp$: $y^C(\infty) \in A$ if C_0 is a face of C and $\phi(y^C(\infty)) = y^{C'}(\infty)$, where C' is the projection of C onto $V(C_0)^\perp$. Note that $V(C)^\perp = V(C')^\perp \cap V(C_0)^\perp$ is the subspace of $V(C_0)^\perp$ orthogonal to $V(C')$ and so the points in V corresponding to $y^C(\infty)$ and $y^{C'}(\infty)$ are the same.

Now one proves that the set $O^*_{(k-1),i+1}$ is open in $\Delta^*_{(k-1),i+1}(V)$. There is an open subset W of $V(C_0)^\perp \cup \Delta^*(V(C_0)^\perp)$ such that $\phi(O^*_{(k-1),i} \cap A) = W \cap \phi(A) = W \cap \Delta^*(V(C_0)^\perp)$. By the remarks preceding this proposition, the image of $O^*_{(k-1),i} + W$ under the quotient map is open in $\Delta^*_{(k-1),i+1}(V)$. As a result, it is enough to show that one may take W to be the subset of $V(C_0)^\perp \cup \Delta^*(V(C_0)^\perp$ that corresponds to $O^*_{(k-1),i+1}$.

This set W equals

$$\phi(O^*_{(k-1),i} \cap A) \cup \{y \in V(C_0)^\perp \mid y^{C_0}(\infty) \in O^*\}$$

$$= \{y^{C'}(\infty) \mid y^C(\infty) \in O^*\} \cup \{y \in V(C_0)^\perp \mid y^{C_0}(\infty) \in O^*\},$$

where C_0 is a face of C and C' is the projection onto $V(C_0)^\perp$ of C.

Since $\{y \in V(C_0)^\perp \mid y^{C_0}(\infty) \in O^*\}$ is an open subset of $V(C_0)^\perp$, the set W is open if $y^C(\infty) \in O^* \cap A$ implies that there is an open subset $U \subset W$ of $V(C_0)^\perp \cup \Delta^*(V(C_0)^\perp)$ containing $y^{C'}(\infty)$. To verify this one uses the following lemma.

A.15. Lemma. *Let $C_0 \subset \overline{C}$, where C_0 and C are two cones of the polyhedral cone decomposition of V.*

Denote by C' the projection of C onto $V(C_0)^\perp$. If $u \in 2c + C$, $c \in C$, let u_0 denote its projection onto $V(C_o)^\perp$. Denote by D the open ball in $V(C) \cap V(C_0)^\perp$ about u_0 of radius $\delta = \delta(c)$. Then, if $c_0 \in C_0$, there is an integer n such that for $n \geq n(c, c_0)$,

$$nc_0 + C_0 + D \subset c + C.$$

Continuation of the proof of Proposition A.14.

Let $y^C(\infty) \in O^* \cap A$. Then $\phi(y^C(\infty)) = y^{C'}(\infty) \in \phi(O^* \cap A)$. Since this is an open set in $\Delta^*(V(C_0)^\perp)$, there exist $c' \in C'$ and $\varepsilon > 0$ such that $\{c' + C' + B^\perp(y;\varepsilon)\}^* \cap \Delta^*(V(C_0)^\perp) \subset \phi(O^* \cap A)$, where $y \in V$ is the common point corresponding to $y^C(\infty)$ and $y^{C'}(\infty)$. If $c \in C$ projects onto c', then $\{c + C + B^\perp(y;\varepsilon)\}$ projects onto $\{c' + C' + B^\perp(y;\varepsilon)\}$ —recall that $V(C)^\perp = V(C')^\perp \cap V(C_0)^\perp$ is the subspace of $V(C_0)^\perp$ orthogonal to $V(C')$. Since $y^C(\infty) \in O^*$, it follows that, for some n, $\{nc + C + B^\perp(y;\varepsilon/n) \subset O$. Let $T \overset{\text{def}}{=} \{2nc + C + B^\perp(y;\varepsilon/2n)$. If $T' = \{2nc' + C' + B^\perp(y;\varepsilon/2n)\}$, then $(T')^* \cap \Delta^*(V(C_0)^\perp) \subset \phi(O^* \cap A)$. Therefore, the open set $(T')^*$, which contains $y^{C'}(\infty)$, is a subset of W provided that, for every $y_0 \in T'$, one has $y_0^{C_0}(\infty) \in O^*$.

Now $y_0 = u_o + v$ is the projection onto $V(C_0)$ of a point $z \in T$, where $z = u + v$, $u \in 2nc + C$, $v \in B^\perp(y;\varepsilon)$. It follows from Lemma A.15 that, if $c_0 \in C_0$ and D is the open ball in $V(C) \cap V(C_0)^\perp$ about u_0 of radius $\delta < \delta(nc)$ then, for $m \geq m(nc, c_0)$, one has $mc_0 + C_0 + D \subset nc + C$. Also, $D + B^\perp(u;\varepsilon) \supset B^\perp(y_0;\eta)$, where the first ball is in $V(C)^\perp$, the second in $V(C_0)^\perp$, and $\eta < \min\{\delta, \varepsilon - \| v - y \|\}$. Hence, $nc_0 + C_0 + B^\perp(y_o;\eta) \subset O$ and so $y_0^{C_o} \in O^*$. □

Proof of Lemma A.15.

Recall that $C \overset{\text{def}}{=} \{v \in V \mid \ell_i(v) > 0, 1 \leq i \leq m,\ \ell_i(v) = 0, m+1 \leq i \leq p\}$, where each ℓ_i is a linear functional on V. Since $V(C) \cap V(C_0)^\perp \oplus V(C_0) = V(C)$, it follows that if ℓ_i, for some i with $1 \leq i \leq m$, vanishes on C_0, then $\ell_i(u_0) = \ell_i(u) > 2\ell_i(c)$.

If $v \in D$ then $\| u_0 - v \| < \delta$ and, so, $\ell_i(v) \geq \ell_i(u_0) - \| \ell_i \| \delta > \ell_i(c)$ if $\| \ell_i \| \delta < \ell_i(c)$.

Choose n so that $n\ell_i(c_0) > \ell_i(c)$ for all i such that ℓ_i does not vanish on C_0. The result follows since $\ell_i > \ell_i(c)$, for $1 \leq i \leq m$, on $nc_0 + C_0 + D$. □

THE TOPOLOGY OF HAUSDORFF CONVERGENCE ON COMPACT SUBSETS

In Bourbaki [B14, Ch. VIII, p. 188], this topology on (G) is introduced via the associated uniform structure. It is also defined in a right invariant manner. To convert it to correspond to a left invariant metric d, one redefines $P(K,V)$ to be the set of ordered pairs (X,Y) such that

$$X \cap K \subset YV^{-1} \text{ and } Y \cap K \subset XV^{-1}.$$

Then if V denotes the ε-ball about e, this is equivalent to stating that

$$X \cap K \subset \{u \mid d(u,Y) < \varepsilon\} \text{ and } Y \cap K \subset \{v \mid d(v,X) < \varepsilon\}.$$

Let $P(K,\varepsilon)$ denote the set of ordered pairs (X,Y) that satisfy the above condition and let

$$P(K,\varepsilon)(X_0) = \{Y \mid (X_0,Y) \in P(K,\varepsilon)\}.$$

This is a basic neighborhood of the closed set X_0 for the topology associated to the uniform structure defined by the $P(K,\varepsilon)$.

A.16. Lemma. *Assume that $C_0 = X_0 \cap K \neq \emptyset$. Then $P(K,\varepsilon) \subset O(C_0, 2\varepsilon)$. Furthermore, if C_0 is a non-void compact subset of X_0 and D_0 denotes the closure of $\{u \mid d(u,C_0) < \varepsilon\}$, then $O(C_0,\varepsilon) \subset P(D_0, 2\varepsilon)$.*

Consequently, the topology of Hausdorff convergence on compact subsets, as defined in § 9.4, is the topology associated to the above uniform structure.

THE EXISTENCE OF N-LEFT INVARIANT SOLUTIONS

The following arguments are due to Karpelevič.

A.17. Lemma. (*See [K3, p. 101, proof of Theorem 9.9.2]*) *Let $n \in N$. If $a \in A^+$ there exists a sequence $(\ell_k) \subset K$ such that $a^k \ell_k a^{-k} \to n$ as $k \to \infty$.*

Proof. First observe that $n = \exp X$ with $X = \sum_{\alpha>0} X_\alpha, X_\alpha \in \mathfrak{g}_\alpha$. Let $U_k = \sum_{\alpha>0} e^{-k\alpha(H)}\{X_\alpha + \theta(X_\alpha)\} \in \mathfrak{k}$ by § 2.1. If $a = \exp H$ then $\alpha(H) > 0$ for all $\alpha > 0$. Let $\ell_k = \exp U_k$. Then $a^k \ell_k a^{-k} = \exp(Ad(a^k)U_k) = \exp(\sum_{\alpha>0}\{X_\alpha = e^{-2k\alpha(H)}\theta(X_\alpha)) \to \exp X = n$ as $k \to \infty$. □

A.18. Proposition. *If $Lu + \lambda u = 0$ has a positive global solution, then it has an N-left invariant positive global solution.*

Proof. Let f denote a positive solution. Denote by f'_k the average of f over the orbits of $a^k K a^{-k}$, i.e., $f'_k(x) = \int_K f(a^k \ell a^{-k} \cdot x) d\ell$ and let $f_k(x) = f'_k(x)/f'_k(o)$. Then, $Lf_k + \lambda f_k = 0$ and $f_k(o) = 1$.

The functions f_k are constant on the orbits of $a^k K a^{-k}$, i.e., $f_k(x) = f_k(a^k \ell a^{-k} \cdot x)$ for any $\ell \in K$. Furthermore, by the compactness of $\mathcal{H}_1^\lambda$, there is a subsequence f_{k_j} that converges uniformly on compact sets to a positive solution f_0 of $Lu + \lambda u = 0$.

It follows from Lemma A.17 that, if $n \in N$ and $x \in X$, then $f_0(n \cdot x) = \lim_j f_{k_j}(a^{k_j} \ell_{k_j} a^{-k_j} \cdot x) = \lim_j f_{k_j}(x) = f_0(x)$. □

Convolution of measures

In the statement of the next lemma and the proof that follows, the notations come from Chapter XI.

Lemma A.19. *Let φ be a compactly supported continuous function. Assume that (μ_n) is a sequence of Radon measures which converges weakly to μ. Then, the sequence of functions $(\mu_n * \varphi)$ converges uniformly on compact sets to $\mu * \varphi$.*

Proof. It suffices to show that, if x_n converges to x, then $(\mu_n * \varphi)(x_n)$ converges to $(\mu * \varphi)(x)$.

Let C denote the support of $\tilde{\varphi}$ and V denote a compact neighbourhood of x. One may assume that the supports of the functions $\delta\tilde{\varphi}_{x_n}$ are all contained in V. Since $\tilde{\varphi}$ is uniformly continuous, there is a non-negative compactly supported function ψ on G, equal to 1 on VC such that $|\delta\tilde{\varphi}_{x_n} - \delta\tilde{\varphi}_x| < \epsilon_n \psi$, where $\lim_n \epsilon_n = 0$.

The Radon measures may be assumed to all be non-negative.

Then

$$\begin{aligned}
|(\mu_n * \varphi)(x_n) - (\mu * \varphi)(x)| &\leq |(\mu_n * \varphi)(x_n) - (\mu_n * \varphi)(x)| \\
&\quad + |(\mu_n * \varphi)(x) - (\mu * \varphi)(x)| \\
&\leq \epsilon_n \mu_n(\psi) + |(\mu_n * \varphi)(x) - (\mu * \varphi)(x)|.
\end{aligned}$$

The result follows. □

A topological lemma

A.20. Lemma. *Let X and Y be locally compact metrizable spaces and denote by ϕ a continuous, proper map of Y onto X. If $\overline{X} = X \cup \partial\overline{X}$ is a compactification of X let $\overline{\phi}$ denote the map of $Y \cup \partial\overline{X}$ (disjoint union) onto $\overline{X}$ defined by*

$$\overline{\phi}(y) = \begin{cases} \phi(y) & \text{if } \; y \in Y \\ y & \text{if } \; y \in \partial\overline{X}. \end{cases}$$

If $A \subset Y$ is open in Y and $B \subset \overline{X}$ is open in $\overline{X}$, let $C = A \cup \overline{\phi}^{-1}(B)$. The collection of sets $C \subset Y \cup \partial\overline{X}$ (disjoint union) of this form is a compact metrizable topology $\mathcal{T}$.

Let $Y \cup_\phi \partial\overline{X}$ denote the resulting topological space. Then,

1. *Y is an open dense subset of $Y \cup_\phi \partial\overline{X}$, and*
2. *a sequence $(y_n) \subset Y \cup_\phi \partial\overline{X}$ converges to $z \in \partial\overline{X}$ if and only if $(\overline{\phi}(y_n))$ converges to $\overline{\phi}(z) = z$.*

Moreover, the topology on the disjoint union $Y \cup \partial\overline{X}$ *is uniquely defined by properties (1) and (2). If* X, Y *are* H*-spaces and* ϕ *is* H*-equivariant, then* $Y \cup_\phi \partial\overline{X}$ *is a* H*-space and* $\overline{\phi}$ *is* H*-equivariant.*

Proof. The proof consists of a sequence of elementary verifications.
Part (a): $\mathcal{T}$ *is a topology and* $\overline{\phi}$ *is continuous.*
If $C_i \in \mathcal{T}$, $i \in I$, and $C_i = A_i \cup \overline{\phi}^{-1}(B_i)$, where A_i and B_i are open in Y and $\overline{X}$ respectively, then $\cup_{i\in I} C_i = A \cup \overline{\phi}^{-1}(B)$, where $A = \cup_{i\in I} A_i$ and $B = \cup_{i\in I} B_i$ are open in Y, $\overline{X}$ respectively.

If $C_i = A_i \cup \overline{\phi}^{-1}(B_i)$, $i = 1, 2$, then $C_1 \cap C_2$ is the union of the sets $A_1 \cap A_2, \overline{\phi}^{-1}(B_1 \cap B_2), A_1 \cap \overline{\phi}^{-1}(B_2), A_2 \cap \overline{\phi}^{-1}(B_1)$. Since $A \cap \overline{\phi}^{-1}(B) = A \cap \phi^{-1}(B \cap X)$, it follows that $C_1 \cap C_2 \in \mathcal{T}$ if each $C_i \in \mathcal{T}$.

The continuity of $\overline{\phi}$ is evident.
Part (b): $\mathcal{T}$ *is Hausdorff and* Y *is a dense open subset of* $Y \cup_\phi \partial\overline{X}$.

Clearly $Y \in \mathcal{T}$. If $z \in \partial\overline{X}$, it has a basis of neighbourhoods of the form $\overline{\phi}^{-1}(B_z)$, where B_z is open in $\overline{X}$ and contains z. It follows that $B_z \cap X \neq \emptyset$ and so $Y \cap \overline{\phi}^{-1}(B_z) = \phi^{-1}(B_z \cap X) \neq \emptyset$. As a result, Y is dense in $Y \cup_\phi \partial\overline{X}$.

Since $Y \in \mathcal{T}$, to verify the Hausdorff property it suffices to show that $y \in Y$ and $z \in \partial\overline{X}$ have disjoint neighbourhoods. The fact that ϕ is proper implies that there exists an open set $A \subset Y$ containing y such that $\overline{A}$ is a compact subset of Y for which $\overline{A} = \phi^{-1}\phi(\overline{A})$. If B_z is an open subset of $\overline{X}$ disjoint from $\phi(\overline{A})$ and containing z, then $A \cap \overline{\phi}^{-1}(B_z) = \emptyset$.
Part (c): *Convergence to a point of* $\partial\overline{X}$.

Since $\overline{\phi}$ is continuous, it suffices to show that if $(y_n) \subset Y$ and $\overline{\phi}(y_n) \to z \in \partial\overline{X}$, then $y_n \to z$. This is obvious, given that the basic neighborhoods of $z \in Y \cup_\phi \partial\overline{X}$ are the sets $\overline{\phi}^{-1}(B_z)$ where $B_z \subset \overline{X}$ is an open set containing z.
Part (d): $\overline{Y}$ *is metrizable and compact.* It is clear from the definition of $\mathcal{T}$ that it has a countable base since this is true of each of the topologies on Y and $\overline{X}$. It is also clear that $\mathcal{T}$ is regular: it suffices to note that the continuity of $\overline{\phi}$ implies that, if $z \in \partial\overline{X}$, every neighborhood of z in $Y \cup_\phi \partial\overline{X}$ contains a closed neighborhood. This shows that $\mathcal{T}$ is metrizable.

To show that $\mathcal{T}$ is compact, consider a sequence (y_n) in $Y \cup_\phi \partial\overline{X} = \overline{Y}$. One may assume that the sequence $(\overline{\phi}(y_n))$ converges in $\overline{X}$. If it converges to $z \in \partial\overline{X}$ then , by Part (c), y_n converges to z.

If, on the other hand, $\overline{\phi}(y_n)$ converges to $x \in X$, it follows that, for some $p > 0$, $\overline{\phi}(y_n) \in X$ for $n \geq p$, as X is open in $\overline{X}$. Hence, $\{x\} \cup \{\overline{\phi}(y_n) \mid n \geq p\}$ is a compact subset of X and its inverse image in Y under the proper map ϕ is compact. As a result, (y_n) has a subsequence that converges in Y. This shows that $\mathcal{T}$ is compact.
Part (e): *The uniqueness of the topology on the disjoint union* $Y \cup \partial\overline{X}$.

Consider a second topology $\mathcal{T}'$ which satisfies the conditions (1) and (2). It is metrizable, by what has been proved. Assume that $(y_n) \subset \overline{Y}$ converges to $y \in \overline{Y}$. If $y \in Y$, the fact that the topologies $\mathcal{T}$ and $\mathcal{T}'$ agree on Y, implies that the sequence (y_n) converges in the $\mathcal{T}'$-topology. If $y \in \partial\overline{X}$, it follows from Part (c) — which is also true for $\mathcal{T}'$ — that (y_n) converges to z with respect to $\mathcal{T}'$. This proves that $\mathcal{T} \supset \mathcal{T}'$. Since $\mathcal{T}$ is compact and $\mathcal{T}'$ is Hausdorff, they coincide.

Part (f): *$Y \cup_\phi \partial\overline{X}$ is an H-space.*

The H-actions on Y and $\partial\overline{X}$ determine a unique H-action on the disjoint union $Y \cup \partial\overline{X}$. The map $\overline{\phi}$ is H-equivariant if ϕ has this property. To check that the action is continuous, consider two sequences $(y_n) \subset \overline{Y}$ and $(g_n) \subset H$ such that $\lim_n y_n = y$ and $\lim_n g_n = g$. If $y \in Y$ it follows that $\lim_n g_n \cdot y_n = g \cdot y$ since Y is an H-space and open in $\overline{Y}$. If $y \in \partial\overline{X}$,it follows from the equivariance of $\overline{\phi}$ and Part (c) that $\lim_n g_n \cdot y_n = g \cdot y$. □

The Laplacian and the Mean-Value Property

The mean-value property of the Laplacian that was used to complete the proof of Corollary 12.9 is a special case of the following general result.

A. 21. Lemma. *Let X be an n-dimensional Riemannian manifold. Let L denote the Laplace–Beltrami operator and m the Riemannian volume measure. Let B_x^r denote the ball of radius r centered at x and let $M^r f(x)$ denote the mean-value $\frac{1}{m(B_x^r)}\int_{B_x^r} f(y)dm(y)$ of a function f on the ball B_x^r.*

Then, if $f \in \mathcal{C}^2(X)$, it follows that, for all $x \in X$,

$$\lim_{r\to 0}\frac{1}{r^2}\left[M^r f(x) - f(x)\right] = \frac{1}{2(n+2)}Lf(x).$$

Proof. Choose an orthonormal basis $Y_1, Y_2, \dots, Y_n$ in the tangent space $T_p(X)$ at p and let N_0 be a neighborhood of $0 \in T_p(X)$ that is mapped diffeomorphically onto a neighborhood N_p of $p \in X$ by the exponential map Exp_p. The corresponding normal (or exponential) coordinates on N_p are the functions y_i such that $y_i(q) = y_i$ if $q = \mathrm{Exp}_p(Y) = \mathrm{Exp}_p(\sum_{i=1}^n y_i Y_i) \in N_p$ (see Helgason [H2, p. 33]). In these coordinates, if $y = (y_1, y_2, \dots, y_n)$ is identified with $Y = \sum_{i=1}^n y_i Y_i$, the metric $(g_{ij}(Y))$ is such that $g_{ij}(Y) = \delta_{ij} + O(\|Y\|^2)$, where $\|Y\|^2$ denotes $\sum_{i=1}^n y_i^2 = \|Y\|^2$. Since distance from 0 in N_0 equals geodesic distance from p in N_p, it follows that, for small r, a Euclidean ball of radius r centered at 0 in $T_p(X)$ is mapped by the exponential map onto a geodesic ball of radius r and center p.

Under the identification of Y with y, the Riemannian metric at p determines a metric on $T_p(X)$that corresponds to the standard Euclidean metric on $\mathbb{R}^n$. Let Δ denote the Laplacian on $T_p(X)$ corresponding to the standard Euclidean Laplacian on $\mathbb{R}^n$. If $F \in \mathcal{C}^2(T_p(X))$ then

$$\Delta F(0) = \sum_{i=1}^n \frac{\partial^2 F}{\partial y_i^2}(0).$$

Since $g_{ij}(Y) = \delta_{ij} + O(\|Y\|^2)$, it follows from the fact that

$$Lf(p) = \frac{1}{\sqrt{g}} \sum_{i=1}^{n} \frac{\partial}{\partial y_i} \{\sqrt{g} g^{ij} \sum_{j=1}^{n} \frac{\partial f}{\partial y_j}\}(p),$$

where $(g^{ij}) = (g_{ij})^{-1}$, that $Lf(p) = \Delta F(0)$ if $f(q) = f(\mathrm{Exp}_p(Y)) = F(Y)$.

Expanding F to second order as

$$F(Y) = F(0) + \sum_{i=1}^{n} y_i \frac{\partial F}{\partial y_i}(0) + \frac{1}{2} \sum_{i,j=1}^{n} y_i y_j \frac{\partial^2 F}{\partial y_i \partial y_j}(0) + o(\|Y\|^2),$$

it follows that, if dY denotes the corresponding Lebesgue measure on $T_p(X)$,

$$\frac{1}{|B_0^r|} \int_{B_0^r} F(Y)dY = F(0) + \frac{1}{2n} \Delta F(0) \frac{1}{|B_0^r|} \int_{B_0^r} \|Y\|^2 dY + o(\|r\|^2),$$

where $|B_0^r|$ is the volume of the ball B_0^r. Since $|B_0^r| = \frac{\omega_n}{n} r^n$, where ω_n is the area of the unit sphere in $\mathbb{R}^n$, it follows that

$$\frac{1}{|B_0^r|} \int_{B_0^r} \|Y\|^2 dY = \frac{n}{n+2} r^2.$$

Hence,

$$\lim_{r \to 0} \frac{1}{r^2} \left[\frac{1}{|B_0^r|} \int_{B_0^r} F(Y)dY - F(0) \right] = \frac{1}{2(n+2)} \Delta F(0).$$

To complete the proof, it suffices to show that

$$\lim_{r \to 0} \frac{1}{r^2} \left[M^r f(p) - \frac{1}{|B_0^r|} \int_{B_0^r} F(Y)dY \right] = 0.$$

Since the volume measure m on N_p corresponds to Lebesgue measure dY on N_0 times the positive smooth density $\sqrt{g}(Y) = 1 + O(\|Y\|^2)$, it follows that

$$\begin{aligned}
M^r f(p) &- \frac{1}{|B_0^r|} \int_{B_0^r} F(Y)dY \\
&= \frac{1}{m(B_p^r)} \int_{B_0^r} f(y)dm(y) - \frac{1}{|B_0^r|} \int_{B_0^r} F(Y)dY \\
&= \frac{1}{m(B_p^r)} \int_{B_0^r} F(Y)\sqrt{g}(Y)dY - \frac{1}{|B_0^r|} \int_{B_0^r} F(Y)dY \\
&= \frac{1}{m(B_p^r)} \int_{B_0^r} F(Y)\{\sqrt{g}(Y) - 1\}dY \\
&\qquad + \left[\frac{1}{m(B_p^r)} - \frac{1}{|B_0^r|} \right] \int_{B_0^r} F(Y)dY.
\end{aligned}$$

Clearly, one may assume that $f(p) = F(0) = 0$. As a result, since $\sqrt{g}(Y) - 1 = O(\|Y\|^2)$, it follows that $F(Y)\{\sqrt{g}(Y) - 1\} = o(\|Y\|^2)$. In addition, $|B_0^r|/m(B_p^r) \to 1$ as $r \to 0$ and, so, the first of the last two integrals is $o(r^2)$. The second term is also $o(r^2)$ since

$$\frac{1}{m(B_p^r)} - \frac{1}{|B_0^r|} = \frac{1}{m(B_p^r)|B_0^r|} \int_{B_0^r} \{1 - \sqrt{g}(Y)\} dY = \frac{O(r^2)}{|B_0^r|}$$

and $\displaystyle\frac{1}{|B_0^r|} \int_{B_0^r} F(Y) dy = o(1)$.

This completes the proof of the lemma. □

APPENDIX B

This appendix continues the examination of the compactifications $\overline{X}_\tau^S$, where τ is an irreducible and faithful representation of G, and $\overline{G \cdot m_Q}$, where m_Q is the unique K-invariant probability on the boundary G/Q, that was begun in Chapters IV and IX. (As in Chapter IV, G is assumed to have no proper compact normal subgroups.) If $Q = P_\tau$ and τ is faithful, these spaces are G-isomorphic compactifications of X (Theorem B. 20) and their structure, which depends only on P_τ, is determined (Theorems B.12 and B.21). The Satake characterization of $\overline{X}_\tau^S$ is obtained as a consequence of this isomorphism (Theorem B.21) and Furstenberg's methods are used to give proofs of the geometrical part of Satake's results [S1].

Furstenberg compactifications

B.1. Definition. Let $Q \supset P$ be a standard parabolic subgroup of G. Denote by $G(Q)$ the largest, normal, connected subgroup of G contained in Q.

B.2. Remarks. (1) $G(Q)$ is the connected component of the kernel, denoted by H, of the action of G on G/Q. Clearly a normal subgroup of G contained in Q acts trivially on G/Q and, so, is contained in H. On the other hand, H is normal in G and contained in the normalizer of Q. Because Q is parabolic, $H \subset Q$. It follows that H is the greatest normal subgroup of G contained in Q. Hence, $G(Q)$ is equal to the connected component of H.

(2) If $P = P^I$, denote by I_0 the union of the components of Δ contained in I. Then $G(P^I) = G^{I_0}$.

If τ is an irreducible representation of G and $Q = P_\tau$, then Proposition 4.19 implies that $Q = P^{I_\tau^\perp}$ and $G(Q) = G^{I^\perp(\tau)}$, where these notations are defined in Chapter IV. Note that Proposition 4.17 implies the tangent representation of τ is faithful if and only if $G(Q) = \{e\}$.

B.3. Lemma. *Assume that $Q = P^I$ is a standard parabolic subgroup of G. Let $e_Q \in G/Q$ denote the origin in G/Q and identify $\overline{N}_I$ with the open dense set $\overline{N}_I \cdot e_Q$ in G/Q. Let η_Q denote a Haar measure on $\overline{N}_I$ and denote by m_Q the K-invariant probability measure on G/Q. Let $u_Q(n) = u_Q(n \cdot e_Q) = \frac{dm_Q}{d\eta_Q}(n \cdot e_Q)$ for $n \in \overline{N}_I$. Then, $u_Q(n)$ is a strictly positive analytic function on $\overline{N}_I$ and for $a \in A$, it follows that*

$$\frac{d(a \cdot m_Q)}{dm_Q}(x) = \frac{u_Q(a^{-1} \cdot x)}{u_Q(x)} \prod_{\alpha \in \Sigma_I^+} e^\alpha(a)$$

where the roots $\alpha \in \Sigma_I^+$ *are taken with their multiplicities and* $x = n \cdot e_Q \in \overline{N}_I \cdot e_Q$.

Proof. Since m_Q and η_Q are defined by non-degenerate differential forms, the function $u_Q(n)$, where $n \in \overline{N}_I$ is the Jacobian of the map $n \to n \cdot e_Q$ with respect to the above differential forms. Hence, u_Q is a strictly positive analytic function. The action $x \to a \cdot x$ of a on $\overline{N}_I \cdot e_Q \subset G/Q$ is identified with the automorphism $n \to ana^{-1}$ of $\overline{N}_I$. The corresponding tangent automorphism of $\overline{\mathfrak{n}}_I$ is diagonal with respect to a basis of $\overline{\mathfrak{n}}_I$, with eigenvalues $e^{-\alpha}(a), \alpha \in \Sigma_I^+$. Hence, $a \cdot \eta_Q = (\prod_{\alpha \in \Sigma_I^+} e^{\alpha}(a))\eta_Q$. Since $m_Q = u_Q \eta_Q$, it follows that $a \cdot m_Q = (u_Q \cdot a^{-1}) a \cdot \eta_Q = (u_Q \circ a^{-1}) \prod_{\alpha \in \Sigma_I^+} e^{\alpha}(a) \eta_Q$. Hence, $\frac{da \cdot m_Q}{dm_Q} = \frac{u_Q \circ a^{-1}}{u_Q} \prod_{\alpha \in \Sigma_I^+} e^{\alpha}(a)$. □

B.4. Proposition. *With the above notations, the stabilizer of* m_Q *is equal to* $KG(Q)$.

Proof. From the remarks above, if $s \in G(Q)$, then s acts trivially on G/Q. Hence, $s \cdot m_Q = m_Q$. It follows that $KG(Q)$ is a group which stabilizes m_Q. Conversely, if $g \cdot m_Q = m_Q$ and $g = kak'$ with $k, k' \in K$ and $a \in \overline{A^+}$, then, $a \cdot m_Q = m_Q$. The formula in Lemma B.3 shows that, for $x = e_Q, u(a^{-1} \cdot e_Q) = u(e_Q)$. Hence, $\prod_{\alpha \in \Sigma_I^+} e^{\alpha}(a) = 1$. Since $a \in \overline{A^+}$ implies that $e^{\alpha}(a) \geq 1$ for any $\alpha \in \Sigma^+$, one has $e^{\alpha}(a) = 1$ if $\alpha \in \Sigma_I^+$. It follows that the action of a on $\overline{N}_I \cdot e_Q$ is trivial. Since $\overline{N}_I \cdot e_Q$ is dense in G/Q, the action of a on G/Q is trivial and, hence, $a \in G(Q)$. As $KG(Q)$ is a group, it follows that $g \in KG(Q)$. □

B.5. Theorem. *Assume that* $G(Q) = \{e\}$. *The map* $gK \to g \cdot m_Q$ *is then an embedding of* $X = G/K$ *into* $\overline{G \cdot m_Q} \subset \mathcal{M}_1(G/Q)$.

Proof. The condition $G(Q) = \{e\}$ and Proposition B.4 imply that the map $gK \to g \cdot m_Q$ is injective. This map is clearly continuous and, in order to show that it is an embedding, it suffices to show that the condition $m_Q = \lim_n g_n \cdot m_Q$ implies that $\lim_n g_n \cdot o = o$. As in the proof of Proposition B.4, if $g_n^{-1} = k_n a_n k_n'$ with $k_n, k_n' \in K, a_n \in \overline{A^+}$, then $\lim_n g_n \cdot m_Q = m_Q$ implies that $\lim_n a_n^{-1} \cdot m_Q = m_Q$. The formula in Lemma B.3 implies that, for a positive and continuous function ϕ on G/Q with compact support contained in $\overline{N}_I \cdot e_Q$, one has

$$(*) \qquad a_n^{-1} \cdot m_Q(\phi) = \prod_{\alpha \in \Sigma_I^+} e^{-\alpha}(a_n) \int \frac{u_Q(a_n \cdot x)}{u_Q(x)} \phi(x) dm_Q(x).$$

Since $a_n \in \overline{A^+}$ and $x \in \overline{N}_I \cdot e_Q$, the point $a_n \cdot x$ belongs to a fixed compact subset of $\overline{N}_I \cdot e_Q$, if x belongs to the support of ϕ. Hence, the integral on the right hand side is bounded when n varies. If $e^{\alpha}(a_n)$ is not bounded for some $\alpha \in \Sigma_I^+$, then $\lim_n a_n^{-1} \cdot m_Q(\phi) = 0$ and $m_Q(\phi) = 0$. As this is impossible, $e^{\alpha}(a_n)$ is bounded for any $\alpha \in \Sigma_I^+$. It follows that there is an $a \in \overline{A^+}$ and a subsequence (a_{n_k}) such that $\lim_k e^\alpha(a_{n_k}) = e^\alpha(a)$ for $\alpha \in \Sigma_I^+$. The formula $(*)$ implies that, for any ϕ with compact support in $\overline{N}_I \cdot e_Q$,

$$\lim_n a_n^{-1} \cdot m_Q(\phi) = \prod_{\alpha\in\Sigma_I^+} e^{-\alpha}(a) \int \frac{u_Q(a\cdot x)}{u_Q(x)} \phi(x) dm_Q(x) \text{ and}$$

$$m_Q(\phi) = \prod_{\alpha\in\Sigma_I^+} e^{-\alpha}(a) \int \frac{u_Q(a\cdot x)}{u_Q(x)} \phi(x) dm_Q(x).$$

In terms of measures on $\overline{N}_I$, this means that

$$m_Q = [\prod_{\alpha\in\Sigma_I^+} e^{-\alpha}(a)] \frac{u_Q \circ a}{u_Q} m_Q = a^{-1} \cdot m_Q.$$

From Proposition B.4, it follows that $a \in G(Q)$. Since $G(Q) = \{e\}$, this implies that $a = e$. Hence, for any $x \in \overline{N}_I \cdot e_Q$, $\lim_n a_n \cdot x = x$. The convergence of the analytic maps a_n of G/Q holds on an open subset of G/Q. Hence, it is valid everywhere. Since the action of G on G/Q is faithful, it follows that $\lim_n a_n = e, \lim g_n \cdot o = o$. □

B.6. Corollary. *Suppose $G(Q) = \{e\}$. Then G/Q is a faithful boundary of G and the compact space $\overline{G \cdot m_Q} \subset \mathcal{M}_1(G/Q)$ is a G-compactification of $X = G/K$.*

Some weight lemmas

The following lemmas correspond to the following lemmas in [S1, Lemmas 5, 6, 7].

B.7. Lemma. *Let τ be an irreducible representation of G and μ_τ its highest weight. For any $\lambda \in \mathfrak{a}^*$ of the form*

$$\lambda = \mu_\tau - \sum_{\alpha\in\Delta} c_\alpha \alpha, \quad c_\alpha \in \mathbb{N},$$

let θ_λ be the support of λ, i.e., $\theta_\lambda = \{\alpha \in \Delta; c_\alpha > 0\}$. Then, if λ is a weight of τ and $\lambda \neq \mu_\tau$, there exists a root $\alpha \in \theta_\lambda$ such that $\lambda + \alpha$ is a weight of τ.

Proof. Assume that for every $\alpha \in \Delta, \lambda + \alpha$ is not a weight of τ. If $X \in \mathfrak{g}_\alpha$, then $\tau(X)V_\lambda \subset V_{\lambda+\alpha}$. Since, by assumption, $\lambda + \alpha$ is not a weight of τ, it follows that $\tau(\mathfrak{g}_\alpha)V_\lambda = 0$. Hence, $\tau(\mathfrak{n})V_\lambda = 0$ because $\cup_{\alpha\in\Delta}\mathfrak{g}_\alpha$ generates the algebra $\mathfrak{n}$.

On the other hand, since $\mathfrak{m} + \mathfrak{a}$ is the centralizer of $\mathfrak{a}$, it follows that $\tau(\mathfrak{m} + \mathfrak{a})V_\lambda \subset V_\lambda$. Hence, $\tau(\mathfrak{m} + \mathfrak{a} + \mathfrak{n})V_\lambda \subset V_\lambda$. It follows that $\tau[(\mathfrak{m} + \mathfrak{a} + \mathfrak{n})_\mathbb{C}]V_\lambda \subset V_\lambda$. As the Borel subalgebra $\hat{\mathfrak{b}}$ of $\mathfrak{g}_\mathbb{C}$ is contained in $(\mathfrak{m} + \mathfrak{a} + \mathfrak{n})_\mathbb{C}$

(see § 4.8) it follows that $\tau(\hat{\mathfrak{b}})V_\lambda \subset V_\lambda$. Then, from Lie's theorem, it follows that V_λ contains a $\tau(\hat{\mathfrak{b}})$-invariant line. As τ is irreducible, this line is unique and $\mathbb{C}v_\tau \subset V_\lambda$. However, this is impossible if $\mu_\tau \neq \lambda$. Hence, there exists $\alpha \in \Delta$ such that $\lambda + \alpha$ is a weight of τ. Formula (4.9) then implies that $\alpha \in \theta_\lambda$. □

B.8. Lemma. *Let λ and θ_λ be as in Lemma B.7. Then, θ_λ is the support of a weight of τ if and only if $\{\mu_\tau\} \cup \theta_\lambda$ is connected.*

Proof. Part (1): Assume that $\lambda = \mu_\tau - \sum_{\alpha\in\Delta} c_\alpha \alpha$ is a weight of τ. Using induction on the positive integer $c(\lambda) = \sum_{\alpha\in\Delta} c_\alpha$ and Lemma B.7, it follows that, for some $\alpha \in \theta_\lambda, \lambda + \alpha$ is a weight and the new sum $c(\lambda+\alpha)$ satisfies $c(\lambda + \alpha) = c(\lambda) - 1$. Hence, if $\theta_{\lambda+\alpha} = \theta_\alpha$, the connectivity of $\mu_\tau \cup \theta_\lambda$ follows from the inductive hypothesis.

If $\theta_{\lambda+\alpha} \neq \theta_\lambda$ and $\theta_\lambda = \theta_{\lambda+\alpha} \cup \{\alpha\}$, where $\alpha \notin \theta_{\lambda+\alpha}$, then α is not orthogonal to $\{\mu_\tau\} \cup \theta_{\lambda+\alpha}$. To see this, note that if α is orthogonal to $\mu_\tau \cup \theta_{\lambda+\alpha}$, the linear form $s_\alpha\lambda = \lambda - 2\frac{\langle\lambda,\alpha\rangle}{\langle\alpha,\alpha\rangle}\alpha = \lambda + 2\alpha = (\lambda+\alpha) + \alpha$ is a weight. Hence, $\alpha \in \theta_{\lambda+\alpha}$, which is a contradiction. The inductive hypothesis says that $\{\mu_\tau\}\cup\theta_{\lambda+\alpha}$ is connected. This also holds for $\{\mu_\tau\}\cup\theta_\lambda$, since $\theta_\lambda = \theta_{\lambda+\alpha} \cup \{\alpha\}$ and, so, α is not orthogonal to $\{\mu_\tau\} \cup \theta_{\lambda+\alpha}$.

Part (2): Assume that $\{\mu_\tau\} \cup \theta$ is connected, where $\theta \subset \Delta$. One shows that there exists a weight λ of τ with $\theta_\lambda = \theta$. Let $\theta = \{b_1, b_2, \cdots, b_m\}$ with $b_i \in \Delta$, $\langle\mu_\tau, b_1\rangle > 0$, $\langle b_i, b_j\rangle < 0$ for some $j < i$ and every $i \leq m$. Then

$$s_{b_1}(\mu_\tau) = \mu_\tau - 2\frac{\langle\mu_\tau, b_1\rangle}{\langle b_1, b_1\rangle} b_1 = \mu_\tau - k_1 b_1,$$

where $k_1 > 0$, is a weight of τ.

In the same way, it follows that $s_{b_i} \cdots s_{b_1}(\mu_\tau) = \mu_\tau - \sum_{j\leq i} k_j b_j$ is a weight of τ for every i and $k_j > 0$ if $j \leq i$. Hence, there exists a weight $\lambda = s_{b_m} \cdots s_{b_1}(\mu_\tau)$ with $\theta_\lambda = \{b_1, b_2, \cdots, b_m\} = \theta$. □

THE STRUCTURE OF $\overline{G \cdot m_Q}$

It is convenient to consider an irreducible representation τ of G such that $Q = P_\tau$ as in Proposition 4.27. Recall that $P_\tau = P^{I_\tau^\perp}$ where $I_\tau^\perp = \{\alpha \in \Delta \mid \langle\mu_\tau, \alpha\rangle = 0\}$. The results will depend only on $I_\tau^\perp$ rather than on τ itself. This is an extension of the results of Chapter IX to the case where τ is non-generic. The action of $g \in G$ on $x \in P(V)$ will be denoted $\tau(g) \cdot x$ or $g \cdot x$ for simplicity.

B.9. Definition. Suppose τ is an irreducible representation of $\mathfrak{g}$ and I is a subset of Δ. Denote by $I_0(\tau)$ the complement of $\{\mu_\tau\}$ in the connected component of μ_τ in $I \cup \{\mu_\tau\}$ containing μ_τ and let $I_0^\perp(\tau) = \{\alpha \in \Delta \mid \langle\alpha, \beta\rangle = 0 \text{ for all } \beta \in I_0(\tau) \cup \{\mu_\tau\}\}$. Let $I^\tau = I_0(\tau) \cup I_0^\perp(\tau)$. A subset I of Δ is said to be **τ-open** if $I = I_0(\tau)$.

B.10. Remark. (1) Proposition 4.17 states that τ is faithful if and only if Δ is τ-open, i.e., $\Delta = \Delta_0(\tau)$ and $\Delta_0^{\perp}(\tau) = \emptyset$.

(2) Consider the decompositions of I and I^τ into connected components. For such a component J, it is either orthogonal to μ_τ, or not, in which case $J \cup \{\mu_\tau\}$ is connected. It follows that $I_0(\tau)$ is the union of the components of I or of I^τ which are not orthogonal to μ_τ. On the other hand $I_0^{\perp}(\tau)$ is the union of the components of I^τ which are orthogonal to μ_τ. As a result, $\mathfrak{g}(I^\tau)$ is the direct product of the ideals $\mathfrak{g}(I_0(\tau))$ and $\mathfrak{g}(I_0^{\perp}(\tau))$ (see Remark 2.14 for the definition of $\mathfrak{g}(I)$). It also follows that $\mathfrak{g}^{I^\tau} = \mathfrak{g}^{I_0(\tau)} + \mathfrak{g}^{I_0^{\perp}(\tau)} = \mathfrak{g}^I + \mathfrak{g}^{I_0^{\perp}(\tau)}$ where $\mathfrak{g}^{I_0(\tau)}, \mathfrak{g}^{I_0^{\perp}(\tau)}, \mathfrak{g}^I$ are ideals of $\mathfrak{g}^{I^\tau}$.

The importance of the concept of a τ-open subset is shown by the following basic lemma, where the scalar product is that used in Proposition 4.32.

B.11. Lemma. *Let $(a_n) \subset \overline{A^+}$ be an I-canonical sequence. Then,*

$$\lim_n \frac{\tau(a_n)}{\|\tau(a_n)\|} = \pi_{I_0(\tau)} = \pi_I = \pi_{I^\tau},$$

where π_J is the orthogonal projection of the subspace V_J generated by the weight subspaces V_λ with $\lambda \in \mathfrak{a}^*$ of the form $\lambda = \mu_\tau - \sum_{\alpha \in J} c_\alpha \alpha$.

Proof. The proof of Lemma 9.50 implies that

$$\lim_n \frac{\tau(a_n)}{\|\tau(a_n)\|} = \pi_I.$$

Denote by V^θ the sum of the weight subspaces V_λ such that $\theta_\lambda = \theta$. Lemma B.8 implies that $V^\theta \neq \{0\}$ if and only if $\theta \cup \{\mu_\tau\}$ is connected. From the definition of I, I^τ, and $I_0(\tau)$, it follows that $\theta \subset I$ and $\theta \cup \{\mu_\tau\}$ is connected if and only if $\theta \subset I_0(\tau)$ and $\theta \cup \{\mu_\tau\}$ is connected. Since $V_I = \oplus_{\theta \subset I} V^\theta$, it follows that $V_I = V_{I_0(\tau)} = V_{I^\tau}$. The lemma follows. □

B.12. Theorem. *Let τ be an irreducible representation of G such that $\dim V_\tau = 1$. If $I \subset \Delta$ let $m_{I,\tau}$ denote the unique K^I-invariant probability measure on $P^I \cdot \bar{v}_\tau \subset G/P_\tau$. Then, the closure of $G \cdot m_\tau$ in $\mathcal{M}_1(G/P_\tau)$ is the disjoint union of the orbits $G \cdot m_{I,\tau}$ where $I \subset \Delta$ is τ-open. Furthermore, $\overline{G \cdot m_\tau} = \bigcup_{\substack{I \subset \Delta \\ \tau-open}} K\overline{A^{I,+}} \cdot m_{I,\tau}$. The stabilizer of $m_{I,\tau}$ is equal to $R^I G^{I_0^{\perp}(\tau)} = R^{I^\tau} G^{I_0^{\perp}(\tau)} = R^{I_0(\tau)} G^{I_0^{\perp}(\tau)}$.*

The proof of this theorem depends on several lemmas.

B.13. Lemma. *Let I be any subset of Δ. The stabilizer of the closed subset $P^I \cdot \bar{v}_\tau$ of G/P_τ is equal to $P^{I^\tau} = P^{I_0(\tau)} G^{I_0^{\perp}(\tau)} = P^I G^{I_0^{\perp}(\tau)}$.*

Proof. Let (a_n) be an I-canonical sequence. Then, $\lim_n a_n \cdot m = m_I$ (see Lemmas 9.48 and 9.49). Consider the canonical map from G/P onto G/P_τ. Clearly, the image of m_I is $m_{I,\tau}$. Hence, $\lim_n \tau(a_n) \cdot m_\tau = m_{I,\tau} = \pi_I \cdot m_\tau$.

Since $P^I \cdot \bar{v}_\tau = G^I \cdot \bar{v}_\tau$ is the support of $m_{I,\tau}$, it follows from Lemma B.11 and the formula $m_{I,\tau} = \pi_I m_\tau$ that

$$P^I \cdot \bar{v}_\tau = P^{I_0(\tau)} \cdot \bar{v}_\tau = P^{I^\tau} \cdot \bar{v}_\tau.$$

Hence, P^{I^τ} stabilizes $P^I \cdot \bar{v}_\tau$. Let H denote the stabilizer of $P^I \cdot \bar{v}_\tau$. Then $H \supset P^{I^\tau}$ is a standard parabolic subgroup. Hence, there exists $J \subset \Delta$ with $P^J = H, J \supset I^\tau$. If $J \neq I^\tau$, then the definition of I^τ implies that $J_0(\tau) \supsetneq I_0(\tau)$. Lemma B.8 implies, therefore, that there exists a weight λ of τ such that $V_\lambda \not\subset V_{I_0(\tau)}$ and, hence, $V_{I_0(\tau)} \subsetneq V_{J_0(\tau)}$. Since the support of $m_{I_0(\tau),\tau}$ (respectively, $m_{J_0(\tau),\tau}$) generates the subspace $V_{I_0(\tau)}$ (respectively, $V_{J_0(\tau)}$), it follows that $P^{I_0(\tau)} \cdot \bar{v}_\tau \neq P^{J_0(\tau)} \cdot \bar{v}_\tau$ and, so, $P^I \cdot \bar{v}_\tau \neq P^J \cdot \bar{v}_\tau$. Hence, P^J cannot stabilize $P^I \cdot \bar{v}_\tau$ if $J \neq I^\tau$. It follows that $H = P^J = P^{I^\tau}$. Remark B.10 (2) implies that $G^{I^\tau} = G^I G^{I_0^\perp(\tau)}$. On the other hand, $P^{I^\tau} = G^{I^\tau} MAN = G^{I_0^\perp(\tau)} G^I MAN = P^I G^{I_0^\perp(\tau)} = G^{I_0^\perp(\tau)} P^I$. As a result, $P^{I^\tau} = P^I G^{I_0^\perp(\tau)} = P^{I_0(\tau)} G^{I_0^\perp(\tau)}$. □

In the following lemma, representations of the semisimple groups G^I are involved. While these semisimple groups, in general, can have proper, connected, compact normal subgroups, the definition of highest weight (see Chapter IV) and formula (4.9) extends immediately to these groups.

B.14. Lemma. *The stabilizer of the subspace V_I is P^{I^τ}. The restriction of τ to G^{I^τ} (respectively, to $G^{I_0(\tau)}$) and V_I defines an irreducible representation τ_I (respectively, τ_I^0) of G^{I^τ} (respectively, of $G^{I_0(\tau)}$). The highest weight of τ_I (respectively, of τ_I^0) is the restriction of the weight μ_τ to $\mathfrak{a}^{I^\tau}$ (respectively, to $\mathfrak{a}^{I_0(\tau)}$) and the connected component of the kernel of τ_I (respectively, of τ_I^0) is $G^{I_0^\perp(\tau)}$ (respectively, $K^{I_0(\tau)} \cap G^{I_0^\perp(\tau)}$). The compact space $P^I \cdot v_\tau$ is a boundary of $G^{I_0(\tau)}$ and the stabilizer of $m_{I,\tau}$ in $G^{I_0(\tau)}$ is $K^{I_0(\tau)}$.*

Proof. Lemmas B.11 and B.13 imply that P^{I^τ} stabilizes the support of $m_{I,\tau} = m_{I^\tau,\tau}$. Since $V_{I^\tau} = V_I$ is generated by the support of m_I, it follows that V_I is stabilized by P^{I^τ} and, so, G^{I^τ} stabilizes V_I. If $v \in V_I$ is such that $\tau_I(H)v = \mu_\tau(H)v$ for any $H \in \mathfrak{a}^{I^\tau}$, the definition of V_I implies that for some weight λ of τ one has $\mu_\tau(H) = \lambda(H) = (\mu_\tau - \sum_{\alpha\in I} c_\alpha \alpha)(H)$ for any $H \in \mathfrak{a}^{I^\tau}$. Since $V_I = V_{I^\tau}$ the restrictions of μ_τ and $\mu_\tau - \sum_{\alpha\in I} c_\alpha\alpha$, $c_\alpha \geq 0$, to $\mathfrak{a}^{I^\tau}$ are equal. By definition of $\mathfrak{a}^{I^\tau}$ this implies that $\sum_{\alpha\in I} c_\alpha\alpha = 0$, $\lambda = \mu_\tau$, and $v \in \mathbb{C}v_\tau$. As a result, the weight subspace of τ_I corresponding to the restriction of μ_τ to $\mathfrak{a}^{I^\tau}$ is $\mathbb{C}v_\tau$. The argument is also valid for $\mathfrak{a}^{I_0(\tau)}$, $G^{I_0(\tau)}$. On the other hand the subspace V_I is generated by $G^{I_0(\tau)}v_\tau$. This follows since $P^{I_0(\tau)} \cdot \bar{v}_\tau = P^I \cdot \bar{v}_\tau$ (see the proof of Lemma B.13) and since $P^{I_0(\tau)} = G^{I_0(\tau)}P$ implies that $P^{I_0(\tau)} \cdot \bar{v}_\tau = G^{I_0(\tau)} \cdot \bar{v}_\tau$. The irreducibility of τ_I is a consequence of these observations, for the following reason. One can decompose τ_I into a direct sum of irreducible representations τ_I^k, $1 \leq k \leq m$, as $V_I = \oplus_{k=1}^m W^k$. The projection v_τ^k of v_τ in W^k, for any $H \in \mathfrak{a}^{I^\tau}$, satisfies

$$\tau_I(H)v_\tau^k = \mu_\tau(H)v_\tau^k.$$

The first observation implies that $v_\tau^k \in \mathbb{C}v_\tau$ and, hence, $v_\tau^k \in W^k$ for some k, say $k = 1, v_\tau^1 \neq 0$ and $v_\tau^k = 0$ for $k \geq 2$. The second observation implies that $G^{I^\tau}v_\tau \subset W^1, W^1 = V_I$. Hence, τ_I is irreducible. The same argument is valid for the restriction of τ to $G^{I_0(\tau)}$ as well as for the restriction τ_I^0 of τ to what may be called the non-compact part $G(I_0(\tau))$ of $G^{I_0(\tau)}$ — its Lie algebra is $\mathfrak{g}(I_0(\tau))$. (If $G_c^{I^\tau}$ denotes the compact part of G^{I^τ}, then $G^{I^\tau} = G_c^{I^\tau}G(I^\tau)$ and the semisimple groups $G_c^{I^\tau}$ and $G(I^\tau)$ commute.) Since $G_c^{I^\tau}$ is contained in any standard parabolic subgroup of G^{I^τ}, and $\mathbb{C}v_\tau$ is the weight subspace corresponding to the highest weight of τ_I, Proposition 4.18 implies that $G_c^{I^\tau}v_\tau \subset \mathbb{C}v_\tau$. Since $G_c^{I^\tau}$ is connected and compact, $G_c^{I^\tau}v_\tau = v_\tau$. The fact that $G_c^{I^\tau}$ commutes with $G(I^\tau)$ and that τ_I is irreducible implies that $G_c^{I^\tau}$ acts trivially on V_I. Hence, the kernel of τ_I contains $G_c^{I^\tau}$. On the other hand, applying Proposition 4.17 to the irreducible representation of $\mathfrak{g}(I_0(\tau))$ defined by V_I and the restriction of the tangent representation of τ to $\mathfrak{g}(I_0(\tau))$, it follows that its kernel is equal to $\mathfrak{g}^{I_0^\perp(\tau)} \cap \mathfrak{g}(I_0(\tau))$. Since $G_c^{I^\tau} \subset G^{I_0^\perp(\tau)}$ and $G^{I_0(\tau)} = G^{I_0^\perp(\tau)}G(I_0(\tau))$ (see the proof of the previous lemma), the connected component of the kernel of the restriction of τ to G^{I^τ} and V_I is $G^{I_0^\perp(\tau)}$. The corresponding result for τ_I^0 follows since $G^{I_0(\tau)} \cap G^{I_0^\perp(\tau)}$ is compact.

As P^{I^τ} stabilizes V_I, the stabilizer of V_I contains P^{I^τ} and, hence, is a standard parabolic subgroup P^J with $J \supset I^\tau$. Assuming that $I^\tau \subsetneq J$, the definition of I^τ and of $I_0(\tau)$ implies that $I_0(\tau) \subsetneq J_0(\tau)$. Hence, Lemma B.11 implies that

$$V_J = V_{J_0(\tau)} \supsetneq V_{I_0(\tau)} = V_I.$$

The first part of the proof implies that the restriction of τ to G^J and V_J is irreducible. Since G^J stabilizes $V_I \subsetneq V_J$, this is a contradiction.

It follows from Lemma B.13 that $P^I \cdot \bar{v}_\tau = P^{I_0(\tau)} \cdot \bar{v}_\tau = G^{I_0(\tau)}P \cdot \bar{v}_\tau = G^{I_0(\tau)} \cdot \bar{v}_\tau = G^{I_0(\tau)}/P_\tau \cap G^{I_0(\tau)}$. Hence, $P^I \cdot \bar{v}_\tau$ is a factor space of the Furstenberg boundary $G^{I_0(\tau)}/P \cap G^{I_0(\tau)}$ and, so, is a boundary. The restriction of τ to $\mathfrak{g}(I_0(\tau))$ and V_I defines, as seen above, an irreducible and faithful representation of $\mathfrak{g}(I_0(\tau))$. Proposition B.4 can be applied to this representation and implies that the stabilizer of $m_{I,\tau} = m_{I_0(\tau),\tau}$ in $G^{I_0(\tau)}$ is equal to $K^{I_0(\tau)}$. As a result, the boundary $P^I \cdot \bar{v}_\tau$ of $G^{I_0(\tau)}$ is faithful. □

B.15. Lemma. *The stabilizer of $m_{I,\tau}$ is $R^I G^{I_0^\perp(\tau)}$.*

If $g \in G$ and I, J are subsets of Δ, the following conditions are equivalent:

(1) *$g \cdot m_{I,\tau} = m_{J,\tau}$; and*
(2) *$I_0(\tau) = J_0(\tau)$, and $g \in R^{I^\tau}G^{I_0^\perp(\tau)} = R^I G^{I_0^\perp(\tau)} = R^{I_0(\tau)}G^{I_0^\perp(\tau)}$.*

Proof. First one proves that the stabilizer H of $m_{I,\tau}$ is $R^{I^\tau}G^{I_0^\perp(\tau)}$. Proposition 9.13 implies that R^{I^τ} stabilizes m_{I^τ} and, hence, its projection $m_{I,\tau}$.

By Lemma B.14, $G^{I_0^\perp(\tau)}$ acts trivially on V_I and, consequently, on the support of $m_{I,\tau}$. It follows that $H \supset R^{I^\tau} G^{I_0^\perp(\tau)}$. Since H stabilizes $P^{I^\tau} \cdot v_\tau$, the support of $m_{I,\tau}$, Lemma B.13 implies that $H \subset P^{I^\tau}$. Since $P^{I^\tau} = G^{I^\tau} R^{I^\tau}$ it suffices to calculate the stabilizer of $m_{I,\tau}$ in G^{I^τ}. Remark B.10 (2) implies that $G^{I_0(\tau)}$ and $G^{I_0^\perp(\tau)}$ are normal in G^{I^τ} with $G^{I^\tau} = G^{I_0(\tau)} G^{I_0^\perp(\tau)}$. From Lemma B.14 it follows that the stabilizer of $m_{I,\tau}$ in $G^{I_0(\tau)}$ is $K^{I_0(\tau)}$. Consequently, $H = R^{I^\tau} K^{I_0(\tau)} G^{I_0^\perp(\tau)} = R^{I^\tau} G^{I_0^\perp(\tau)}$. In the same way one proves that $H = R^{I_0(\tau)} G^{I_0^\perp(\tau)} = R^I G^{I_0^\perp(\tau)}$. Since $R^{I^\tau} G^{I_0^\perp(\tau)}$ stabilizes $m_{I,\tau}$, condition (2) implies condition (1).

Assume that condition (1) holds. Then, the stabilizers of the supports of $g \cdot m_{I,\tau}$ and $m_{J,\tau}$ are equal. Hence, Lemma B.13 implies that $gP^{I^\tau} g^{-1} = P^{J^\tau}$. It follows from Proposition 2.18 that $I^\tau = J^\tau$ and $g \in P^{I^\tau}$. Consequently, $I_0(\tau) = J_0(\tau)$. As a result, condition (1) implies that $g \cdot m_{I,\tau} = m_{I,\tau}$. The first part of the proof shows that $g \in R^{I^\tau} G^{I_0^\perp(\tau)} = R^I G^{I_0^\perp(\tau)} = R^{I_0(\tau)} G^{I_0^\perp(\tau)}$. □

Proof of Theorem B.12.

In view of Proposition B.4, the map $g \to \tau(g) \cdot m_\tau$ factors through the projection $g \to g \cdot o$ of G onto $X = G/K$. As a result, if $g = kak'$ then $\tau(g) \cdot m_\tau = \tau(ka) \cdot m_\tau$ is a function of the point $ka \cdot o \in X$.

To calculate $\overline{G \cdot m_\tau}$ one uses fundamental sequences $(g_n \cdot o)$. Let $g_n \cdot o = k_n a_n \cdot o$ with $k_n \to k$, $a_n \in \overline{A^+}$ and $a_n = a_n^I a_{n,I}$ where $a_n^I \to a^I \in \overline{A^{I,+}}$ and $(a_{n,I})$ is I-canonical. Then, by Lemma B.11,

$$\lim_n \tau(g_n) \cdot m_\tau = \tau(ka^I) \cdot m_{I,\tau} = \tau(ka^I) \cdot m_{I_0(\tau),\tau}.$$

If $a^I = aa'$ with $a \in \overline{A^{I_0(\tau),+}}$ and $a' \in A_{I_0(\tau)}$ it follows that $\lim_n \tau(g_n) \cdot m_\tau = \tau(k)\tau(a)\tau(a') \cdot m_{I_0(\tau),\tau} = \tau(ka) \cdot m_{I_0(\tau),\tau} \in K\overline{A^{I_0(\tau),+}} \cdot m_{I_0(\tau),\tau}$, because $A_{I_0(\tau)}$ stabilizes both $m_{I_0(\tau)} \in \mathcal{M}_1(G/P)$ and $m_{I_0(\tau),\tau}$. Hence, $\overline{G \cdot m_\tau} = \cup_{I^\tau \subset \Delta} K\overline{A^{I^\tau,+}} \cdot m_{I^\tau,\tau}$ is the union of the G-orbits of the measures $m_{I,\tau}$ where I is τ-open. Furthermore, if $\alpha \in I_0(\tau)$, then $\alpha(\log a) = \lim_n \alpha(\log a_{n,I})$.

By Lemma B.15, the above union is disjoint, since $g \cdot m_{I,\tau} = m_{J,\tau}$ with I and J τ-open implies that $I_0(\tau) = J_0(\tau) = I = J, m_{I,\tau} = m_{J,\tau}$, and $g \in R^{I^\tau} G^{I_0^\perp(\tau)} = R^{I_0(\tau)} G^{I_0^\perp(\tau)} = R^I G^{I_0^\perp(\tau)}$.

To conclude, note that the stabilizer of $m_{I,\tau}$ was calculated in Lemma B.15. □

The convergence in $\overline{G \cdot m_\tau}$ is described by the following proposition.

B.16. Proposition. *Let $I \subset \Delta$ be a subset of Δ. If $g_n \in G$ and $(g_n \cdot o) = (k_n a_n \cdot o)$ is I-fundamental, the sequence $(g_n \cdot m_\tau)$ converges to $ka^{I_0(\tau)} \cdot m_{I_0(\tau),\tau}$, where $k_n \to k$ and $a^{I_0(\tau)} \in \overline{A^{I_0(\tau),+}}$ is such that $\alpha(\log a^{I_0(\tau)}) = \lim_n \alpha(\log a_n)$ for $\alpha \in I_0(\tau)$. The parameters (k, a, I) of a realization $ka \cdot$*

$m_{I,\tau}$, with $k \in K, a \in \overline{A^{I,+}}$, and I a τ-open subset of Δ, are unique modulo an element of $(K^{I^\tau} \cap aK^{I^\tau}a^{-1})M$. In other words:

(1) *$g_1 \cdot m_{I_1,\tau} = g_2 \cdot m_{I_2,\tau}$ implies that $I_1 = I_2 = I$ and $g_2^{-1}g_1 \in R^I G^{I_0^\perp(\tau)}$.*

(2) *$k_1a_1 \cdot m_{I_1,\tau} = k_2a_2 \cdot m_{I_2,\tau}$ implies that $I_1 = I_2 = I, a^{I_1} = a^{I_2} = a$ and $k_1^{-1}k_2 \in (K^{I^\tau} \cap aK^{I^\tau}a^{-1})M$.*

Proof. Except for the uniqueness assertion, these assertions were proved in the course of proving Theorem B.12.

If I_1, I_2 are τ-open and $k_1a_1 \cdot m_{I_1,\tau} = k_2a_2 \cdot m_{I_2,\tau}$, it follows from Theorem B.12 that $I_1 = I_2 = I, a_2^{-1}k_2^{-1}k_1a_1 \in R^I G^{I_0^\perp(\tau)}$, and $k_2^{-1}k_1 \in K \cap P^{I^\tau} = K^{I^\tau}M$. Hence, $a_2^{-1}k_2^{-1}k_1a_1 \in G^{I^\tau}M \cap R^I G^{I_0^\perp(\tau)} = K^{I^\tau}G^{I_0^\perp(\tau)}$. Since $G^{I_0^\perp(\tau)}$ commutes with $A^I \subset G^I$, it follows that $a_1 = k^{-1}a_2\ell$ with $k, \ell \in K^{I^\tau}$. As in Proposition 9.16, it follows that $a_1 = a_2 = a$ and $k_2^{-1}k_1 \in (K^{I^\tau} \cap aK^{I^\tau}a^{-1})M$. □

The closure $\overline{A^+ \cdot o}$ of $\overline{A^+ \cdot m_\tau}$ in $\overline{G \cdot m_\tau}$ is characterized by the following proposition.

B.17. Proposition. *The closure $\overline{A^+ \cdot m_\tau}$ in $\overline{G \cdot m_\tau}$ is the disjoint union of the subsets $\overline{A^{I,+}} \cdot m_{I,\tau}$ with I τ-open in Δ.*

A sequence $(a_n \cdot m_{I,\tau}) \subset \overline{A^{I,+}} \cdot m_{I,\tau}$ $(I, J$ being $\tau -$ open) converges to $a \cdot m_{J,\tau} \in \overline{A^{J,+}} \cdot m_{J,\tau}$ if and only if

(1) *$J \subset I$,*

(2) *for any $\alpha \in J$, $\lim_n \alpha(\log a_n) = \alpha(\log a)$,*

(3) *for any τ-open subset L of Δ, with $J \subsetneq L \subset I$, there exists at least one $\alpha \in L \setminus J$ such that $\lim_n \alpha(\log a_n) = +\infty$.*

Proof. From the theorem one knows that the closure $\overline{A^+ \cdot m_\tau} \subset \overline{G \cdot m_\tau}$ can be calculated with I-fundamental sequences, where $I \subset \Delta$ is τ-open. If $a_n = a_n^I a_{n,I}$, with $(a_{n,I})$ I-canonical and $a_n^I \in \overline{A^{I,+}}$ such that $\lim_n a_n^I = a^I$, then $\lim_n a_n \cdot m_\tau = a^I \cdot m_{I,\tau} \in \overline{A^{I,+}} \cdot m_{I,\tau}$. It follows from this that $\overline{A^+ \cdot m_\tau} = \cup_{I\tau-\text{open}} \overline{A^{I,+}} \cdot m_{I_0(\tau),\tau} = \cup_{I\tau-\text{open}} \overline{A^{I,+}} \cdot m_{I,\tau}$ and from Theorem B.12 that the union is disjoint.

To prove the second statement, assume first that $(a_n) \subset \overline{A^{I,+}}$ is a fundamental sequence (and calculate the limit of $a_n \cdot m_{I,\tau}$). Hence, for some $I' \subset I$, $a_n = a_{n,I'}^I a_n^{I'}$, with $(a_{n,I'}^I)$ an I'-canonical sequence in $\overline{A^{I',+}}$ (i.e, $\log(a_{n,I'}^I) \subset C_{I'}^I$ and goes to infinity through that cone) and $a_n^{I'} \in \overline{A^{I',+}}$ converging to $a^{I'} \in \overline{A^{I,+}}$. Let (b_k) denote an I-canonical sequence such that $\lim_k b_k \cdot m_\tau = m_{I,\tau}$. One can approximate arbitrarily closely $a_n \cdot m_{I,\tau}$ by $a_nb_{k_n} \cdot m_\tau$ with some $b_{k_n} = b'_n$. Then, $\lim_n a_n \cdot m_{I,\tau} = \lim_n a_nb'_n \cdot m_\tau = \lim_n a_{n,I'}^I b'_n a_n^{I'} \cdot m_\tau$. Since $\alpha \in \Delta \setminus I'$ implies that $\lim_n \alpha(\log a_{n,I'}^I b'_n) = +\infty$, and $\alpha \in I'$, where $I' \subset I$ implies that $\alpha(\log(a_{n,I'}^I b'_n)) = 0$, it follows that $(a_{n,I'}^I b'_n)$ is I'-canonical. Hence, $\lim_n a_n \cdot m_{I,\tau} = \lim_n a_n^{I'} a_{n,I'}^I b'_n \cdot m_\tau = a^{I'} \cdot m_{I',\tau} = a^{I'} \cdot m_{I'_0(\tau),\tau} = a' \cdot m_{I'_0(\tau),\tau}$, where $a' \in \overline{A^{I'_0(\tau),+}}$.

Assume now that $\lim_n a_n \cdot m_{I,\tau} = a \cdot m_{J,\tau}$ with I and J τ-open. From the above argument, it follows that $a \cdot m_{J,\tau} = a^{I'} \cdot m_{I',\tau} = a' \cdot m_{I_0'(\tau),\tau}$, with $a' \in \overline{A^{I_0'(\tau),+}}$. By Proposition B.16, $J = J_0(\tau) = I_0'(\tau) \subset I$ and $a' = a$. Conditions (1) follows. Furthermore, for any $\alpha \in J = I_0'(\tau)$, $\lim_n \alpha(\log a_n) = \alpha(\log a') = \alpha(\log a)$. Conditions (2) and (3) follow since $J = I_0'(\tau)$.

To finish the proof, note that if (a_n) satisfies the conditions (1), (2), (3) of the proposition and is supposed to be I'-fundamental, then $\alpha(\log a) = \alpha(\log a')$ with $a' \in \overline{A^{I_0'(\tau),+}}$ for every $\alpha \in J$. Furthermore, condition (3) implies that $I_0'(\tau) = J$. As a result, $a' = a$. Hence, $a_n \cdot m_{I,\tau}$ converges to $a' \cdot m_{J,\tau} = a \cdot m_{J,\tau}$. □

THE G-ISOMORPHISM OF $\overline{X}_{\tau}^{S}$ AND $\overline{G \cdot m_{\tau}}$

B.18. Lemma. *Let τ be an irreducible representation of $G, g \in G$, and let I and J be two τ-open subsets of Δ. Then the condition $\tau(g)\pi_I\tau(g)^* = \lambda\pi_J, \lambda \in \mathbb{R}^*$, is equivalent to $I = J$, and $g \in R^I G^{I_0^{\perp}(\tau)}$.*

Proof. Lemma B.13 shows that $P^{I_0(\tau)} \subset P^{I^\tau}$ preserves $V_I = Im\pi_I$. From Lemma 9.13, it follows that $A_{I_0(\tau)}N_{I_0(\tau)}$ acts trivially on the support of $m_{I_0(\tau)}$ and, hence, on the support of $m_{I,\tau}$. This implies that $A_{I_0(\tau)}N_{I_0(\tau)}$ acts on V_I by homotheties. Lemma B.14 implies that $G^{I_0^{\perp}(\tau)}$ acts trivially on V_I and $K^{I^\tau}M$ acts on V_I by orthogonal transformations. Hence, $R^I G^{I_0^{\perp}(\tau)}$ preserves V_I and acts on V_I by similitudes.

Since, if $g \in R^I G^{I_0^{\perp}(\tau)}$, $\tau(g)\pi_I\tau(g^*)$ is an hermitian operator with range $\tau(g)^{-1}(V_I)$, it suffices to consider the action of $\tau(g)^{-1}$ on V_I in order to calculate $\tau(g)\pi_I\tau(g)^*$. Since this action reduces to similitudes one concludes that $\tau(g)\pi_I\tau(g)^* = \lambda\pi_I$, with $\lambda > 0$, if $g \in R^I G^{I_0^{\perp}(\tau)}$.

Conversely, if $\tau(g)\pi_I\tau(g)^* = \lambda\pi_J$, then $\tau^\wedge(g)\pi_I^\wedge\tau^\wedge(g)^* = \lambda^{p_\tau}\pi_J^\wedge$. Since the support of the measure $\pi_I^\wedge\tau^\wedge(g)^* m_\tau$ is equal to the support of $m_{I,\tau} = \pi_{I,\tau}^\wedge m_\tau$, the stabilizers of the supports of $\tau^\wedge(g)m_{I,\tau}$ and $m_{J,\tau}$ agree. From Lemma B.13, it follows that $gP^{I^\tau}g^{-1} = P^{J^\tau}$. Hence, $I^\tau = J^\tau, I = J$, and $g \in P^{I^\tau}$.

From the first argument one can now assume that $g \in G^{I_0(\tau)}$ and, then, make use of the Cartan decomposition for this group. If $g = kak'$ with $k, k' \in K^{I_0(\tau)}, a \in \overline{A^{I_0(\tau),+}}$, since $\tau^\wedge(kak')\pi_I^\wedge\tau(kak')^* = \pi_I^\wedge$ it follows that

$$\tau^\wedge(ka^2) \cdot m_{I,\tau} = m_{I,\tau}.$$

Hence, from Theorem B.12, it follows that $ka^2 \in R^I G^{I_0^{\perp}(\tau)}, k \in K^{I^\tau}, a \in \overline{A^{I_0(\tau),+}}, a \in A^{I_0^{\perp}(\tau)}$, and $a = e$, $g \in K^{I_0(\tau)}$. This proves the last assertion. □

B.19. Lemma. *If $g \in G$ and $u \in \mathrm{End}V$, define $g \cdot \bar{u} = \overline{g \cdot u}$, where $\bar{u} \in P(\mathrm{End}V)$ corresponds to $u \in \mathrm{End}V$ and $g \cdot u = gug^*$. The space $\overline{X}_{\tau}^{S}$ is* *the disjoint union of the G-orbits of the points $\bar{\pi}_I \in \overline{X}^S_\tau$ corresponding to the projections π_I with I τ-open, i.e.,*

$$\overline{X}^S_\tau = \cup_{I\tau-\text{open}} G \cdot \bar{\pi}_I.$$

Proof. The disjointness of the G-orbits follows from Lemma B.18, because $\tau(g)\pi_I\tau(g)^* = \lambda\pi_J$ implies that $I = J$, $\pi_I = \pi_J$.

In order to calculate $\overline{X}^S_\tau$, one uses fundamental sequences in X. If $g_n \cdot o = k_n a_n \cdot o$, with $k_n \to k$ and $a_n = a^I_n a_{n,I}$, where $a^I_n \in \overline{A^{I,+}}$ converges to a^I and $(a_{n,I})$ is an I-canonical sequence, it follows that $\tau(g_n)\tau(g_n)^* = \tau(k_n(a^I_n)^2(a_{n,I})^2)\tau(k_n^{-1})$.

From Lemma 9.50 and the definition of π_J, the sequence $\frac{\tau((a_{n,I})^2)}{\|\tau((a_{n,I})^2)\|}$ converges to $\pi_{I_0(\tau)}$. Hence, it follows that:

$$\lim_n \frac{\tau(g_n)\tau(g_n)^*}{\|\tau(a_{n,I})^2\|} = \tau(k(a^I)^2)\pi_{I_0(\tau)}\tau(k^{-1}) = \lambda^2\tau(ka')\pi_{I_0(\tau)}\tau(ka')^*,$$

where a' is the projection of a^I on $\overline{A^{I_0(\tau),+}}$ and $\tau_I(a^I) = \lambda\tau(a')$. It follows that $\overline{X}^S_\tau = \cup_{I\tau-\text{open}} G \cdot \pi_I$. □

B.20. Theorem. *If τ is faithful, the compactifications $\overline{X}^S_\tau$ and $\overline{G \cdot m_\tau}$ are G-isomorphic.*

Proof. Because τ is faithful, the two compact spaces $\bar{X}^S_\tau$ and $\overline{G \cdot m_\tau}$ are G-compactifications of X (Corollary B.6 and Remark B.2). Proposition B.16 and Lemma B.17 show that, as G-spaces they are both the disjoint union of the same homogeneous spaces $G/R^I G^{I_0^\perp(\tau)}$ with I $\tau-$ open . Clearly, a bijective G-map ϕ is defined by setting $\phi(m_{I,\tau}) = \pi_I$ if I is τ-open. The criterion of convergence in Proposition B.16 for fundamental sequences and Lemma B.19 imply that this map is continuous. Since ϕ is bijective and $\overline{G \cdot m_\tau}$ is compact, ϕ is a G-isomorphism. □

B.21. Theorem. *Let τ be an irreducible and faithful representation of G. The space $Y = \overline{X}^S_\tau$ is characterized by the following conditions:*

1. *Y is a G-compactification of X.*
2. *For any τ-open subset I of Δ , there exists a P^{I^τ}-equivariant embedding ϕ_I of X^I into Y. Define $X^I_o = \phi_I(X^I)$ and let $o_I = \phi_I(o) \in X^I_o, X^\Delta_o = X, o_\Delta = o$.*
3. *$Y = \cup_{I\tau-\text{open}} G \cdot X^I_o$ and if $g \cdot X^I_o \cap X^J_o \neq \emptyset$ for $g \in G$, with I, J both τ-open, then $I = J$, and $g \in P^{I^\tau}$.*
4. *The closure of $A^+ \cdot o$ in Y is equal to $\cup_{I\tau-\text{open}} \overline{A^{I,+}} \cdot o_I$. If I and J are τ-open, a sequence $(a^I_n \cdot o_I) \subset \overline{A^{I,+}} \cdot o_I$ converges to $a^J \cdot o_J \in \overline{A^{J,+}} \cdot o_J$ if and only if (a) $J \subset I$, (b) for any $\alpha \in J, \lim_n \alpha(\log a^J_n) = \alpha(\log a^J)$, (c) for any τ-open subset L of Δ such that $J \subsetneq L \subset \Delta$, there exists at least one $\alpha \in L \setminus J$ such that $\lim_n \alpha(\log a_n) = +\infty$.*

Proof. The argument follows the proof of Theorem 9.56, with $\overline{X}_\tau^S = \overline{G\cdot m_\tau}$ replacing $\overline{X}^{SF} = \overline{G\cdot m}$ and Theorem B.20 playing the role of Corollary 9.53.

Part (1): If $Y = \overline{G\cdot m_\tau}$, the first condition is stated in Theorem B.5. Recalling that o is the origin in X^I, set $\phi_I(p\cdot o) = p\cdot m_{I,\tau} \in \overline{G\cdot m_\tau}$ if $p \in P^{I^\tau}$ where I is τ-open. Note that, since $I_0(\tau)$ is orthogonal to $I_0^\perp(\tau)$, $G^{I_0^\perp(\tau)}$ commutes with $G(I_0(\tau)) = G(I)$ (see Remark B.10). Hence, $G^{I_0^\perp(\tau)}$ stabilizes $X^I = G^I/K^I$ and acts trivially on X^I. Because the stabilizers of o and $m_{I,\tau}$ in P^{I^τ} are both equal to $R^I G^{I_0^\perp(\tau)}$ (see Lemma B.14), the map ϕ_I is well-defined, P^{I^τ}-equivariant, and maps X^I continuously onto $G^I\cdot m_{I,\tau} \subset \mathcal{M}_1(P^{I^\tau}/P^{I^\tau}\cap P_\tau) = \mathcal{M}_1(P^{I^\tau}/P^{I^\tau}\cap P_\tau)$.

On the other hand, $\mathcal{M}_1(P^I/P^I\cap P_\tau)$ is embedded in $\mathcal{M}_1(G/P_\tau)$ and Theorem B.5 implies that ϕ_I is an embedding because I is τ-open. Theorem B.12. implies that $\overline{G\cdot m_\tau} = \cup_{I\tau-\text{open}} G\cdot m_{I,\tau}$ and that the orbits $G\cdot m_{I,\tau}$ are disjoint. The same theorem implies that the stabilizer of $G^I\cdot m_{I,\tau}$ is equal to P^{I^τ}, since G^I is transitive on $X_o^I = G^I\cdot m_{I,\tau}$. Finally, condition (4) follows directly from Proposition B.17.

Part (2): Conditions (1), (2), and (3) determine the structure of Y as a G-set: Y is the disjoint union of the G-orbits $G/R^I G^{I_0^\perp(\tau)}$. On the other hand, since $G = KP^I$, it follows that $G\cdot X_o^I = K\overline{A^{I,+}}\cdot o_I$. Hence, every point of Y can be written in the form $y = ka^I\cdot o_I$, for some τ-open subset I of Δ, with $k\in K$ and $a^I \in \overline{A^{I,+}}$. Define a map ϕ from Y to $\overline{X}_\tau^S$ by setting $\phi(y) = g\cdot m_{I,\tau}$ if $y = g\cdot o_I$. Because the stabilizers in G of o_I and $m_{I,\tau}$ are both equal to $R^I G^{I_0^\perp(\tau)}$, the map ϕ is well-defined and is a G-equivariant bijection of Y onto $\overline{G\cdot m_\tau}$. In order to show the continuity of ϕ, it suffices to show that if $\lim_n y_n = y$, $y_n \in Y$, then $\lim_n \phi(y_n) = \phi(y)$. By the compactness of Y, one can extract subsequences and, hence, assume that y_n is of the form $y_n = k_n a_n^I\cdot o_I$, with $k_n \to k\in K$, $a_n^I \in \overline{A^{I,+}}$, and I a τ-open subset of Δ. Because of condition (1), it follows that $\lim a_n^I\cdot o_I = k^{-1}\cdot y$. Condition (4) then implies that $k^{-1}\cdot y = a^J\cdot o_J$ for some τ-open subset J of I, where $a^J \in \overline{A^{J,+}}$, with $\lim_n \alpha(\log a_n^I) = \alpha(\log a^J)$ if $\alpha\in J$. Furthermore, for any τ-open subset L of Δ, where $J \subsetneq L$, one has that $\lim_n \alpha(\log a_n) = +\infty$ for some $\alpha \in L\setminus J$. Therefore, $\phi(y_n) = k_n a_n^I\cdot\phi(o_I) = k_n a_n^I\cdot m_{I,\tau}$. The stated properties of a_n^I, a^J and Proposition B.17 imply that $\lim_n a_n^I\cdot m_{I,\tau} = a^J\cdot m_{J,\tau}$. Hence, $\lim_n \phi(y_n) = ka^J\cdot m_{J,\tau} = ka^J\cdot\phi(o_J) = \phi(ka^J\cdot o_J) = \phi(y)$. $\square$

BIBLIOGRAPHY

[A1]. A. Abels, *Distal automorphism groups of Lie groups*, J. reine angew. Math. **329** (1981), 82–87.

[A2]. A. Ancona, *Negatively curved manifolds, elliptic operators, and the Martin boundary*, Ann. of Math. **125** (1987), 495–536.

[A3]. A. Ancona, *Sur les fonctions propres positives des variétés de Cartan–Hadamard*, Comm. Math. Helv. **64** (1989), 62–83.

[A4]. M. Anderson and R Schoen, *Positive harmonic functions on complete manifolds of negative curvature*, Ann. of Math. **121** (1985), 429–461.

[A5]. J.-P. Anker, *La forme exacte de l'estimation fondamentale de Harish-Chandra*, C.R. Acad. Sci. Paris, (Sér. I) **305** (1987), 371–374.

[A6]. J.-P. Anker & L. Ji, *Comportement exact du noyau de la chaleur et de la fonction de Green sur les espaces symétriques non compacts*, C. R. Acad. Sci. Paris, (Sér. I) **326** (1998), 153–156.

[A7]. A. Ash, D. Mumford, M. Rapaport, and Y. Tai, *Smooth Compactifications of Locally Symmetric Varieties*, Math Sci Press, Brookline, Massachusetts, 1975.

[B1]. M. Babillot, *Potential at infinity on symmetric spaces and Martin boundary* in *Harmonic analysis and discrete potential theory*, ed. M.Picardello, Plenum, New York, 1992.

[B2]. M. Babillot, *Asymptotics of Green functions on a class of solvable groups*, Potential Analysis **8** (1998), 69–100.

[B3]. W. Ballmann, *On the Dirichlet problem at infinity for manifolds of nonpositive curvature*, Forum Math. **1** (1989), 201–213.

[B4]. W. Ballmann, M. Gromov, and V. Schroeder, *Manifolds of nonpositive curvature*, Birkhäuser, Boston, 1985.

[B5]. W. Ballmann, *The Martin boundary of certain Hadamard manifolds*, preprint, 1994.

[B6]. W. Ballmann and F. Ledrappier, *The Poisson boundary for rank one manifolds and their cocompact lattices*, Forum Math. **6** (1994), 301–313.

[B7]. R.F. Bass and K. Burdzy, *The Martin boundary in non-Lipschitz domains*, Trans. AMS **337** (1993), 361–378.

[B8]. A. Borel, *Introduction aux groupes arithmétiques*, Hermann, Paris, 1965.

[B9]. A. Borel and J. Tits, *Groupes réductifs*, Publ. Math. IHES **27** (1965), 55–150.

[B10]. A. Borel and J. Tits, *Elements unipotents et sous-groupes paraboliques de groupes reductifs I*, Invent. Math. **12** (1971), 95–104.

[B11]. A. Borel and J.P. Serre, *Cohomologie d'immeubles et de groupes S-arithmétiques*, Topology **15** (1976), 211–232.

[B12]. P. Bougerol, *Comportement à l'infini du noyau potentiel sur un espace symétrique* in *Colloque J. Deny*, Lecture Notes in Mathematics **1096** (1984), Springer-Verlag, Berlin.

[B13]. N. Bourbaki, *Éleḿents de Mathématique, Livre III, Topologie générale*, Hermann, Paris, 1951.

[B14]. N. Bourbaki, *Éleḿents de Mathématique, Livre VI, Intégration*, Hermann, Paris, 1963.

[B15]. K.S. Brown, *Buildings*, Springer-Verlag, New York, 1989.

[B16]. F. Bruhat and J. Tits, *Groupes réductifs sur un corps local I*, Publ. Math. IHES **41** (1972).

[B17]. F. Bruhat and J. Tits, *Groupes réductifs sur un corps local II*, Publ. Math. IHES **60** (1984).

[B18]. K. Burns and R. Spatzier, *On the topological Tits buildings and their classifications*, IHES **65** (1987), 5–34.

[C1]. P. Cartier, *Géometrie et analyse sur les arbres*, Sém. Bourbaki 24e année exp 407, 71–72.

[C2]. G. Choquet, *Le théorème de représentation intégrale dans les ensembles convexes compacts*, Annales de l'Institut Fourier (Grenoble) **10** (1960), 333–344.

[C3]. C. Constantinescu and A. Cornea, *Ideale Ränder Riemannscher Flächen*, Springer-Verlag, Berlin, 1963.

[C4]. J. P. Conze and Y. Guivarc'h, *Remarques sur la distalité dans les espaces vectoriels*, C.R. Acad. Sci. Paris, (Sér. A) **278** (1974), 1083–1086.

[C5]. J. P. Conze and Y. Guivarc'h, *Propriété de droite fixe et fonctions propres des opérateurs de convolution*, Lecture Notes in Math. **404** (1974), Springer-Verlag, Berlin.

[C6]. M.C. Cranston and T.S. Salisbury, *Martin boundaries of sectorial domains*, Ark. Math. **31** (1993), 27–49.

[D1]. E. Damek, *Pointwise estimates for the Poisson kernel on NA-groups by the Ancona method*, Ann. Fac. Sci. Toulouse (6) **5** (1966), 421-441.

[D2]. Y. Derriennic and Y. Guivarc'h, *Théorème de renouvellement pour les groupes non moyennables*, C.R.A.S., Paris **277** (1973), 613–615.

[D3]. E.B. Dynkin, *Markov process and problems in analysis*, Proc. of ICM at Stockholm, 1962, pp. 36–58 (A.M.S. Translation ser.2, **31**, pp.1–24).

[D4]. E.B. Dynkin, *Brownian motion in certain symmetric spaces and non-negative eigenfunctions of the Laplace–Beltrami operator*, A.M.S. Translations Series 2 **72** (1968), 203–228.

[E1]. P. Eberlein and B. O'Neill, *Visibility manifolds*, Pac. J. Math. **46** (1973), 45–110.

[E2]. L. Elie, *Comportement asymptotique du noyau potentiel sur les groupes de Lie*, Annales de l'ENS **15** (1982), 257–364.

[E3]. P. Eymard and N. Lohoué, *Sur la racine carrée du noyau de Poisson dans les espaces symétriques*, Annales de l'ENS **8** (1975), 257–364.

[F1]. H. Freudenthal, *Linear Lie Groups*, Academic Press, 1969.

[F2]. W. Fulton and J. Harris, *Representation Theory*, Graduate Texts in Mathematics **129** (1991), Springer-Verlag, New York.

[F3]. H. Furstenberg, *A Poisson formula for semi-simple Lie groups*, Ann. Math. **77** (1963), 335–386.

[F4]. H. Furstenberg, *Translation invariant cones of functions*, Bull. A.M.S. **71** (1965), 272–326.

[F5]. H. Furstenberg, *Boundary theory and stochastic processes on homogeneous spaces*, Proc. Symp. Pure Math. **XXVI** (1972), 193–229.

[G1]. R. Gangolli and V.S. Varadarajan, *Harmonic analysis of spherical functions on real reductive groups*, Ergebnisse der Mathematik, **101** (1988), Springer-Verlag, Berlin.

[G2]. P. Gerardin, *Harmonic functions on reductive split groups in "Operator algebras and group representations", Vol. I*, Pitman, Boston, 1984.

[G3]. D.Gilbarg and N.S. Trudinger, *Elliptic partial differential equations of second order*, Springer-Verlag, Berlin, 1983.

[G4]. S. Giulini and W. Woess, *The Martin compactification of the cartesian product of two hyperbolic spaces*, J. für die reine u. angw. Math. **444** (1993), 17–28.

[G5]. S. Glasner, *Proximal flows*, Lecture Notes in Mathematics **517** (1976), Springer-Verlag, Berlin.

[G6]. R. Godement, *Une generalisation du théorème de la moyenne pour les fonctions harmoniques*, C.R.A.S Paris **234** (1952), 2137–2139.

[G7]. I. Goldsheid and Y. Guivarc'h, *Zariski closure and the dimension of the Gaussian law of the product of random matrices*, Probab. Theory Related Fields **105** (996), 109–142.

[G8]. I. Goldsheid and G.A.Margulis, *Lyapunov exponents of a product of random matrices*, Russian Math. Surveys **44** (1989), 11–71.

[G9]. F. Greenleaf, *Invariant means in topological groups*, Van Nostrand, New York, 1969.

[G10]. F. Guimier, *Simplicité du spectre de Liapounoff d'un produit de matrices aléatoires sur un corps ultramétrique*, C.R. Acad. Sci. Paris, (Sér. I) **309** (1989), 885–888.

[G11]. Y. Guivarc'h, M.Keane and B.Roynette, *Marches aléatoires sur les groupes de Lie*, Lecture Notes in Math. **624** (1977), Springer-Verlag, Berlin.

[G12]. Y. Guivarc'h, *Théorème quotients pour les marches aléatoires*, Astérisque **74** (1980), 15–38.

[G13]. Y. Guivarc'h, *Loi des grands nombres et rayon spectral d'une marche aléatoire sur un groupe de Lie*, Astérisque **74** (1980), 47–98.

[G14]. Y. Guivarc'h, *Sur la représentation intégrale des fonctions propres du Laplacien dans un espace symétrique*, Bull. Sci. Math. 2ème série **108** (1984), 373–392.

[G15]. Y. Guivarc'h and A. Raugi, *Frontière de Furstenberg, propriétés de contraction et théorèmes de convergence*, Z. Wahr. **89** (1985), 187–242.

[G16]. Y. Guivarc'h and J.C.Taylor, *The Martin compactification of the polydisc at the bottom of the positive spectrum*, Colloquium Mathematicum **LX/LXI** (1990), 537–546.

[G17]. Y. Guivarc'h, L. Ji and J.C.Taylor, *Compactifications of symmetric spaces*, C.R. Acad. Sci. Paris, (Sér. I) **317** (1993), 1103–1108.

[G18]. Y. Guivarc'h, *Compactifications of symmetric spaces and positive eigenfunctions of the Laplacian*, Séminaires de l'Universite de Rennes (1994).

[H1]. Harish-Chandra, *On a lemma of F. Bruhat*, J. Math. Pures et Appl. **35** (1956), 203–210.

[H2]. S. Helgason, *Differential geometry, Lie groups, and Symmetric spaces*, Academic press, New York, 1978.

[H3]. S. Helgason, *Groups and Geometric Analysis*, Academic Press, Orlando, Florida, 1984.

[H4]. S. Helgason, *Geometric Analysis on Symmetric Spaces*, Mathematical Surveys and Monographs **39** (1994), American Mathematical Society, Providence, Rhode Island.

[H5]. G. Hochschild, *The structure of Lie groups*, Holden–Day, Inc., San Francisco, 1965.

[H6]. J. Humphreys, *Introduction to Lie algebras and Representation Theory*, Graduate Texts in Mathematics **9** (1972), Springer-Verlag, New York.

[H6]. R. A. Hunt and R. L. Wheeden, *Positive harmonic functions on Lipschitz domains*, Trans. AMS **147** (1970), 507–527.

[J1]. L. Ji, *Compactifications of symmetric spaces and locally symmetric spaces* in *Geometric complex analysis,* ed. J. Noguchi et al, World Scientific, 1996.

[J2]. L. Ji, *Satake and Martin compactifications of symmetric spaces are topological balls*, Mathematical Research Letters **4** (1997), 79–89.

[J3]. L. Ji and R. MacPherson, *Geometry of compactifications of locally symmetric spaces*, preprint 1997.

[K1]. V.A. Kaimanovič and A.M. Vershik, *Random walks on discrete groups and entropy*, Ann. of Prob. **11** (1983), 457–490.

[K2]. V.A. Kaimanovič, *Boundaries of invariant Markov operators: the identification problem*, Lecture Note series of London Math Soc. **228** (1966), 127–176.

[K3]. F.I. Karpelevič, *The geometry of geodesics and the eigenfunctions of the Beltrami–Laplace operator on symmetric spaces*, Trans. Moscow Math. Soc. **14** (1965), 51–199.

[K4]. A. Koranyi, *Harmonic functions on symmetric spaces* in *Symmetric spaces,* ed. W. M. Boothby and G.Weiss, Marcel Dekker, New York, 1972.

[K5]. A. Koranyi, *Poisson integrals and boundary components of symmetric spaces*, Invent. Math. **34** (1976), 19–35.

[K6]. A. Koranyi, *Compactification of symmetric spaces and harmonic functions*, Lecture Notes in Mathematics **739** (1979), Springer-Verlag, Berlin.

[K7]. A. Koranyi, *A survey of harmonic functions on symmetric space*, Proceeding of Symposia in Pure mathematics **XXXV** Part I (1979), 323–344.

[L1]. M. Liao, *Brownian motion and the canonical stochastic flow on a symmetric space*, Trans. A.M.S. **341** (1994), 253–274.

[L2]. V. Losert, *On the structure of groups with polynomial growth*, Math. Zeit. **195** (1987), 109–117.

[L3]. T.J. Lyons, B. MacGibbon and J.C.Taylor, *Projection theorems for hitting probabilities and a theorem of Littlewood*, Jour. of Fun. Anal. **59** (1984), 470–489.

[M1]. I.G. Macdonald, *Spherical functions on a group of p-adic type*, Univ. Madras, Madras, 1971.

[M2]. M.P. Malliavin and P. Malliavin, *Factorisation et lois limites de la diffusion horizontale audessus d'un espace Riemannien symétrique*, Lecture Notes in Mathematics **404** (1974), Springer-Verlag, Berlin.

[M3]. G.A. Margulis, *Positive harmonic functions on nilpotent groups*, Soviet Math Doklady **7** (1966), 241–243.

[M4]. G.A. Margulis, *Discrete subgroups of semi-simple Lie groups*, Ergebnisse der Mathematik 3. Folge **17** (1989), Springer-Verlag, Berlin.

[M5]. R.S. Martin, *Minimal positive harmonic functions*, Trans. Amer. Math. Soc. **49** (1941), 137–172.

[M6]. R. Melrose, *Geometric Scattering Theory*, Cambridge Univ Press, New York, 1995.

[M7]. P.A. Meyer, *Probability and Potentials*, Blaisdell Publishing Company, Waltham, Mass., 1966.

[M8]. C.C. Moore, *Compactifications of symmetric spaces I*, Amer. J. Math. **86** (1964), 201–218.

[M9]. C.C. Moore, *Amenable subgroups of semi-simple groups and proximal flows*, Israël J. Math. **84** (1979), 121–138.

[M10]. G.Mostow, *Strong rigidity of locally symmetric spaces,* Ann. Math. Studies 78, Princeton University Press, Princeton, N.J., 1973.

[N]. E. Nolde, *Nonnegative eigenfunctions of Beltrami–Laplace operator on symmetric spaces with complex semi-simple Lie group* (in Russian), Usp. Math. Nauk. **21** (1966), 260–261.

[O1]. M.A. Olshanetsky, *Martin boundary for the Beltrami–Laplace operator on a Riemannian space of non-positive curvature* (in Russian), Usp. Math. Nauk. **24** (1969), 189–190.

[O2]. M.A. Olshanetsky, *Martin boundaries for real semisimple Lie groups*, Jour. Fun. Anal. **126** (1994), 169–216.

[O3]. A.L. Onishchik and E.B. Vinberg, *Lie groups and algebraic groups*, Springer-Verlag, Berlin, 1990.

[O4]. A. Orihara, *On random ellipsoid*, J. Fac. Sci. Univ. Tokyo Sect. IA Math **17** (1970), 73–85.

[P1]. M. Picardello and W. Woess, *Martin boundaries of random walks: ends of trees and groups*, Trans. of A.M.S. **302** (1987), 185–205.

[P2]. M. Prat, *Étude asymptotique et convergence angulaire du mouvement brownien sur une variété à courbure negative*, C.R. Acad. Sci. Paris, (Sér. A) **280** (1975), 1539–1542.

[P3]. I.I.Pyatetskii-Shapiro, *Automorphic functions and the geometry of classical domains*, Gordon and Breach Science Publishers, New York, 1969.

[R1]. A. Raugi, *Fonctions harmoniques sur les groupes localement compacts à base denombrable*, Bull. Soc. Math. France Mem. **54** (1977), 5–118.

[R2]. A. Raugi, *Fonctions harmoniques positives sur certains groupes de Lie resolubles connexes*, Bull. Soc. Math. France **124** (1996), 649–684.

[R3]. D. Revuz, *Markov chains*, North-Holland Pub. Co., Amsterdam, 1975.

[S1]. I. Satake, *On representations and compactifications of symmetric Riemannian spaces*, Ann. of Math. **71** (1960), 77–110.

[S2]. I. Satake, *On compactifications of the quotient spaces for arithmetically defined discontinuous groups*, Ann. of Math. **72** (1960), 555–580.

[S3]. J. Serrin, *On the Harnack inequality for linear elliptic equations*, J. Anal. Math. **4** (1954–1956), 292–308.

[S4]. D. Sullivan, *Related aspects of positivity in Riemannian geometry*, J. Diff. Geom. **25** (1987), 327–351.

[S5]. M.G. Šur, *The Martin boundary for a linear elliptic second-order operator*, A.M.S. Translations Series 2 **56** (1966), 19–36.

[T1]. J.C. Taylor, *The Martin boundaries of equivalent sheaves*, Ann. Inst. Fourier **XX (7)** (1970), 433–456.

[T2]. J.C. Taylor, *Brownian motion on a symmetric space of non-compact type: asymptotic behavior in polar coordinates*, Can. J. Math. **43** (1991), 1–21.

[T3]. J.C. Taylor, *The Martin compactification of a symmetric space of non-compact type at the bottom of the positive spectrum: an introduction* in *Potential theory,* ed. by Masanori Kishi, Walter de Gruyter, Berlin, 1992.

[T4]. J.C. Taylor, *Compactifications determined by a polyhedral cone decomposition of* $\mathbb{R}^n$ in *Harmonic Analysis and Discrete Potential Theory,* ed. M.A.Picardello, Plenum Press, New York, 1992.

[T5]. J. C. Taylor, *Martin compactifications,* in *Boundaries and Lie groups* ed. J.C. Taylor, CRM-AMS, 1998.

[T6]. J. Tits, *Buildings of spherical type and finite BN-pairs*, Lecture Notes in Mathematics **386** (1974), Springer-Verlag, Berlin.

[T7]. J. Tits, *On Buildings and Their Applications*, Proc. ICM, Vancouver (1974), 209–220.

[T8]. V. Tutubalin, *Some theorems of the type of the strong law of large numbers*, Theory of Probability and its Applications **14** (1969), 313–319.

[T9]. V. Tutubalin, *A central limit theorem for products of random matrices and some of its applications*, Symposia Math. vol 21 (1977), 101–116.

[V]. N.T. Varopoulos, L. Saloff-Coste, and T. Coulhon, *Analysis and geometry on groups*, Cambridge tracts in Mathematics **100** (1992), Cambridge University Press, Cambridge.

[W1]. G. Warner, *Harmonic analysis on semi-simple Lie groups I*, Springer-Verlag, New York, 1972.

[W2]. G. Watson, *A treatise on the theory of Bessel functions*, Cambridge University Press, Cambridge, 1962.

[Y]. S.T. Yau, *Open Problems in differential geometry in Differential geometry*, Proc. Symp. in Pure Math. **54** Part 1 (1993), 1–28.

[Z]. R.J. Zimmer, *Ergodic theory and semisimple groups*, Birkhauser, Boston, 1984.

LIST OF SYMBOLS

INDEX

Progress in Mathematics

Edited by:

Progress in Mathematics is a series of books intended for professional mathematicians and scientists, encompassing all areas of pure mathematics. This distinguished series, which began in 1979, includes authored monographs and edited collections of papers on important research developments as well as expositions of particular subject areas.

We encourage preparation of manuscripts in some form of TEX for delivery in camera-ready copy which leads to rapid publication, or in electronic form for interfacing with laser printers or typesetters.

Proposals should be sent directly to the editors or to: Birkhäuser Boston, 675 Massachusetts Avenue, Cambridge, MA 02139, U. S. A.

100 TAYLOR. Pseudodifferential Operators and Nonlinear PDE
101 BARKER/SALLY (eds). Harmonic Analysis on Reductive Groups
102 DAVID (ed). Séminaire de Théorie des Nombres, Paris 1989-90
103 ANGER/PORTENIER. Radon Integrals
104 ADAMS/BARBASCH/VOGAN. The Langlands Classification and Irreducible Characters for Real Reductive Groups
105 TIRAO/WALLACH (eds). New Developments in Lie Theory and Their Applications
106 BUSER. Geometry and Spectra of Compact Riemann Surfaces
107 BRYLINSKI. Loop Spaces, Characteristic Classes and Geometric Quantization
108 DAVID (ed). Séminaire de Théorie des Nombres, Paris 1990-91
109 EYSSETTE/GALLIGO (eds). Computational Algebraic Geometry
110 LUSZTIG. Introduction to Quantum Groups
111 SCHWARZ. Morse Homology
112 DONG/LEPOWSKY. Generalized Vertex Algebras and Relative Vertex Operators
113 MOEGLIN/WALDSPURGER. Décomposition spectrale et séries d'Eisenstein
114 BERENSTEIN/GAY/VIDRAS/YGER. Residue Currents and Bezout Identities
115 BABELON/CARTIER/KOSMANN-SCHWARZBACH (eds). Integrable Systems, The Verdier Memorial Conference: Actes du Colloque International de Luminy
116 DAVID (ed). Séminaire de Théorie des Nombres, Paris 1991-92
117 AUDIN/LAFONTAINE (eds). Holomorphic Curves in Symplectic Geometry
118 VAISMAN. Lectures on the Geometry of Poisson Manifolds
119 JOSEPH/ MEURAT/MIGNON/PRUM/ RENTSCHLER (eds). First European Congress of Mathematics, July, 1992, Vol. I
120 JOSEPH/ MEURAT/MIGNON/PRUM/ RENTSCHLER (eds). First European Congress of Mathematics, July, 1992, Vol. II
121 JOSEPH/ MEURAT/MIGNON/PRUM/ RENTSCHLER (eds). First European Congress of Mathematics, July, 1992, Vol. III (Round Tables)
122 GUILLEMIN. Moment Maps and Combinatorial Invariants of T^n-spaces

123 BRYLINSKI/BRYLINSKI/GUILLEMIN/KAC (eds). Lie Theory and Geometry: In Honor of Bertram Kostant
124 AEBISCHER/BORER/KALIN/LEUENBERGER/REIMANN (eds). Symplectic Geometry
125 LUBOTZKY. Discrete Groups, Expanding Graphs and Invariant Measures
126 RIESEL. Prime Numbers and Computer Methods for Factorization
127 HÖRMANDER. Notions of Convexity
128 SCHMIDT. Dynamical Systems of Algebraic Origin
129 DIJKGRAAF/FABER/VAN DER GEER (eds). The Moduli Space of Curves
130 DUISTERMAAT. Fourier Integral Operators
131 GINDIKIN/LEPOWSKY/WILSON (eds). Functional Analysis on the Eve of the 21st Century. In Honor of the Eightieth Birthday of I. M. Gelfand, Vol. 1
132 GINDIKIN/LEPOWSKY/WILSON (eds.) Functional Analysis on the Eve of the 21st Century. In Honor of the Eightieth Birthday of I. M. Gelfand, Vol. 2
133 HOFER/TAUBES/WEINSTEIN/ZEHNDER (eds). The Floer Memorial Volume
134 CAMPILLO LOPEZ/NARVAEZ MACARRO (eds) Algebraic Geometry and Singularities
135 AMREIN/BOUTET DE MONVEL/GEORGESCU. C_0-Groups, Commutator Methods and Spectral Theory of N-Body Hamiltonians
136 BROTO/CASACUBERTA/MISLIN (eds). Algebraic Topology: New Trends in Localization and Periodicity
137 VIGNERAS. Représentations l-modulaires d'un groupe réductif p-adique avec $l \neq p$
138 BERNDT/DIAMOND/HILDEBRAND (eds). Analytic Number Theory, Vol. 1 In Honor of Heini Halberstam
139 BERNDT/DIAMOND/HILDEBRAND (eds). Analytic Number Theory, Vol. 2 In Honor of Heini Halberstam
140 KNAPP. Lie Groups Beyond an Introduction
141 CABANES (eds). Finite Reductive Groups: Related Structures and Representations
142 MONK. Cardinal Invariants on Boolean Algebras
143 GONZALEZ-VEGA/RECIO (eds). Algorithms in Algebraic Geometry and Applications
144 BELLAÏCHE/RISLER (eds). Sub-Riemannian Geometry
145 ALBERT/BROUZET/DUFOUR (eds). Integrable Systems and Foliations Feuilletages et Systèmes Intégrables
146 JARDINE. Generalized Etale Cohomology
147 DIBIASE. Fatou TypeTheorems. Maximal Functions and Approach Regions
148 HUANG. Two-Dimensional Conformal Geometry and Vertex Operator Algebras
149 SOURIAU. Structure of Dynamical Systems. A Symplectic View of Physics
150 SHIOTA. Geometry of Subanalytic and Semialgebraic Sets
151 HUMMEL. Gromov's Compactness Theorem For Pseudo-holomorphic Curves
152 GROMOV. Metric Structures for Riemannian and Non-Riemannian Spaces
153 BUESCU. Exotic Attractors: From Liapunov Stability to Riddled Basins
154 BÖTTCHER/KARLOVICH. Carleson Curves, Muckenhoupt Weights, and Toeplitz Operators
155 DRAGOMIR/ORNEA. Locally Conformal Kähler Geometry
156 GUIVARC'H/JI/TAYLOR. Compactifications of Symmetric Spaces
157 MURTY/MURTY. Non-vanishing of L-functions and Applications
158 TIRAO/VOGAN/WOLF (eds). Geometry and Representation Theory of Real and p-adic Groups
159 THANGAVELU. Harmonic Analysis on the Heisenberg Group
160 KASHIWARA/MATSUO/SAITO/SATAKE (eds). Topological Field Theory, Primitive Forms and Related Topics
161 SAGAN/STANLEY (eds). Mathematical Essays in honor of Gian-Carlo Rota